Lecture Notes in Artificial Intelligence 7639

Subseries of Lecture Notes in Computer Science

LNAI Series Editors

Randy Goebel
University of Alberta, Edmonton, Canada
Yuzuru Tanaka
Hokkaido University, Sapporo, Japan
Wolfgang Wahlster
DFKI and Saarland University, Saarbrücken, Germany

LNAI Founding Series Editor

Joerg Siekmann
DFKI and Saarland University, Saarbrücken, Germany

Monica Palmirani   Ugo Pagallo
Pompeu Casanovas   Giovanni Sartor (Eds.)

AI Approaches to the Complexity of Legal Systems

Models and Ethical Challenges
for Legal Systems,
Legal Language and Legal Ontologies,
Argumentation and Software Agents

International Workshop AICOL-III
Held as Part of the 25th IVR Congress
Frankfurt am Main, Germany, August 15-16, 2011
Revised Selected Papers

 Springer

Series Editors

Randy Goebel, University of Alberta, Edmonton, Canada
Jörg Siekmann, University of Saarland, Saarbrücken, Germany
Wolfgang Wahlster, DFKI and University of Saarland, Saarbrücken, Germany

Volume Editors

Monica Palmirani
University of Bologna, CIRSFID, Italy
E-mail: monica.palmirani@unibo.it

Ugo Pagallo
University of Torino, Italy
E-mail: ugo.pagallo@unito.it

Pompeu Casanovas
Universitat Autònoma de Barcelona, Bellaterra, Spain
E-mail: pompeu.casanovas@uab.es

Giovanni Sartor
European University Institute and CIRSFID, Bologna, Italy
E-mail: giovanni.sartor@eui.eu

ISSN 0302-9743   e-ISSN 1611-3349
ISBN 978-3-642-35730-5   e-ISBN 978-3-642-35731-2
DOI 10.1007/978-3-642-35731-2
Springer Heidelberg Dordrecht London New York

Library of Congress Control Number: Applied for

CR Subject Classification (1998): I.2, H.4, H.3, H.5, C.2, J.1, K.4.1-2

LNCS Sublibrary: SL 7 – Artificial Intelligence

Typesetting: Camera-ready by author, data conversion by Scientific Publishing Services, Chennai, India

Printed on acid-free paper

Springer is part of Springer Science+Business Media (www.springer.com)

Preface

AI Approaches to the Complexity of Legal Systems, or AICOL, for short, was first organized as a thematic workshop of the 24th World Congress of Philosophy of Law and Social Philosophy (IVR), held in Beijing, China, during September 15–20, 2009. This led to a successful second edition of the workshop, organized as part of JURIX-09 (Rotterdam, The Netherlands, November 16–18). And now this book collects the contributions to the workshop's third edition, which took place as part of the 25th IVR congress, held in Frankfurt, Germany, during August 15–20, 2011.

Work in artificial intelligence and law has been particularly fruitful over the last decade. Besides providing advanced computer applications for the legal domain, with the development of knowledge-based systems and intelligent information retrieval among other things, research in AI and law has yielded innovative interdisciplinary models for understanding legal systems and legal reasoning. These models—highly significant for the philosophy of law and legal theory—include logical frameworks for defeasible legal reasoning and dialectical argumentation, logics for normative positions, theories of case-based reasoning, and computable models of legal concepts.

Today there is a strong need not only to bring research in AI and law to bear on legal theory, but also to foster mutual feedback and interaction among the different lines of research in AI and law. In fact, when different branches develop at a fast pace, we are at risk of squandering an opportunity to exchange knowledge and methodologies.

This is particularly so in multiagent systems and in social-network analysis, which share concepts and objects of study, and yet any overlap between them tends to be merely superficial in practice and theory alike. Multisystem and multilingual ontologies provide an important opportunity to integrate different trends of research in AI and law, including comparative legal studies. Complexity theory, graph theory, game theory, and any other contributions from the mathematical disciplines can help both to formalize the dynamics of legal systems and to capture relations among norms. Cognitive science can help the modeling of legal ontology by taking into account not only the formal features of law but also social behavior, psychology, and cultural factors.

This book is thus meant to support scholars in different areas of science in sharing knowledge and methodological approaches. This is done by highlighting similarities as well as differences among these approaches, and the contributions accordingly seek to capture this interdisciplinary aspect by laying out the scientific ground common to all of the disciplines in question, without any exclusive focus on what the state of the art is in each of these disciplines.

In keeping with this overarching purpose, the discussion is organized into six main parts devoted to each of the six topics addressed in the workshop:

- Models for the Legal System
- Ethics and the Regulation of ICT
- Legal Knowledge Management
- Legal Information for Open Access
- Software Agent Systems in the Legal Domain
- Legal Language and Legal Ontology

In the first part—Models for the Legal System—Sartor presents a new approach to the analysis of compliance with normative systems by taking into account different individual attitudes, ranging from self-interest to altruism, as well as an array of social and moral reasons for action. Araszkiewicz presents a coherence-based model of legal argumentation (CMLA) for assessing the doctrine of consistent interpretation developed by the European Court of Justice. New models for the legal system contribute to the state of the art in both ICT and legal theory, since they support the coherent and harmonized development of new technologies.

In the second part—Ethics and the Regulation of ICT—Pagallo discusses the impact of robotics on contemporary legal systems, looking in particular at some legal challenges the information revolution is posing for criminal law, contractual obligation, and tort law. Three new possible theories of robot liability and responsibility are presented, with strong implications for interaction between humans and artificial systems, thus also considering how such interaction can make for added complexity in the legal system. Similarly, the research conducted by Weng and Zhao on networked robots addresses the legal implications of combining unstructured physical environments with virtual ones, discussing the attendant risks as well as the safety and liability issues arising in connection with the use and behavior of such neworked robots. The authors argue that we can and should inject core ethical principles into robot technology. Moral issues are an emerging concern, not as a discipline per se but as an element to be integrated into the study of law and ICT in a new complex dimension, a world lying between cyberspace and reality. In this line of thought, Bourcier and De Filippi discuss the complexity of cloud computing and how to manage that complexity through policy. Cloud computing is based on a new business paradigm applied to an already mature technology: the outcome is a completely new legal landscape. Broker servers play a key role in negotiating the best strategy, resolving disputes, and providing the best connection services for each customer profile, while taking privacy and security issues into account. The contribution envisages a new paradigm where cloud-computing nodes are regulated by third-party certification authorities guaranteeing that end users can count on services affording transparency, privacy, and security, including protection from cybercrimes and an anti-corruption policy. This is another scenario where intelligent agents can be designed by building into them rules and principles of moral action.

The third part—Legal Knowledge Management—is focused on the ways in which computational applications can be implemented on a bottom-up approach, offering empirical evidence on which basis to sustain theoretical models. The contribution by Tiscornia et al. looks at the case law of the Italian High Court and of selected administrative and lower courts for the purpose of explaining the criteria of legal argumentation used in balancing competing legal rights and values: the authors apply natural-language tools to a sampling of 300 cases in an effort to understand the underlying approach to legal argumentation in treating a range of topics, with a view to helping legal practitioners go about their work. Winkles and Ruyter investigate the role of citations in the case law of the Supreme Court of The Netherlands for the purpose of detecting semantic information concerning the quality of the case law in the top-ten list: they look at 376 cases and 15,053 citations, and the outcome is visualized in a graph allowing legal practitioners to better understand the different ways cases relate to one another. Palmirani and Ceci present a contribution intended to promote the use of OWL2.0 properties for modeling and capturing judicial arguments as set out in case-law texts marked up in Akoma Ntoso: they combine three levels of ontology (argumentation, core, and domain ontology), focusing on the last of these to illustrate their working methodology. The research in this third part relies importantly on natural language tools in detecting, extracting, and qualifying legal knowledge to support future applications based on the Semantic Web. Boella et al. present a paper where legal semantics contribute to improving Web services. The authors introduce the Eunomos software, an advanced management system for legal terminology that helps expert users keep abreast of relevant law on any given topic. In the effort to formalize rules on top of the semantic level, Francesconi presents research where RDF/OWL is used to describe legal provisions and their interrelationships. More to the point, he presents an implementation of Hohfeldian relations and illustrates the approach by walking us through an example.

The fourth part—Legal Information for Open Access—presents research intended to develop new legal-information systems incorporating legal models, formalized legal knowledge, and ethical policies. Francesconi and Peruginelli investigate open access phenomena as an outgrowth of the digitization process, addressing important priorities in the production and dissemination of knowledge. They focus in particular on a project to build an open digital archive on the Web for legal informatics in the new digital era, considering that the dissemination of knowledge must be in the service of scholars and scientists, and not the other way around. In the same vein, Casanovas and Plaza propose an open access model for the content and publications put out by legal information institutes/by the Legal Information Institute of (LII) of the Cornell Law School, discussing some moral and legal issues that cannot be ignored in dealing with privacy and intellectual property.

In the fifth part—Software Agent Systems in the Legal Domain—we consider how these software tools can be designed in such a way as to embody legal principles and values, and how their behavior can be adjusted accordingly. Smith et al. offer a technical solution for combining normal and non-normal logics for dealing with the idea of collective trust. Laukyte discusses the different ways that software agents for multi-agent systems are conceived in law, AI, and software engineering, pointing out how the concept is narrowly defined in each of these three areas of practice. She thus introduces the idea of software agents as boundary objects, a sociological approach on which the three communities in question can find common ground and interact in developing an adequate model for MASs. Vincent and Zeleznikow discuss knowledge representation and work out an information system designed to support judges in sentencing: they describe the nature of sentencing in the Australian state of Victoria and the corresponding method of judicial decision making, while also considering argumentation in relation to procedure and to cognitive decision making models. Boer and Van Engers present a model-based diagnosis of the complex social systems in which large government bodies operate: their purpose is to identify areas and instances in which agents may play a problematic role in multi-agent systems.

The final part of the book—Legal Language and Legal Ontology—considers techniques for natural-language processing as a bridge between text and semantic Web annotations and ontologies. There is still much work to be done in this area in closing the gap between the legal terminology for specific legal concepts and the corresponding ontology classes. This has been attempted using FrameNet, a highly formalized tool that accordingly lends itself to this sort of endeavor. Palmirani et al. build on this approach in a novel way by using NLP tools to qualify normative modificatory provisions in legal texts marked up using the NormeInRete XML standard: they take a specific class of modificatory provisions (suspension of a norm's efficacy) and subject it to linguistic and legal analysis to show how such knowledge can be formalized through a linguistic tool such as FrameNet and then used by a semantic interpreter. Bertoli and Chishman also use a FrameNet database, but for semantic tagging and for developing a multilingual lexicon. The authors describe the initial steps in the development of a lexicographic project aimed at building a legal frame-based lexicon for Brazilian legal language. Myška et al. take a different approach in an effort to simplify legal language and make possible a better understanding of what the law says, so as to minimize noncompliance. They investigate two possible approaches intended to make legal language simpler and easier to understand for nonlawyers. However, a case study on the Creative Commons computerized system suggests that, in this case, simplifying the legal language does not necessarily reduce the level of uncertainty in the law. Very much driven by the same goals are Fernández-Barrera and Casanovas, who proceed on the basis of legal-domain semantics to provide simplified tools that citizens can use to query the case law pertaining to consumer rights. Their research was conducted as part of the ONTOMEDIA project, aimed at designing a semantic platform enabling users and professional mediators to meet in a community-driven Web portal.

June 2012, Bologna

Pompeu Casanovas\
Ugo Pagallo\
Monica Palmirani\
Giovanni Sartor

Organization

Organizing Committee

Danièle Bourcier	CERSA-CNRS, Paris, France
Pompeu Casanovas	UAB Institute of Law and Technology, Barcelona, Spain
Monica Palmirani	CIRSFID - University of Bologna, Italy
Ugo Pagallo	University of Turin, Italy
Giovanni Sartor	European University Insitute and University of Bologna, Italy

Program Committee

Gianmaria Ajani	University of Turin, Italy
Kevin Ashley	University of Pittsburgh, USA
Guido Boella	University of Turin, Italy
Joost Breuker	Leibniz Institute, Amsterdam University, The Netherlands
Tom Bruce	University of Cornell, USA
Núria Casellas	UAB Institute of Law and Technology, Barcelona, Spain
Cristiano Castelfranchi	ISTC - CNR, Italy
Tom van Engers	Leibniz Institute, Amsterdam University, The Netherlands
Enrico Francesconi	ITTIG-CNR Florence, Italy
Michael Genesereth	Stanford University, USA
Thomas Gordon	Fraunhofer Institute for Open Communications Systems
Guido Governatori	NICTA, Australia
Rinke Hoekstra	Leibniz Institute, Amsterdam University, The Netherlands
Arno Lodder	Vrije University, The Netherlands
Pablo Noriega	IIIA-CSIC, Barcelona, Spain
Marta Poblet	ICREA, UAB Institute of Law and Technology, Barcelona, Spain
Henry Prakken	Universiteit Utrecht, Groningen University, The Netherlands
Piercarlo Rossi	Facolta' di Economia, Università del Piemonte Orientale, Italy
Barry Smith	University of Buffalo, USA

Table of Contents

Part I: Models for the Legal System

Compliance with Normative Systems 1
  Giovanni Sartor

Coherence-Based Account of the Doctrine of Consistent
Interpretation 33
  Michał Araszkiewicz

Part II: Ethics and the Regulation of ICT

Three Roads to Complexity, AI and the Law of Robots:
On Crimes, Contracts, and Torts.................................. 48
  Ugo Pagallo

The Legal Challenges of Networked Robotics: From the Safety
Intelligence Perspective 61
  Yueh-Hsuan Weng and Sophie Ting Hong Zhao

Cloud Computing: New Research Perspectives for Computers
and Law 73
  Daniele Bourcier and Primavera De Filippi

Part III: Legal Knowledge Management

Balancing Rights and Values in the Italian Courts: A Benchmark for a
Quantitative Analysis 93
  Tommaso Agnoloni, Maria-Teresa Sagri, and Daniela Tiscornia

Survival of the Fittest: Network Analysis of Dutch Supreme Court
Cases.. 106
  Radboud Winkels and Jelle de Ruyter

Ontology Framework for Judgment Modelling 116
  Marcello Ceci and Monica Palmirani

Eunomos, a Legal Document and Knowledge Management System
to Build Legal Services 131
  *Guido Boella, Llio Humphreys, Marco Martin, Piercarlo Rossi, and
  Leendert van der Torre*

Axioms on a Semantic Model for Legislation for Accessing
and Reasoning over Normative Provisions   147
    Enrico Francesconi

Part IV: Legal Information for Open Access

An Open Access Policy for Legal Informatics Dissemination
and Sharing   162
    Enrico Francesconi and Ginevra Peruginelli

Advancing an Open Access Publication Model for Legal Information
Institutes   171
    Pompeu Casanovas and Enric Plaza

Part V: Software Agent Systems in the Legal Domain

Combinations of Normal and Non-normal Modal Logics for Modeling
Collective Trust in Normative MAS   189
    Clara Smith, Agustín Ambrossio, Leandro Mendoza, and
    Antonino Rotolo

Software Agents as Boundary Objects   204
    Migle Laukyte

Argumentation and Intuitive Decision Making: Criminal Sentencing
and Sentence Indication   217
    Andrew Vincent

Application of Model-Based Diagnosis to Multi-Agent Systems
Representing Public Administration   235
    Alexander Boer and Tom van Engers

Part VI: Legal Language and Legal Ontology

Semantic Annotation of Legal Texts through a FrameNet-Based
Approach   245
    Marcello Ceci, Leonardo Lesmo, Alessandro Mazzei,
    Monica Palmirani, and Daniele P. Radicioni

Developing a Frame-Based Lexicon for the Brazilian Legal Language:
The Case of the Criminal_Process Frame   256
    Anderson Bertoldi and Rove Luiza de Oliveira Chishman

Creative Commons and Grand Challenge to Make Legal Language
Simple..   271
    Matěj Myška, Terezie Smejkalová, Jaromír Šavelka, and Martin Škop

From User Needs to Expert Knowledge: Mapping Laymen Queries
with Ontologies in the Domain of Consumer Mediation   286
    Meritxell Fernández-Barrera and Pompeu Casanovas

Author Index...   309

Compliance with Normative Systems*

Giovanni Sartor

University of Bologna, Law Faculty-CIRSFID and European University
Institute of Florence

Abstract. I will argue that the cognitive attitudes and operations involved in compliance with normative systems are usually different from
those involved in complying with isolated social norms. While isolated
norms must be stored in the memory of the agents endorsing them, this
does not happen with regard to large normative systems. In the latter
case, the agent adopts a general policy-based intention to comply with
the normative system as a whole, an intention that provides an abstract
motivation for specific acts of compliance, once the agent has established
that these acts are obligatory according the system. I will show how the
endorsement of such a policy can be based on different individual attitudes, ranging from self-interest to altruistic, social or moral motivations.
Finally, I will analyse how a normative system may both constrain powers
and extend them, relying on this abstract motivation of its addressees.

1 Introduction

I will here address a challenge to mentalistic theories of norms, i.e., the views that
a norm's existence results from the norm itself being the content of appropriate
mental states of the concerned agents (such as the shared belief that the norm is
binding, and the goal or intention to comply with it). This challenge results from
the fact that we follow not only shared social norms, but also complex normative
systems: while shared social norms are represented in the mind of the concerned
agents, large normative systems direct people's thoughts and actions without
becoming, as a whole, mental objects for individuals.[1] We are often faced with
systems of this kind in our daily life (the legal system, but also the prescriptions
of an institutionalised religion, or the regulations of a company, a condominium,
a regulated market, a teaching institution, a sociotechnical infrastructure such
as an airport or a harbour, etc.). All norms of such a system cannot be stored
in one's memory since they exceed human capacities (at least for the largest
normative systems, such as a municipal law, containing many thousands, even
millions, of rules) and moreover such norms persistently change as a consequence

---

* Parts of this paper have been published in In Paglieri, F., Tummolini, L., Falcone,
R., and Miceli, M., editors, The Goals of Cognition. Essays in Honor of Cristiano
Castelfranchi. College Publications, London.
[1] The term *agent* is here used as in AI, to mean an entity endowed with cognitive
capacities and capable of autonomous action; it is not used in the legal-economical
sense of someone delegated to act on behalf of another.

M. Palmirani et al. (Eds.): AICOL Workshops 2011, LNAI 7639, pp. 1–32, 2012.

of intervening facts (such as the adoption of new regulations, new decisions interpreting, them, etc.). For instance, while each of us has some knowledge of a few rules of our legal system (the ones corresponding to shared moral rules, such as the prohibition of killing, or most frequently encountered, such as certain traffic rules, or governing one's particular activity, such as rules on software copyright for a computer programmer), generally the common citizen has a very vague idea of the content the law of his of her country, especially in technical domains such as tax law, land planning law, environmental law, etc.

When referring to a large normative system N an agent usually does not immediately find an answer to the question "What ought I to do?" (as it usually happens when applying a shared social norm). One rather needs asks oneself (or the appropriate expert) "What does N require from me?", i.e., "What ought I do to according to N?" The answer to this question ("I ought to do action A according to N") does not have, by itself, a motivating force for the agent. The concerned agent may well refuse to take into account the system's requests (for instance one may ask oneself what a certain religion requires from oneself, without having the slightest intention to follow the prescriptions of that religion, whatever they may be).

I will suggest that the motivation to perform a particular action qualified as obligatory by a normative system results from a general intention to comply with the system as a whole. The latter attitude provides an abstract motivation for specific acts of compliance, once the addressee has established that certain actions are obligatory according to the system. I will show how the endorsement of such an intention can be based on different individual attitudes. Finally, I will analyse how a normative systems may both constrain social powers and extend them, relying on this abstract attitude of its addressees.

2 Preliminary Notions: Actions, Obligations, Norms

For analysing compliance, we need some basic notions. First, a way of expressing action and obligations is required. For actions I will use the simple E operator of Pörn (1977), though other action logics would be appropriate as well for this discussion of compliance (on the E operator see also Sergot, 2001, for a different approach to action, see for instance, Horty, 2001).

Definition 1 (Actions). *Let proposition $E_j S$ describe agent j's positive action consisting in the production of state of affairs S, where "S" is any proposition. Thus $E_j S$ means "j brings it about that S". Similarly, let $\neg E_j S$ describe the negative action (the omission) consisting in not bringing about that S. Thus $\neg E_j S$ means "j omits to bring about that S" or "j does not bring it about that S". When the distinction between positive and negative action is not relevant, let us use $\mathcal{A}_j$ to cover both. Let $\overline{\mathcal{A}_j}$ denote the complement of $\mathcal{A}_j$ ($\overline{\mathcal{A}_j}$ stands for $\neg E_j S$ if $\mathcal{A}_j = E_j S$; it stands for $E_j S$ if $\mathcal{A}_j = \neg E_j S$).*

For simplicity when an agent brings about its own action, I will not repeat the agent's name in the action's result. Thus, for expressing the idea that

John smokes (*John* brings it about that he smokes, meaning that *John* does the action of smoking) rather than writing $E_{John}Smoke(John)$, I will write $E_{John}Smoke$.

This notion of an action does not involve intentionality (an aspect which is involved in the notion of an action as a goal-directed behaviour in Conte and Castelfranchi, 1995). I prefer to stick to this minimal understanding of agency since compliance with normative systems usually prescinds from an action's intentionality: holding the required behaviour is usually sufficient for compliance. Intentions may instead be relevant for the consequences of violations (where intention may be required, or negligence, for certain normative consequences to take place), an aspect that I am not considering here.

As an example of an action-proposition, consider the following

$$E_{John}Damaged(Tom) \tag{1}$$

which means "*John* brings it about that *Tom* is damaged", or more simply "*John* damages *Tom*" while the following

$$\neg E_{John}Damaged(Tom) \tag{2}$$

means "*John* does not bring it about is about that *Tom* is damaged", or more simply "*John* does not damage *Tom*". I do not need to discuss here the logic of E, which is a classical modal logic (if A and B are logically equivalent, then $E_x A \to E_x B$), including inference rule

$$\frac{A}{\neg E_x A} \tag{3}$$

and axiom schema

$$E_x S \to S \tag{4}$$

Inference rule (8) says that one cannot realise what is a logical theorem (a necessary truth). For instance since $A \lor \neg A$ is a necessary truth, being a theorem of propositional logic, *Tom* cannot be said to bring it about (it would hold independently of his action).

Axiom schema (4) says that that if the state of affairs S is realised though an action, then it is the case that S. For instance the fact that *Tom* makes it so that *Ann* suffers damage, obviously entails that *Ann* suffers damage:

$$E_{Tom}Damaged(Ann) \to Damaged(Ann) \tag{5}$$

Definition 2 (Obligations and Prohibitions). *Let $\emptyset$ denote obligation. $OE_j S$ means "it is obligatory hat j brings it about that S". Similarly $O\neg E_j S$ means "it is obligatory that j does not bring about that S", or "it is forbidden that j brings about that S".*

For instance, the following means "it is obligatory that *John* makes it so that *Tom* is compensated", or more simply, "it is obligatory that *John* compensates *Tom*",

$$OE_{John}Compensated(Tom) \tag{6}$$

while the following means "it is obligatory that *John* does not makes it so that *Tom* is damaged", or more simply, "it is forbidden that *John* damages *Tom*".

$$O\neg E_{John} Damages(Tom) \tag{7}$$

I will not specify here a particular deontic logic, since the following considerations may apply to different deontic logics. The reader may assume, for instance, standard deontic logic, as characterised in Føllesdal and Hilpinen, 1971, but my preference would to to a simpler deontic logic, limited to the substitution of logically equivalent formulas inside the deontic operator, namely, the schema:

$$\frac{\mathcal{A} \leftrightarrow \mathcal{B}}{O\mathcal{A} \leftrightarrow O\mathcal{B}} \tag{8}$$

Permission can be defined as usually as the negation of a prohibition:

$$P\mathcal{A} \stackrel{\text{def}}{=} \neg O\overline{\mathcal{A}}$$

To keep the language as simple as possible, I shall not address how a deontic language can be enriched through Hohfeldian concepts (for a logical analysis, Sartor, 2006), and how this this extension can be useful for addressing compliance (Siena et al., 2009). While I am making use of the E action logic, I consider that the ideas on compliance here developed are generally compatible also with approaches to deontic reasoning based on different logics for action.

Definition 3 (Norms). *I represent norms as defeasible conditionals*

$$[A \stackrel{n}{\Rightarrow} B] \tag{9}$$

where A is a proposition and B is any kind of normative qualification, deontic or non deontic, and $\stackrel{n}{\Rightarrow}$ expresses normative conditionality, namely the link between an antecedent (possibly empty) and the normative consequent that is generated by that antecedent. A norm including variables stands for the set of all of its ground instances.

I take normative conditionals to be non truth-functional, but to allow for (defeasible) modus ponens. Note that the conditional $A \stackrel{n}{\Rightarrow} B$ is not a statement of fact, but can rather be viewed as rule, according to which consequent B is produced (it holds, according to the normative system being considered) when the antecedent A holds. Here is an example of two deontic norms, the first stating that it is forbidden to cause damage to others, and the second that who causes a damage to another has the obligation to compensate the latter (in the following when obvious I drop the requirement $x \neq y$):

$$[x \neq y \stackrel{n}{\Rightarrow} O\neg E_x Damaged(y)]$$
$$[x \neq y \wedge E_x Damaged(y) \stackrel{n}{\Rightarrow} OE_x Compensated(y)] \tag{10}$$

The following is an example of a constitutive norm, saying that if we injure a person (make so that someone is injured), we cause damage to that person (injuring counts as damaging):

$$[E_x Injured(y) \overset{n}{\Rightarrow} E_x Damaged(y)] \tag{11}$$

Also concerning the normative conditional $\overset{n}{\Rightarrow}$, I will not provide a full logical account. I will just require that it enables defeasible detachment (*modus ponens*), i.e., that from A and norm $A \overset{n}{\Rightarrow} B$, the conclusion B can be inferred.

$$\{A, A \overset{n}{\Rightarrow} B\} \hspace{0.3em}\vdash\hspace{-0.9em}\sim\hspace{0.3em} B \tag{12}$$

I use the symbol $\vdash\!\!\!\sim$ for non-monotonic derivability, assuming that normative conditionals are inherently defeasible, but in this paper I will not discuss defeasibility and its logical treatment (see Prakken and Sartor, 2003, Sartor, 2011).

Note that I do not distinguish deontic conditionals and constitutive or counts-as conditionals (Searle, 1995, Jones and Sergot, 1996, Grossi et al., 2008), assuming that the same inferences apply to both (on normative conditionality, see Sartor, 2005; on the connection between deontic and constitutive conditionality, see Boella and van der Torre, 2006). The following example shows how from a conditional and an instance of its antecedent we can defeasibly derive an instance of the conditional's consequent.

$$\{E_{Tom}Damaged(John), E_x Damaged(y) \overset{n}{\Rightarrow} OE_x Compensated(y)\} \hspace{0.3em}\vdash\!\!\!\sim\hspace{0.3em} \tag{13}$$
$$OE_{Tom}Compensated(John)$$

3 Relativised Normative Statements (In Particular Obligations)

In addressing compliance we have to connect a normative system N (a set of norms) and (the propositions describing) the factual circumstances C relevant to N's application. Here I am only interested in the obligations and the institutional facts that are generated by norms in N, when applied to facts in C. Thus we can assume that C contains (or entails) all factual literals (atomic sentences or negations of them) which are true in the real or hypothetical situation (the world) in which the norms have to be applied, without considering how the truth of such literals can be established. For simplicity's sake we can limit C to the factual literals that are relevant to the application of norms in N, matching literals in the antecedent of a norm in C. When the considered factual circumstances are those that hold in the real world (rather than in a merely possibly situation), i.e., they are the truths relevant to the application of N in the case at hand, I shall denote them through the expression $T(N)$.

I will now introduce relativised normative statements, expressing that a proposition (in particular, an obligation) holds with regard to a normative system.

Definition 4 (Relativised Normative Statements). *We say that any proposition B holds relatively to normative system N and circumstances C, and write $[B]_{N,C}$ iff $N \cup C \hspace{0.2em}\vdash\!\!\!\sim\hspace{0.2em} B$*

$$[B]_{N,C} \overset{\text{def}}{=} N \cup C \hspace{0.3em}\vdash\!\!\!\sim\hspace{0.3em} B \tag{14}$$

In particular when the proposition which is affirmed to hold is an obligation $O\mathcal{A}_x$, we abbreviate the corresponding normative statements $[O\mathcal{A}_x]_{N,C}$ as $\mathbb{O}_{N,C}\mathcal{A}_x$.

Definition 5 (Relativised Obligation-Statements). *We say that it is obligatory relatively to N and C that x does $\mathcal{A}$, and write $\mathbb{O}_{N,C}\mathcal{A}_x$, to express that $N \cup C \mathrel{\vdash\!\!\!\sim} O\mathcal{A}x$:*

$$\mathbb{O}_{N,C}\mathcal{A}_x \overset{\text{def}}{=} N \cup C \mathrel{\vdash\!\!\!\sim} O\mathcal{A}_x \tag{15}$$

According to Definition (5), a relativised obligation statement does not express a norm, but it expresses an assertion about the implications of norms (normative systems) and circumstances (in the terminology of Alchourrón, 1969 and Alchourrón and Bulygin, 1971 such assertions are called "normative propositions").

When we are referring to the true relevant circumstances of the real world— i.e., to the set of truths relevant to the application of N, denoted as $T(N)$—, rather than to circumstances of hypothetical situations, we simply write $[B]_N$, or $\mathbb{O}_N\mathcal{A}_x$.

$$[B]_N \overset{\text{def}}{=} N \cup T(N) \mathrel{\vdash\!\!\!\sim} B$$
$$\mathbb{O}_N\mathcal{A}_x \overset{\text{def}}{=} N \cup T(N) \mathrel{\vdash\!\!\!\sim} O\mathcal{A}_x \tag{16}$$

For instance, let us consider the following example, where N_1 includes a simplified version of the three norms above, and C_1 is limited to the fact that *John* injured *Tom*:

Example 1.

$$C_1 = \{E_{John}Injured(Tom)\}$$
$$N_1 = \{[E_x Injured(y) \overset{n}{\Rightarrow} E_x Damaged(y)];$$
$$[O\neg E_x Damaged(y)]; \tag{17}$$
$$[E_x Damaged(y) \overset{n}{\Rightarrow} OE_x Compensated(y)]\}$$

It is easy to see that the following inferences holds on the basis of example (1):

$$(C_1 \cup N_1) \mathrel{\vdash\!\!\!\sim} E_{John}Damaged(Tom)$$
$$(C_1 \cup N_1) \mathrel{\vdash\!\!\!\sim} O\neg E_{John}Damaged(Tom) \tag{18}$$
$$(C_1 \cup N_1) \mathrel{\vdash\!\!\!\sim} OE_{John}Compensated(Tom)$$

Therefore, we can say that, relatively to N_1 and C_1, *John* has damaged *Tom*, it is obligatory that John does not damage Tom, and it is obligatory that *John* compensates *Tom*:

$$[E_{John}Damaged(Tom)]_{N_1,C_1} \wedge \mathbb{O}_{N_1,C_1}\neg E_{John}Damaged(Tom)\wedge$$
$$\mathbb{O}_{N_1,C_1}E_{John}Compensated(Tom) \tag{19}$$

If *John* has really injured *Tom* (and no other relevant circumstances obtain, such as exception excluding the application of the norms at issue), i.e., if $C_1 = T(N_1)$, we can simply say that according to N_1, *John* has damaged *Tom*, he ought not to damage him, and he ought to compensate him:

$$[E_{John}Damaged(Tom)]_{N_1} \wedge \mathbb{O}_{N_1}\neg E_{John}Damaged(Tom)\wedge$$
$$\mathbb{O}_{N_1}E_{John}Compensated(Tom) \tag{20}$$

Here is another small example. The first norm in N_2 says that places open to the public are (count as) public places. The second says that if one is in a public place then one is forbidden to smoke.

Example 2.

$$C_2 = \{OpenToPublic(LectureRoom); in(John, LectureRoom); E_{john}smoke\}$$
$$N_2 = \{[OpenToPublic(y) \overset{n}{\Rightarrow} PublicPlace(y)];$$
$$[PublicPlace(y) \wedge in(x,y) \overset{n}{\Rightarrow} O\neg E_x Smoke]\}$$

$$(21)$$

We can say then say that according to N_2 given circumstances C_2 it is obligatory that *John* does not smoke ($\mathbb{O}_{N_2,C_2}\neg E_{Tom}Smoke$), and that *John* violates this obligation ($Violated_{N_2,C_2})O\neg E_{John}smoke)$).

The extent of the set of action obligatory according to normative system N depends on the content of N, but also on the deontic logic we have adopted for N. For instance, if we adopt standard deontic logic for N, then if $N \mathrel{|\!\sim} O\mathcal{A}$ and $\mathcal{A} \to \mathcal{B}$, then $N \mathrel{|\!\sim} O\mathcal{B}$. This will not hold if instead we adopt the minimal deontic logic we described above (which requires that $\mathcal{A} \leftrightarrow \mathcal{B}$).

We can however recover the extent of obligations according to standard deontic logic, by defining a broader notion of a relativised obligation. For instance, following the idea of a logic of satisfaction, we could say that action $\mathcal{A}$ is weakly obligatory, relatively to a normative system N, if $\mathcal{A}$ is entailed by actions that are obligatory relatively to the system.

The language of relativised obligation allows us to say that according to different normative systems different obligations hold. For instance, given that Canon law contains a universal norm prohibiting the use of contraception as well as a constitutive rule saying any action meant to make a sex act unfruitful counts as artificial contraception, and given that taking the pill in order to prevent pregnancy is meant to make subsequent sex acts unfruitful, we can conclude that according to the Canon law a woman, say Ann, is forbidden to take the pill in order to prevent pregnancy. Similarly, given that Islamic law contains a norm that prohibits receiving interest on loans of money, we can say that according to Islamic law John is forbidden to receive interest on loans of money .

A notion of relativised permission can be provided that corresponds to the above analysis of an obligation.

Definition 6 (Relativised Permission). *Let us say that it is permissible relatively to N and C that x does $\mathcal{A}$, and write $\mathbb{P}_{N,C}\mathcal{A}_x$ iff N and C entail $P\mathcal{A}_x$:*

$$\mathbb{P}_{N,C}\mathcal{A}_x \overset{def}{=} N \cup C \mathrel{|\!\sim} P\mathcal{A}_x \qquad (22)$$

Note that according to this definition, saying that an action $E_x S$ is permissible relatively to normative system N and circumstances C ($\mathbb{P}_{N,C}E_x S$) does not amount to saying that it is not the case that $E_x S$ is forbidden relatively to the same system and circumstances ($\neg\mathbb{O}_{N,C}\neg E_x S$). Proposition $\mathbb{P}_{N,C}E_x S$ is not equivalent to $\neg\mathbb{O}_{N,C}\neg E_x S$, since the former holds when $N \cup C$ entails $PE_x S$, while the latter holds when $N \cup C$ does not entail $O\neg E_x S$ (see Alchourrón, 1969, Alchourrón and Bulygin, 1971).

4   Compliance

With the help of the notions introduced in the previous section, we can now address compliance. The issue of compliance can arise in very different context, as the following examples shows:

- Mary is appointed to a professorship. She signs a contract stating her commitment to comply with the University regulations.
- John enters a PhD program. He is directed to the booklet containing the regulations he has to comply with.
- Linda is appointed as a judge. She takes an oath to respect the Constitution and the laws of her country.
- Adolf Eichman enters the SS. He takes an oath of obedience to death to Adolph Hitler and the superiors he has designated.
- Antony enters the Franciscan order. He promises to respect the body of regulations known as "The Rule of St. Francis" as well as the law of the Catholic Church.
- Mary, a shop-owner, receives a threats by gangsters belonging to a mafia organisation. She chooses to comply with all rules imposed by that organisation (monthly protection money, code of silence, etc.) to avoid problems with the bad guys.
- A digital agent enters and electronic marketplace. It commits to respect all rules of the marketplace.

In all these contexts the agent has taken the commitment (adopted the intention to) comply with a certain normative system.

We can distinguish different notions of compliance. The first notion is behavioural compliance, which simply consist in behaving is such a way as to fulfil an obligation.[2]

Definition 7 (Behavioural Compliance). *An agent x behaviourally complies with an obligation $O\mathcal{A}_x$ of a normative system N, iff the obligation holds according to N and x's behaviour counts as $\mathcal{A}$ according to N, i.e., iff*

$$\mathbb{O}_N \mathcal{A}_x \wedge [\mathcal{A}_x]_N) \tag{23}$$

For instance, if a non-smoker does not smoke in a public office, ignoring that there is a prohibition to do so (she does not know about the prohibition in Example (2) above), she will still behaviourally comply with that prohibition. She will do that even if she is taking a siesta, and therefore is not aware that she is not smoking. On the basis of this notion of behavioural compliance we can develop the idea of conscious compliance, which consists in complying with a an obligation, while being aware that it is entailed by a certain normative system.

---

[2] As above, I will often omit to make explicit reference to the circumstances in which a normative set N is to be applied, assuming that an implicit reference is made to $T(N)$, the true circumstances relevant to the application of N.

Definition 8 (Conscious Compliance). *An agent* x *consciously complies with an obligation* OE_xS *of a normative system* N, *iff* x *behaviourally complies with the obligation, believes that the normative system entails that obligation, and is aware of doing the required action, i.e., iff*

$$\mathbb{O}_N\mathcal{A}_x \wedge [\mathcal{A}_x]_N \wedge Bel_x(\mathbb{O}_N\mathcal{A}_x) \wedge Bel_x([\mathcal{A}_x]_N) \tag{24}$$

By assimilating knowledge and true belief we can say that x *consciously complies with an obligation* OA_x *according to* N *iff* x *knows that, according to systems* N, x *is doing an action which is obligatory:*

$$Knows_x([\mathcal{A}_x]_N) \wedge Knows_x(\mathbb{O}_N\mathcal{A}_x) \tag{25}$$

Many instances of compliance with norms in a legal system are unconscious: the concerned agent is not aware that the law prescribes a certain behaviour, but behaves correspondingly, either motivated by moral or social norms or by any other factors (self interest, altruism, etc.).

5 Intentional Compliance

Here I am interested with acts of compliance motivated by the (belief) in the existence of an obligation relatively to a normative system. First of all the action considered must be intentional, namely motivated by the intention to perform it, and moreover such an intention must be motivated by the awareness of the obligation.

Definition 9 (Intentional Compliance). *An agent* x *intentionally complies with an obligation according to* N, *when* N *entails that obligation* $(\mathbb{O}_N\mathcal{A}_x)$, *and* x's *belief that this is the case* $(Bel_x(\mathbb{O}_N\mathcal{A}_x))$ *motivates* x *to intend to hold the prescribed behaviour* $(Int_x\mathcal{A}_x)$, *which, in its turns motivates* x *to hold that behaviour* $(\mathcal{A}_x)$.

$$\mathbb{O}_N\mathcal{A}_x \wedge (Bel_x(\mathbb{O}_N\mathcal{A}_x) \vartriangleright^m Int_x\mathcal{A}_x) \wedge (Int_x\mathcal{A}_x \vartriangleright^m [\mathcal{A}_x]_N) \tag{26}$$

where $\vartriangleright^m$ *denotes motivation, understood as mental causation.*

This definition would require refinements, linked to the difficulties inherent to the notion of motivation, which I cannot address here. Let me just state that I take $M_1 \vartriangleright^m M_2$ to be true, when both M_1 and M_2 are true and the fact that the agent instantiated M_1 was the reason why the agent subsequently instantiated M_2. With regard to the notion of an intention, I assume that the unconditioned intention to perform an action or omission consists in having the chosen goal to perform the action (for a discussion of the connection between goal and intentions, and for the proposal of a refined formalisation, see Castelfranchi and Paglieri, 2007):

$$Int_x\mathcal{A} = CGoal_x\mathcal{A}_x \tag{27}$$

For my purpose (and given that I do not need to distinguish actions and omissions) this simple notion of an intention will suffice.

Observe that one can perform an action wrongly believing that one is under an obligation. This is the situation where there is no obligation to behave in a certain way, but the agent believes that such an obligation exists.

$$\neg \mathbb{O}_N \mathcal{A}_x \wedge (Bel_x(\mathbb{O}_N \mathcal{A}_x) \triangleright^m Int_x \mathcal{A}_x) \wedge (Int_x \mathcal{A}_x \triangleright^m \mathcal{A}_x) \qquad (28)$$

Let us now consider again example (1), and assume that $C_1 = T(N)$ (C_1 contains all true circumstances relevant to the application of N). Given that

$$\mathbb{O}_N E_{John} Compensated(Tom) \qquad (29)$$

we can say that *John* behaviourally complies with that obligation if *John* performs the obligatory action:

$$E_{John} Compensated(Tom) \qquad (30)$$

We can say that *John* intentionally complies if he has the obligation, and his awareness of having the obligation leads him to intend to perform the obligatory action, which leads him to perform it:

$$Bel_{John}(\mathbb{O}_N E_{John} Compensated(Tom)) \triangleright^m Int_{John} E_{John} Compensated(Tom) \wedge$$
$$Int_{John} E_{John} Compensated(Tom) \triangleright^m E_{John} Compensated(Tom)$$
$$(31)$$

6   Compliance with a Normative System

So fare we have been considering compliance with a single obligation established by a normative systems. Now we need to consider compliance with a whole normative system, possibly including thousands of obligations (as any modern legal system).

Definition 10 (Compliance with a Normative Systems). *An agent x complies with a normative system N, iff x complies with all obligations established by N. In other words, x complies with N, iff x performs every action $[\mathcal{A}_x]_i$ which is obligatory according to N:*

$$Complies_x(N) \stackrel{def}{=} [\mathcal{A}_x]_1 \wedge \ldots \wedge [\mathcal{A}_x]_n \qquad (32)$$

where $[\mathcal{A}_x]_1 \wedge \ldots \wedge [\mathcal{A}_x]_n$ is the conjunction of every action or omission $[\mathcal{A}_x]_i$ such that $\mathbb{O}_N[\mathcal{A}_x]_i$, i.e., such that $N \cup T(N) \mathrel{\vdash\!\sim} O[\mathcal{A}_x]_i$.[3]

Complying with the whole of a normative system N (rather than with a single obligation) can be the object of a deliberation, on the basis of which an agent j adopts the corresponding goal, i.e., the goal of j's own compliance, which

---

[3] To avoid infinite conjunction of redundant action propositions, we may add the requirement and for each such $[\mathcal{A}_x]_i$ there must exist an instance of a norm in N, whose conclusion is $[\mathbb{O}\mathcal{A}_x]_i$.

becomes j's intention to comply. Thus a consequentialist agent j, given that for him the utility of complying is higher than the utility of non-complying

$$u_j(Complies_j(N)) > u_j(\neg Complies_j(N)) \qquad (33)$$

would assume that the utility of performing the action of complying (making so that he complies) is higher than the utility of omitting to do so

$$u_x(E_j Complies_j(N)) > u_j(\neg E_j Complies_j(N)) \qquad (34)$$

Consequently, given the principle stated in definition (14) we can conclude that an agent j believing in Proposition (33) will adopt the intention achieve compliance.

$$Int(E_j Complies_j(N)) \qquad (35)$$

However, it seems to me that the representation of the intention to comply in formula (35) above (namely, as an agent's intention to achieve a state of affairs where every obligation of that agent is fulfilled) fails to capture the usual state of mind of of an agent who has decided to comply with a normative system. In fact, an agent usually cannot have a precise mental representation of the state of affairs of full compliance, as specified in definition (10), since the agent ignores the norms in the system, and therefore cannot know what needs to be done to achieve full compliance. For instance, we all know that our country has a legal system, some of us know a few criteria for identifying the norms belonging to that system, but none of us knows all or most norms it contains. How can we intend to realise a state of affair of which we are not aware?

This objection be countered by conditionalising the actions to be performed to achieve compliance. Even if we cannot know what actions are obligatory, we can still intend to performs any action which happens to be obligatory. This is expressed by the following definition.

Definition 11 (Compliance with a Normative Systems (Conditionalised Version)). *An agent x complies with a normative system N, iff x complies with all obligations established by N. In other words, x complies with N, iff whenever an action or omission by x, denoted as $[A_x]_i$, is obligatory according to N, x performs it:*

$$Complies_x(N) \overset{\text{def}}{=} \bigwedge_{i \in [1..n]} (\mathbb{O}_N [A_x]_i \rightarrow [A_x]_i \qquad (36)$$

where $\bigwedge_{i \in [1..n]} (\mathbb{O}_N [A_x]_i \rightarrow [A_x]_i)$ stands for the conjunction of all formulas having the form $\mathbb{O}_N [A_x]_y \rightarrow [A_x]_y$, one per each of x's action $[A_x]_i$ prescribed by one of the norms of N.

Also this representation, however, seems inadequate to me. Firstly, we do not know what antecedents of the conditionals included in the big conjunction will turn out to be true, and thus to what actions we are committing ourselves. Can we as rational agent intend, without qualifications, to bring about full realisation of an open set of demands whose content is unknown to us?

Secondly, even we could have a mental representation of the state of full compliance, we should know that this state of affairs is unlikely to happen: given the high number of obligations arising from the system, and the fact that we is not aware of many of them, we will most likely violate some of them, even though we are doing are best. How can one intend to realise a state of affairs being aware that most likely this state of affairs will not take place?

Thirdly, an agent committed to compliance should maintain its motivation even when the agent has failed to comply with one obligation , and even when when the agent deliberately chooses not to comply with one particular norm. But full compliance is an all or nothing state of affairs, which becomes impossible once one obligation is violated.

7   Policy-Based Intention to Comply

It seems to me that rather than committing itself to achieving full compliance, a reasonable agent could consider adopting a general compliance policy, namely the policy of intending to perform any action which is obligatory according to N. Thus the intention to comply will appear to be a policy-based intention, namely, an intention to act in a certain way under conditions characterised in a general way, so that they may be instantiated in different specific circumstances (on such policy-based intentions, see Bratman, 1987, 87-92 and Bratman, 1989, 451 ff., for a formalisation in defeasible logic see Governatori et al., 2009, for some considerations, see also Sartor, 2005, 31-40). According to this policy, the agent will comply whenever the conditions are met, giving a separate and independent relevance to each opportunity for compliance: the agent may fail to comply in one occasion (when the agent ignores that the conditions are met, or when overriding reasons exist defeating the application of the policy), but still keep a defeasible commitment to the policy and be governed by it in other occasions.

Definition 12 (Policy-Based Intentions). *Let us represent policy-based intentions in the form:*

$$S \overset{i}{\Rightarrow} Int_j \mathcal{A}_j \tag{37}$$

where $\overset{i}{\Rightarrow}$ is a non-truthfunctional connective (similar to $\overset{n}{\Rightarrow}$ for norms), meaning that the state of affairs S (the belief that it holds) triggers agent j's intention to do action $\mathcal{A}_j$.

I assume that also that a modus ponens-like inference applies to $\overset{i}{\Rightarrow}$, so that:

$$\{S, S \overset{i}{\Rightarrow} Int_j \mathcal{A}_j\} \hspace{0.3em} \vdash \hspace{0.3em} Int_j(\mathcal{A}_j) \tag{38}$$

In fact, intentions often take a conditional form, which supports detachment. For instance, Tom, given that today is a working day and that he intends to work today if it is a working day, can conclude with the intention to work today.

$$\{workingDay(today), workingDay(today) \overset{i}{\Rightarrow} Int_{Tom} E_{Tom} Work(today)\} \hspace{0.3em} \vdash$$
$$Int_{Tom} E_{Tom} Work(today)$$

$$\tag{39}$$

Conditional intentions can have an abstract form, which enables multiple instantiations. For instance, agent Tom may have the following policy-based intention to work on any working day x:

$$workingDay(x) \overset{i}{\Rightarrow} Int_{Tom}E_{Tom}Work(x) \qquad (40)$$

A general conditioned intention stands for the set of all of its ground instances, such as:

$$workingDay(Tomorrow) \overset{i}{\Rightarrow} Int_{Tom}E_{Tom}Work(Tomorrow) \qquad (41)$$

so that from such a general intention, given a specific fact matching its antecedent, like

$$workingDay(Tomorrow) \qquad (42)$$

Tom can infer the corresponding instance of the conclusion:

$$Int_{Tom}(E_{Tom}Work(Tomorrow)) \qquad (43)$$

However, Tom does not need to store in his mind all of such ground instances (that he intends to work today if today is a working that, that he intends to work tomorrow if tomorrow is a working day, that he intends to work the day after tomorrow ...). he just needs the policy-based intention expressed in abstract terms, and can use it for specific inferences when needed.

Let us now consider how the commitment to comply with a normative system can be modelled as a policy-based intention.

Definition 13 (Policy-Based Intention to Comply). *An agent j's commitment to comply with normative system N can be understood as the agent's j conditioned intention to do any action $\mathcal{A}_j$ that is obligatory according to N:*

$$\mathbb{O}_N\mathcal{A}_j \overset{i}{\Rightarrow} Int_j\mathcal{A}_j \qquad (44)$$

Assume, for instance that Tom, while being in a place open to the public, is considering the implications of the normative system N_2 of example (2) (which says that places open to the public count as public spaces, and that it is forbidden to smoke in public places). Then Tom can establish that he is forbidden to smoke according to N_2:

$$\mathbb{O}_{N_2}(\neg E_{Tom}Smoke) \qquad (45)$$

Assume also that Tom has adopted the following policy-based intention to comply with N_2:

$$\mathbb{O}_{N_2}\mathcal{A}_{Tom} \overset{i}{\Rightarrow} Int_{Tom}\mathcal{A}_{Tom} \qquad (46)$$

one of whose grounds instances is:

$$\mathbb{O}_{N_2}(\neg E_{Tom}Smoke) \overset{i}{\Rightarrow} Int_{Tom}(\neg E_{Tom}Smoke) \qquad (47)$$

From (45) and (47) Tom can conclude that he intends to abstain from smoking:

$$Int_{Tom}(\neg E_{Tom}Smoke) \qquad (48)$$

As this example shows, the meaning of the policy-based intention to comply consists in its inferential role: it works in the agent's mind as defeasible rule, allowing the derivation of an instance of its conclusion given (the belief in) an instance of its antecedent. Its peculiarity in comparison to other inference policies is that its conclusion is an intention to be implemented, rather than a proposition to be believed. In conclusion, we have found two ways to understand the commitment (intention) to comply with a normative system N by an agent j:

- j's intention to realise the state of affairs where all obligations directed to j are satisfied through its action $(Int_j(E_jComplies_j(N))$
- j's endorsement of the policy according to which j intends to comply with any N-obligation directed to itself $(\mathbb{O}_N\mathcal{A}_j \overset{i}{\Rightarrow} Int_j\mathcal{A}_j)$

It seems to me that there is only a one-way dependency between these two intentions. Adopting the latter policy-based intention is the most obvious way to realise (at least to some extent) the state of affairs of one's compliance. However, the converse does not hold: j may adopt the compliance policy, even when j does not intend to realise full compliance, knowing that it is not possible to achieve it. Moreover, such a policy may be limited by specific exceptions, whose detection would prevent the application of the policy (and would take j further away from full compliance), as I shall argue in section (9).

8 Compliance by Different Kinds of Agents

Compliance is neutral: the choice to comply with a normative system may result from the most different attitudes and goals. It is even doubtful whether in many cases a choice is involved in the adoption of the attitude to comply. When one lives in a certain community one tends to adopt the norms which are endorsed and followed in that community without the need of a specific act of choice. Correspondingly, when we know that our community has a normative system, but we don't know what rules belong to that system, we tend to adopt a general policy to comply with whatever rules will belong to that system, i.e., the policy-based intention above described. This happens in the communities in which we participate without an explicit choice (such as a country, a local community, a family, etc.), but also in those organisations that we enter by choice (a university, a company, a sport club, etc.), where a compliant attitude appears as a natural implication of one's choice to join a certain group or activity, rather than as a separate independent choice. Different explanations can be provided for the unreflected adoption of a determination to comply. For instance, it has been affirmed that humans are naturally endowed with the attitude of "docility", meant as " the propensity to behave in socially approved ways and to refrain from behaving in ways that are disapproved", and attitude that may have an evolutionary explanation since it "enhances human fitness tremendously by allowing children to enjoy a long period of dependence, and to acquire effective skills through learning" (Simon, 1983, 64). So, it seems that humans living within a certain organisation or community would "naturally" desire to be included and approved, and consequently adopt the goal (the intention) to comply with the norms of that organisation or community.

This fact, however, does not exclude that one's intention to comply may be the result of a deliberate choice. Such a choice may provide the motivation for compliance even when one has no desire to be involved in a certain organisation or community. For instance a prisoner in concentration camp may choose to comply with the regulation of the camp, for fear of sanctions linked to non-compliance. He may also criticise those who do not comply (rather than approving of their courage), for fear of retaliation.

In other cases, a conscious deliberation to comply may support an existing insufficient commitment to do so. For instance a rebellious teenager may accept that he should comply with the school regulations (or with the law more generally) when convinced that non-compliance can easily get him into trouble.

Even people already having a certain propensity to comply may engage in a deliberation on whether to comply or not, when critically assessing whether they should or not maintain this attitude.

Different agents may have different ways of approaching the deliberation on whether they should comply with a norm or a normative system. For our purposes it is sufficient to focus on a broad category of agents, *consequentialist choosers*, namely, agents choosing their actions on the basis on an assessment of the consequences of such actions, an assessments determined by the expected utility (differential benefit) the agent expects as a result of the action. Here the notion of "result of an action" is understood is a very broad way, including the fact of adopting the action itself, as well as the further consequences of this fact (for a broad notion of consequentialism, see Pettit, 1997).

I will distinguish two aspects involved in the assessment of a choice by an agent:

- the utility of action $\mathcal{A}_x$ according to agent x, denoted by $u_x\mathcal{A}_x$, i.e., the measure of the net desirability of that choice, according to x'assessment,
- the impact of an action $\mathcal{A}_x$ on the well being of a subject y according to x, denoted by $w_y\mathcal{A}_x$, i.e., the measure of how much $\mathcal{A}_x$ advances or diminishes y's well-being, according to x's assessment.

Let us first characterise the general idea of a consequentialist chooser.

Definition 14 (Consequentialist Chooser). *A consequentialist chooser x will intend to do an action $\mathcal{A}_x$ whenever x believes that the expected utility of doing that action is superior to the utility of not doing it:*

$$Bel_x(u_x(\mathcal{A}_x) > u_x(\overline{\mathcal{A}_x})) \rightarrow Int_x\mathcal{A}_x \qquad (49)$$

Let us now distinguish different kinds of consequentialist choosers:

- Self-centred (egoistic). For a self-centred chooser x, the utility of a choice is equal to the choice's impact on x's own well-being: $u_x(\mathcal{A}_x) = w_x(\mathcal{A}_x)$.
- Altruistic. For an altruistic chooser x the utility of a choice corresponds to its impact on the wellbeing of a set of agents, possibly including also (but not only) x: $u_x(\mathcal{A}_x) = w_{y_1}(\mathcal{A}_x) + \cdots + w_{y_m}(\mathcal{A}_x)$, where $y_1 \ldots y_m$ are the agents x considers relevant to its choice.

– Communitarian. For a communitarian chooser x, the utility of a choice corresponds to its impact on the wellbeing of x's community: $u_x(\mathcal{A}_x) = w_g(\mathcal{A}_x)$, where g is the community x cares about.

– Utilitarian. For a utilitarian chooser x, the utility of a choice corresponds to the sum of its impacts on the wellbeing of each human being $u_x(\mathcal{A}_x) = w_{y_1}(\mathcal{A}_x) + \ldots w_{y_n}(\mathcal{A}_x)$ where $y_1 \ldots y_n$ are all human beings (by "utilitarianism", I mean the idea that the "standard of what is right in conduct, is not the agent's own happiness, but that of all concerned", Mill, 1991, Ch. 2).

Clearly, different kinds of consequentialist choosers will take different actions in the same situation. For instance when an action positively affects x's welfare, but negatively affects relevant others to a larger extent, a self-centred agent will do it, but an altruistic (or utilitarian) agent will not. However all consequentialist choosers act with the purpose of increasing utility, and consequently, they should address the issue of endorsing the general policy-based intention of fulfilling any obligation established by a certain normative system, i.e., the intention to comply as expressed in by formula (44) above, in the following way. Assume that j believes that a higher utility will be obtained by adopting the policy to comply rather that by not having this policy (to express that a policy-based intention is considered as a whole in j's reasoning about it, I enclose it in square brackets):

$$u_j([\mathbb{O}_N\mathcal{A}_j \overset{i}{\Rightarrow} Int_j\mathcal{A}_j]) > u_j(\neg[\mathbb{O}_N\mathcal{A}_j \overset{i}{\Rightarrow} Int_j\mathcal{A}_j]) \tag{50}$$

According to (50), making so that j has (acquires or maintains) the policy-based intention to comply is better than omitting to do that:

$$u_j(E_j[\mathbb{O}_N\mathcal{A}_j \overset{i}{\Rightarrow} Int_j\mathcal{A}_j]) > u_j(\neg E_j[\mathbb{O}_N\mathcal{A}_j \overset{i}{\Rightarrow} Int_j\mathcal{A}_j]) \tag{51}$$

From (51), j can conclude that it intends to acquire (bring it about that it has) the intention-based policy to comply:

$$Int_j E_j[\mathbb{O}_N\mathcal{A}_j \overset{i}{\Rightarrow} Int_j(\mathcal{A}_j]) \tag{52}$$

Executing such an action, i.e., achieving that intention, would consist in adopting the policy-based intention to comply, namely, being ready to form the intention to perform an action $\mathcal{A}_j$ whenever (j believes that) this action is obligatory according to N.

For my purposes I do not need to engage in a discussion of the logic of meta-intention. It is sufficient to assume that a rational agent x, having the intention to perform the action consisting in adopting a (conditioned or unconditioned) intention INT_x will perform such a mental action and acquire INT_x, according to the following schema:

$$Int_x E_x(INT_x) \rightarrow E_x INT_x \tag{53}$$

Given that actions are successful by formula (4) above, performing $E_x INT_x$ entails acquiring INT_x, i.e., in our example, adopting the policy-based intention to comply.

Various refinements and extensions of the consequentialist model of agency are indeed possible: intermediate positions could be distinguished (as when one is moderately altruistic, giving some importance to the well-being of others, but less importance than to one's own well being) or egalitarian-prioritarian elements may be introduced (so that the differential welfare or certain people is more significant than that of others). The bounds of rationality could also be considered, and the ways in which the social environment influences attitudes and choices. Finally, the analysis of compliance could also go beyond consequentialist reasoning, extending to cases where compliance follows from a deontological ethics (for a discussion of deontology and consequentialism, see Baron et al., 1997) or from a religious faith. All these refinements and extensions of the model here proposed are beyond the scope of this contribution, where I will limit my analysis to the simplistic typology of consequentialist reasoners just proposed.

9   Non-compliance

An agent may also choose not to comply or to be indifferent to compliance. We can distinguish different ideas in this regard.

Firstly, the agent may be completely *indifferent* to compliance. In this case, for any obligation $O\mathcal{A}_j$, the fact that the obligation is prescribed by N is no motivation for j to perform. From j's perspective, the N-obligatoriness of an action is no reason to (intend to) do it (I write $A \not\Rightarrow B$ to mean that the conditional $A \Rightarrow B$ does not hold, is not applicable):

$$\mathbb{O}_N\mathcal{A}_j \not\overset{i}{\Rightarrow} Int_j\mathcal{A}_j \tag{54}$$

Secondly, the j may be *diabolic*, as far as N is concerned (in the sense of wanting to violate N's obligations just for the sake of doing it). For such a j, the very fact that an action $\mathcal{A}_x$ is obligatory according to N provides a motivation to violate N. In other terms, j has adopted the policy of doing the contrary of anything obligatory according to N:

$$\mathbb{O}_N\mathcal{A}_j \overset{i}{\Rightarrow} Int_j\overline{\mathcal{A}_j} \tag{55}$$

Thirdly, j's commitment to compliance may be *limited*, since j together with the compliance policy also adopts one or more exception-policies to it, namely, rules stating that the compliance policy does not hold under certain conditions (such rules would be undercutters, in the model of Pollock, 1995, see also Prakken and Sartor, 1997 and Prakken, 2010). Different defeasible compliers may recognise different exceptions.

An *opportunistic* complier j (the bad man, see Holmes, 1897) makes an exception to the compliance policy whenever j comes to believe that by violating an obligation it will get a higher personal advantage (well-being) than complying with it. Thus j would adopt the following reasoning policy, which blocks the defeasible compliance policy of formula (44) above whenever the utility of non-compliance exceeds that of compliance: when the utility of doing $\mathcal{A}_j$ is inferior to the utility of not doing it, then the obligatoriness of $\mathcal{A}_j$ does not provide a (defeasibly sufficient) reason to have the intention to do it.

$$w_j \mathcal{A}_j < w_j \overline{\mathcal{A}_j} \overset{i}{\Rightarrow} (\mathbb{O}_N \mathcal{A}_j \not\overset{i}{\Rightarrow} Int_j \mathcal{A}_j) \tag{56}$$

Note that the opportunistic complier is not uncommitted toward compliance: j still has the defeasible commitment to comply expressed by formula (44) above, but this commitment is overridden by the belief that non-compliance (in a particular case) would get j a better outcome.

Effective sanctions could neutralise in many cases the opportunistic complier's exception, by making it so that that for any action $\mathcal{A}_j$, j's expected utility of non-compliance (once that the risk of sanctions is also taken into account) is inferior to the utility of compliance. This however depends of the expected impact of the sanction on j, namely, on the amount of the punishment and its probability, which should outweigh the advantage that $\overline{\mathcal{A}_j}$ would provide if there were no sanction.

Not all exceptions to the compliance policy are determined by self-interest. For instance, if Ann believes in some versions of natural law, or in some doctrine supporting civil disobedience, she would make an exception to her policy to comply with N, whenever she believes the N requires her to do an action $\mathcal{A}_{Ann}$ which is (unbearably) unjust. Thus Ann would adopt the following policy, according to which when an action of her is unjust, then its obligatoriness according to N is not a defeasibly sufficient reason for intending to do it:

$$Unjust(\mathcal{A}_{Ann}) \overset{i}{\Rightarrow} (\mathbb{O}_N \mathcal{A}_{Ann} \not\overset{i}{\Rightarrow} Int_{Ann} \mathcal{A}_{Ann}) \tag{57}$$

Other kinds of exceptions could be distinguished. For instance an act utilitarian agent would make an exception to the compliance policy whenever it considers that complying causes more harm than good to humanity. Similarly a corruptible agent would make an exception to the compliance policy when by non-complying the agent would get a substantial differential personal advantage (the amount required for leaning toward non-compliance, being inversely proportional to the corruptibility).

Note that according to this construction of compliance, there is no direct clash (no-balancing) between one's conditioned intention to comply and the reason for holding a different behaviour. Rather the agent needs to consider whether such reasons instantiate an undercutter for the agent's intention to comply, i.e, whether they exclude the applicability of the compliance-policy to the situation under scrutiny. Such exception may also be introduced when an agent j is aware of its cognitive limitations.

10 Endorsement of Norms and Commitment to Comply

Research on social norms has recently addressed social processes through which norms are shared in a community, namely, the interlinked processes of the social emergence of norms and of their immergence in the mind of the concerned agents (Andrighetto et al., 2007). Besides considering the spontaneous emergence of shared customary rules, also the psycho-social process involved in compliance with authoritative orders has been studied (Conte and Castelfranchi, 1999). However, I think that a further step is required to adequately capture the reasoning involved in the application of complex norm-systems.

Let us consider for instance a municipal tax law, such as the Italian one (which is a section of the larger Italian legal system). First of all, very few people have precise knowledge of a large set of rules from Italian tax law, and nobody's mind contains all of Italian tax law. It would be difficult to claim that such rules have "immerged" (and are stored) in the minds of Italian citizens since most the latter do not know (and have never known) most of those rules. What citizens share is only the ability to identify somehow the law in force in their country as distinguished from other laws (foreign or ancient laws) and a general commitment (in many case a very qualified one), to comply with this law and possibly some criteria to identify its main contents. Citizens also have some ideas on the implications of this law that are most important to them (e.g., that the law requires them to pay the income tax every year, that VAT has to be paid on purchases, etc.), but are unable to determine such implications with precision (on the distinction between identifying the law and determining its content, see Jori, 2011).

Usually common citizens usually approach tax issues with the help of tax experts, who give them some indications of what obligations follow from tax law under specific real or hypothetical cases, what sanctions may follow from violating such obligations, what line of actions are most advantageous with regard to tax-law effects. On the basis of this fragmentary information, law-abiding people will determine how to comply with tax law. Let us try to analyse the reasoning process involved in applying this kind of normative information (and more generally all complex normative systems, such as advanced legal systems).

Let us assume that Tom has a general commitment to comply the normative systems L (the law), which includes many tax regulation (without knowing what it the precise content of L). In other words, he endorses the policy based intention to perform any action that is obligatory according to L (the law):

$$\mathbb{O}_L \mathcal{A}_{Tom} \overset{i}{\Rightarrow} Int_j \mathcal{A}_{Tom} \tag{58}$$

Tom is now wondering whether he should pay income tax on the capital gains he obtained by selling his house. Being committed to comply with the law, but not knowing what the law requires from him, Tom asks the tax expert Ann for advise. Assume that the Ann remembers that there is a rule in the tax code that establishes the requirement to pay income taxes on capital gains, but vaguely remembers that there are exceptions to it. This prompts Ann to look for exceptions, and she finds indeed one matching houses. This exception says (in a simplified form) that capital gains from the sale of houses purchased more than 5 year before the sale and inhabited by the seller are exempted from income tax.

Assume that Ann's inquiry has let that to conclude that the legal system L contains certain norms:

$$L \supseteq \{SellsHouse(x) \overset{n}{\Rightarrow} OE_x PayIncomeTaxOnSale;$$
$$BoughtMoreThan5YearsBefore(x) \wedge HasInhabitedHouse(x) \overset{n}{\Rightarrow} \quad (59)$$
$$\neg(SellsHouse(x) \overset{n}{\Rightarrow} OE_x PayIncomeTaxOnSale)\}$$

where the second norms in (59) says that under the indicated conditions the first one does not hold (is not applicable).

Ann then asks *Tom* whether at the time of the sale more that 5 years had elapsed from *Tom*'s purchase, and whether he has been living in the house. Assume that *Tom* replies positively to the first question and negatively to the second one. Then *Ann* says: "Dear, *Tom*, unfortunately you are legally bound to pay income tax on your gains". In fact, by combining the law L with these factual circumstances (let us assume these circumstances are the only relevant ones), *Ann* can see that the following inference holds:

$$L \cup \{\neg HasInhabitedHouse(Tom)\} \sim OE_{Tom} PayIncomeTaxOnSale \quad (60)$$

so that she can infer what she tells her client:

$$\mathbb{O}_L E_{Tom} PayIncomeTaxOnSale \quad (61)$$

If *Tom* asks for an explanation, *Ann* would probably answer by saying that whenever one was has not lived in the house one sells, then according to the law one has the obligation to pay income tax:

$$SellsHouse(x) \wedge \neg HasInhabitedSoldHouse(x) \rightarrow \mathbb{O}_L E_{(x)} PayIncomeTaxOnSale \quad (62)$$

Note that formula (62) does not express a norm of L (there is no norm in L which has exactly that content, see formula (59)). More generally (62) is no norm at all, but rather is a general conditional statement about L, namely the statement that in case that the seller has not inhabited the sold house, then L entails that the seller has to pay taxes on capital gains. Similarly, if *Ann* were contacted by *Tom* before making the sale, she would tell him: "Since you have not inhabited the house, you will have to pay income tax on your capital gain".

I think that this example may suffice to show that norms included in large normative systems operate differently from social norms. When we learn social norms we permanently store them in our memory, as the content of appropriate normative beliefs and goals, so that they can directly govern our behaviour. On the contrary, we do not learn and store in our memory most norms included in a large normative systems. We rather possess some ideas about the existence of such a system and the ways to identify its content. When needed, we collect some fragmentary information about the system and combine this information with the relevant facts, both tasks being often delegated to experts. On the basis of this information we can conclude that the system requires us to perform certain actions. By combining such conclusions with our general commitment to comply with the system we adopt intentions to perform such actions.

11   Compliance by Officers

Certain normative systems have officers (typically judges) charged with ensuring compliance, in particular by sanctioning non-compliance.

For each obligation OA_x, let us denote with $Punished(x)$, the situation where x is punished. Let us assume for simplicity's sake that a single compliance officer, a judge named Jud, who is responsible for for ascertaining and repressing all violations of N, and that the punishment is the same for all violations. In other words, let us assume that N contains a norm stating than whenever an obligation OA_x is violated, the obligatory action not having been accomplished, it is Jud's obligation to punish x:

$$OA_x \wedge \overline{A_x} \overset{n}{\Rightarrow} OE_{Jud}Punished(x) \tag{63}$$

Note that for rule in formula (63) above to fire, OA_x must be derivable from N itself in combination with the facts of the case (thus we do not need to substitute OA_x with the metalevel normative proposition $\mathbb{O}A_x$).

This representation is a simplification with regard to complex normative systems, where we have multiple interlocked rules determining who is in charge for each kind of violation, and we need to distinguish officers charged with providing evidence of violations, officers charged with establishing whether a violation has taken place and order a sanction, officers charged with carrying out the sanction. In fact officers in a complex normative system have a shared task which is more complex that simple punishment, a task which may possibly be characterised as the development and maintenance of their normative system. To accomplish this task they need to coordinate, to some extent, their activities consisting in creating, modifying, interpreting and applying the norms in the system (see Shapiro, 2002 who sees this activity as a shared cooperative activity in the sense of Bratman, 1992). For the purposes of this paper, however, a simplistic analysis will suffice.

Thus compliance by Jud (in its role as law enforcer) could be expressed as follows:

$$CompliesWith_{Jud}(N) \overset{\text{def}}{=} \forall(x)(\mathbb{O}_N E_{Jud}Punished(x) \to E_{Jud}Punished(x)) \tag{64}$$

Jud's commitment to a policy-based intention to comply could be expressed as the intention to punish anybody it has the obligation to punish (namely, anybody who violated a norms):

$$\mathbb{O}E_{Jud}Punished(x) \overset{i}{\Rightarrow} Int_{Jud}E_{Jud}Punished(x) \tag{65}$$

Following the reasoning in Sections 8 and 9 above we may consider the various conditions under which different judges, having different concerns, could adopt policy (65): they could adopt it out self-interest (to advance their career, have a good reputation, etc.), altruism, communal interests, moral commitments, and any mixtures of these and other motivations. Moreover, they may subject this policy to various limitations. For instance, a corruptible judge will not apply the policy when there is great advantage to be gained through non-compliance.

12   Spreading Compliance

I will not examine here the social determinants of compliance and the social factors that encourage or discourage compliance, which would require me to address the many issues dealing with the theory of social norms (see, for instance, Conte and Castelfranchi, 2006, Bicchieri, 2011) and their connection to legal systems.

I will just observe that the expected utility of x complying with N often depends on how many other agents will comply with N, as officers or as private individuals. As the number of compliers of a normative systems N increases usually both the individual and social differential benefit of compliance (as compared to non-compliance) increases: in a context of compliance, legal and social sanctions for non-compliance are more likely to take place, compliance is more likely to have a socially beneficial effect, compliance can be viewed as an exercise in reciprocity. This explains why in a context of increased (decreased) compliance, individuals are usually more (less) motivated to comply: so both compliance and non-compliance tend to spread in the community. For most people there is a threshold of compliance-frequency that makes the utility of compliance positive, so that they would choose to comply when the threshold is overcome. However, this threshold may be different for different people (both officers or common fellows) who may be differently motivated (by the individual, social, or communal benefit of compliance).

Thus, compliance by agents who are sufficiently motivated only where there is a higher compliance frequency may depend on whether there is a sufficient number of other agents sufficiently motivated at lower compliance levels, who can bootstrap the process.

Clearly a more complex picture could be developed though a more accurate and diversified representation of motivations for compliance (see for instance Bénabou and Tirole, 2006), which may indeed lead different agents to different choices. For instance one may comply with norms that others do not comply with, in order to better advertise one's commitment to the common good, or to get a confirmation of one's morality; a Kantian agent should be unmoved by the non-compliance by others to a norms the agent approves of; a "myopic" agent would only care about the compliance by the nearest neighbours, etc.

I cannot here address all the many issues concerning the spreading of normative attitudes (see for instance Andrighetto et al., 2007). One general consideration, however, is that people will have normative expectations about the compliance by others, and this expectation will be strengthened insofar as other people as a matter of fact do comply. Here not only one's belief in the value of having a certain normative system, but also reciprocity is at issue, as well as the fact that people will make their own choices (and take risks) on the basis of the factual expectation that others will comply.

13 The Morality of Compliance

When a normative system N is generally complied with and enforced, there will be usually a general attitude of viewing compliance as morally obligatory. This may indeed support the adoption of the intention to comply.

Let us assume that an agent's morality M (the set of moral norms the agent endorses) contains a norm stating the obligatoriness of whatever is obligatory (for any agent x) relatively to a certain normative system N:

$$(\mathbb{O}_N \mathcal{A}_x \overset{n}{\Rightarrow} O \mathcal{A}_x) \in M \tag{66}$$

If an agent j believes in proposition (66), and that $\mathcal{A}_j$ is really obligatory according to N (i.e., that $\mathbb{O}_N \mathcal{A}_j$), j will conclude that the obligation to do $\mathcal{A}_j$ is entailed by morality:

$$M \cup T(M) \mathrel{\vdash\mkern-9mu\sim} O \mathcal{A}_x \tag{67}$$

Thus, j will view action $\mathcal{A}_j$ as morally obligatory (according to definition (5), i.e., j will believe that

$$\mathbb{O}_M \mathcal{A}_j \tag{68}$$

Rule (66), when applied to the norms governing a political organisation (typically a state) expresses the idea of the political obligation, namely, the moral obligation to obey the law.

The obligation to comply may be qualified by exceptions (e.g., one may argue that it is not morally obligatory to comply with norms enjoining a serious violation of human rights, or which are blatantly unjust or absurd) especially when non-compliance is done in public to convey a political message urging resistance or change, so that it may qualify as civil disobedience.

The idea that that there is a moral obligation to obey a normative system N can contribute to compliance with N, as long as the concerned agent j is committed to do what is required by morality (as identified by j itself), i.e., as long as j endorses the following policy:

$$\mathbb{O}_M \mathcal{A}_j \overset{i}{\Rightarrow} Int_x \mathcal{A}_j \tag{69}$$

Thus j, believing that it has the obligation to do action $\mathcal{A}_j$ according to N (i.e, $\mathbb{O}_N \mathcal{A}_j$) can use moral rule (66) to conclude that it has a moral obligation to do $\mathcal{A}_j$ (i.e., $\mathbb{O}_M \mathcal{A}_j$) and consequently use policy (69) to adopt the intention of doing $\mathcal{A}_j$. Those who endorse rule (66) will also tend to extend their moral condemnation to the violators of norms in N.

In conclusion, moral beliefs may ground or reinforce the endorsement of a policy to comply, but this is not always the case, since the adoption of such a policy may also follows from self-interest or other motivations, as shown above.

14 Compliance and Social Power

The model of compliance here proposed can be related to the theory of power and influence proposed in Castelfranchi (2003). The basic idea I will use is that an agent j influences another agent k when j makes it so that k adopts j's goals, and that influence is a most important mechanism for social power.

Let us assume a state of generalised compliance, so that all (or most) addressees of normative system N have adopted a policy-based intention to comply with N, according to the model indicated in formula (44) above. I will argue that under these conditions a normative systems can be an efficient machine not only for limiting, but also for producing influence and power.

Obviously, normative systems can limit social influence. For instance, assume that John is physically stronger than Tom. If there were no legal system prohibiting the use of violence, John could influence Tom and induce him to (intend to) accomplish what John likes (working for John, paying John for protection, etc.), by threatening to use violence against Tom. However, this is no longer possible (or at least more difficult) when there is an effective legal system N which prohibits using violence against others. If John himself is rigorously committed to the policy to comply with N, then John will adopt the intention to abstain from prohibited actions, and therefore also from violence. In case John is not committed to compliance (or is only defeasibly committed to it, with his self-interest providing for an exception), the compliance of others (and in particular of the enforcement officers) will make it so that the criminal behaviour is prevented of at least made less attractive by the prospect of punishment. This should prevent the threat or make it not credible. Therefore John will not use the threat, or at least Tom will not be influenceable through it.

Let us now examine how normative systems, rather than limiting social influence, can extend it. We need to consider that what obligations are generated by N depends on two factors: the norms in N and the true relevant factual circumstances $T(N)$. This means that N can work as an input-output machine. The input consists in changes in $T(N)$ (the creation of new relevant facts), and the output consists changes in the obligations entailed by N. The input can produce the output in two ways: (a) by providing (or removing) facts that produce obligations according to the norms in N, or (b) by changing the norms in N, these changes having an impact on the obligations derivable from N. In this section I will consider the the first way of changing N's obligations, and in the following I will address the latter.

For instance a normative system can make orders binding (for instance, the orders of a military commander to a soldier, or of an employer or manager to a worker, by making obligatory for the addressee of an order to comply with it. This idea could also be expressed by using the notion of institutional (norm-based) power (Jones and Sergot, 1996, Gelati et al., 2002a, Sartor, 2006,Hage, 2011b, Hage, 2011a, Tummolini and Castelfranchi, 2006), but here a simpler representation will be provided, without expressly formalising the concept of institutional power. Assume the system N contains a rule according to which *Ann* has the obligation to do whatever action $\mathcal{A}_{Ann}$ is ordered by her manager *Tom* (for simplicity I do not consider the limitation of such an obligation in modern legal systems, where the order must pertain to the execution of the work, and respect the worker's rights and dignity):

$$E_{Tom}Order(\mathcal{A}_{Ann}) \overset{n}{\Rightarrow} \mathbb{O}\mathcal{A}_{Ann} \tag{70}$$

Assume that Tom does indeed order Ann to do something (for instance, to draft the minutes of a meeting):

$$E_{Tom}Order(E_{Ann}DraftMinutes) \tag{71}$$

so that this action-proposition becomes one the true relevant facts

$$(E_{Tom}Order(E_{Ann}DraftMinutes)) \in T(N)) \tag{72}$$

Given that N contains rule (70) and $T(N)$ contains fact (71) the following holds:

$$N \cup T(N) \mathrel{\vdash\mkern-9mu\sim} OE_{Ann}DraftMinutes \tag{73}$$

so that we can say that according to N it is indeed obligatory that Ann drafts the minutes

$$\mathbb{O}_N E_{Ann}DraftMinutes \tag{74}$$

Assume that Ann has adopted the general compliance policy of formula (44) above relatively to normative system N, so that she intends to do whatever action of her is obligatory according to N:

$$\mathbb{O}_N\mathcal{A}_{Ann} \overset{i}{\Rightarrow} Int_{Ann}\mathcal{A}_{Ann} \tag{75}$$

Policy-based intention (75) and normative proposition (74) entail that Ann will adopt the intention to draft the minutes

$$Int_{Ann}E_{Ann}DraftMinutes \tag{76}$$

Thus, given that Ann is committed to comply with N, Tom can influence her. By ordering any action, he modifies $T(N)$ and makes it so that $N \cup T(N)$ entails the obligatoriness of that action, which makes it so that Ann adopts the intention of doing that action. Note that this power by Tom does not depend on his personal qualities (Ann may dislike Tom or believe that he an incapable idiot), it only depends on the content of the normative system, on the relevant facts, and on Ann's commitment to policy-based intention (75).

A normative system N can also provide individuals with the possibility of binding themselves, i.e., of undertaking obligations according to N, or more generally of creating any normative positions concerning themselves. For this purpose it is sufficient that N contains the following rule:

$$[E_x Promise(\mathcal{A}_x) \overset{n}{\Rightarrow} O\mathcal{A}_x] \tag{77}$$

meaning that whenever an x promises to do $\mathcal{A}$ then x has the obligation to do $\mathcal{A}$.

A complied with (and protected through sanctions or other means of social pressure) normative system containing the rule in (77) enables agents to create credible commitment for themselves (given the costs of non-compliance), on the basis of which others can act (e.g., I promise to give 1,000 euros to the person who will bring back to me my lost dog), or can be induced to take similar commitments, as in contracts (on a more general approach to contract, which views them as means to create not just obligations but any kind of normative positions see Gelati et al., 2002b, Sartor, 2006, Hage, 2011b). For example, assume the following: 1) system N contains the rule in (77), 2) I promised that I will give 1000 euros to the best law student of this year; 3) Ann is this year's best law student. It follows that according to N, I have the obligation to give 1000 euros to Ann.

15   The Machine of the Law

Let us now consider how an agent (a legislator) can have the ability to introduce new norms in N. For this purpose, we need to assume that N is a dynamic normative system (Kelsen, 1967), including meta-rules determining what new noms will belong to N. For simplicity I shall leave temporal aspects implicit even though they are essential in an adequate account of normative dynamics (see Governatori et al., 2007). So, let us assume that N includes a meta-norm saying that whatever norm ϕ is issued by the legislator Leg is included in N (ϕ is a variable ranging over norms):

$$[E_{Leg}Issued(\phi) \overset{n}{\Rightarrow} \phi \in N] \tag{78}$$

I cannot here develop the analysis of the dynamics of normative systems, which would require a discussion on how to model defeasibility and time (see for instance Governatori et al., 2006). Thus, for our purposes it is sufficient to characterise N as the minimal set satisfying the following equality:

$$N = \{[E_{Leg}Issued(\phi) \overset{n}{\Rightarrow} \phi \in N]\} \cup \{\psi : N \cup T(N) \mathrel{|\!\sim} \psi \in N\} \tag{79}$$

According to equation (79), N is defined as containing the meta-norm of (78) (which would work as the "constitution" in a logical sense of N, following Kelsen, 1967) plus every other norm that is qualified as being in N according to N itself. i.e., any norm ψ (ψ is a variable ranging over norms such that N entails the proposition that ψ is contained in N (for a presentation of this idea, see Sartor, 2009, on modelling legal systems through metanorms, se also Yoshino, 1995, Yoshino, 1997 and Hernandez Marín and Sartor, 1999).

Alternatively we could assume that that the content of equality (79) is rephrased by a fundamental norm, which is not does not belong to N, but constitutes the ultimate ground for membership to N (as a Kelsenian Grundnorm, or as a Hartian rule of recognition, see Hart, 1994).

$$([E_{Leg}Issued(\phi) \overset{n}{\Rightarrow} \phi \in N] \in N) \wedge ((N \cup T(N) \mathrel{|\!\sim} \psi \in N) \overset{n}{\Rightarrow} \psi \in N) \tag{80}$$

The two-pronged norm in (80), let us call it *Fundamental*, states the norm empowering the legislator is in N, and that all norms are in N, whose membership to N is entailed by N itself.[4] Then N can be defined as the minimal set of the norms whose legality is entailed by $Fundamental$, together with the relevant facts.

$$N = \{\phi : (T(N) \cup Fundamental) \mathrel{\vdash\mkern-5mu\sim} \phi \in N\} \tag{81}$$

Given this background (i.e., either equation (79) or (81)), let us assume that legislator accomplishes the action of issuing a new norm, for instance, a norm prohibiting any agent x to smoke:

$$E_{Leg}Issued(O\neg E_x Smoke) \tag{82}$$

The accomplishment of the action described in this formula is a fact, which is added to the true factual circumstance $T(N)$. With this addition, the following holds according to the rule of formula (78) above (when useful for clarity, I bracket norms included in meta-linguistic expression):

$$N \cup T(N) \mathrel{\vdash\mkern-5mu\sim} [O\neg E_x Smoke] \in N) \tag{83}$$

Consequently N contains norm $O\neg E_x Smoke$, according to formula (79):

$$[O\neg E_x Smoke] \in N \tag{84}$$

Since it now holds that

$$N \cup T(N) \mathrel{\vdash\mkern-5mu\sim} O\neg E_{Tom} Smoke \tag{85}$$

so that we can say that now smoking is forbidden to Tom according to N:

$$\mathbb{O}_N \neg E_{Tom} Smoke \tag{86}$$

The legislator can use the power provided by formula (78) above to put a judge in charge of punishing violators. To achieve this result, the legislator just has to perform the action of issuing a norm to that effect, namely a norm saying that the judge Jud should punish any agent who violates a norm in N, i.e., any agent who does the opposite of what is obligatory for that agent:

$$E_{Leg}Issued(O\mathcal{A}_x \wedge \overline{\mathcal{A}_x} \overset{n}{\Rightarrow} OE_{Jud}Punished(x)) \tag{87}$$

As a consequence of this legislative action, N now contains the issued norm

$$[O\mathcal{A}_x \wedge \overline{\mathcal{A}_x} \overset{n}{\Rightarrow} OE_{Jud}Punished(x)] \in N \tag{88}$$

with means that Jud has, according to N, the obligation to punish any violator.

---

[4] The rule in (80) can also be rephrased as having a single conclusion (using variables in a very liberal way):

$$((\phi = [E_{Leg}Issued(\phi) \overset{n}{\Rightarrow} \phi \in N]) \vee (N \cup T(N) \mathrel{\vdash\mkern-5mu\sim} \phi \in N)) \overset{n}{\Rightarrow} \phi \in N$$

Assume now that both *Ann* and *Jud* have the policy-based intention to comply with N, and that *Ann* views non-compliance as immoral. Then, on the basis of the statement of the legislator, *Ann* will adopt the intention not to smoke, *Jud* will adopt the intention to punish smokers in public places, and *Ann* would believe that anyone who smokes in a public place behaves immorally.

The legislator can also confer to another agent, the administrator *Admin*, the ability to insert new norms in N (delegated legislation) by enacting such norms (while respecting certain legal constraints on *Admin*'s legislative action):

$$[E_{Leg}Issued(E_{Admin}Issued(\phi) \land E_{Admin}RespectConstraints(\phi) \overset{n}{\Rightarrow} \phi \in N]$$
(89)

As a consequence of the action described in formula (89) and the characterisation of N in (79), the norm empowering *Admin* is now contained in N:

$$[E_{Admin}Issued(\phi) \land E_{Admin}RespectConstraints(\phi) \overset{n}{\Rightarrow} \phi \in N] \in N \qquad (90)$$

Consequently whatever new norm ϕ is issued by *Admin*, respecting the relative constraints (concerning the content of ϕ or the procedure for its creation), that norm will be inputted in N. In this way, the legislator transfers to *Admin* the legislator's ability to influence people's behaviour, by exploiting their commitment to compliance.

Not only the generalised commitment to comply with N provides the legislator (and its delegatees) with the possibility to influence the behaviour of compliers and judges. It also provides those who are able to influence the legislator with the ability to influence the behaviour of all others. Assume for instance that *Tom* is the leader of the party having the majority in the legislative assembly. Then *Tom* can make it so that the legislator adopts the intention to introduce (or repeal) a norm $B \overset{n}{\Rightarrow} A$, to make it so that the population intends to do (and does) action A under circumstances B.

A normative system supported by a generally endorsed policy-based intention to comply can thus work as an input-output machine, empowering those who can control its input: by providing appropriate normative and factual inputs, they can obtain corresponding intentions and actions and so implement their aims. As Karl Olivecrona put it "[t]he purpose of the lawgivers is to influence the actions of men, but this can only be done through influencing their minds" (Olivecrona, 1971, 21-2, Spaak, 2009). Thus legislators (and those able to influence them) can use the "machinery of the law" for reaching their social, political (and sometimes personal) purposes (see Pattaro, 2009, Pattaro, 2005). Normative systems, in a way, precede certain social powers, and provide for their foundation. The extent of norm-systems based powers may indeed be very large, which explains why developed legal systems contain constitutional limitations and controls over the exercise of such powers (such as democratic procedures for electing the legislative body, judicial review over legislation and administration, more generally, an institutional system of "checks and balances").

16   Conclusion

I have first considered how obligations can be relative to a particular normative systems, and I have provided a meta-logical representation of this idea. Then I have analysed the intention to comply with a normative system, affirming that that the commitment to comply must be understood as a policy-based intention. I have then considered why consequential choosers may come to this determination, as simple addressees or enforcement officers. Finally I have developed some considerations on how compliance can spread and how it can both restrain and provide power.

The study of compliance with normative systems involves various aspects I could not address here. First of all there is the issue of interpretation, i.e., of determining the content of the normative system to be complied with, on the basis of the available materials (texts, cases, practices, values, etc.), a problem that legal theorists have been discussing for centuries, and on whose epistemological-methodological nature the debate is still on-going. Other important issues concern modelling contrary to duty obligations (and other technical aspects of deontic logic), taking into account cooperation between the involved agents and dependencies and trust relationships between them, addressing negotiation and argumentation regarding how to comply and the consequences of violations, considering how a shared awareness of each one's intention to comply and a shared belief in a duty to comply can contribute to compliance.

Finally, the model here presented provides a minimal understanding of the internal point of view towards a normative system (i.e., the point of view of an agent that has chosen to use that system as a guide to its own behaviour). The analysis of such a point of view can be developed by adding further requirements, which may or may not apply with regard to particular normative systems or addressees of them: a social or conventional dimension (expected compliance by others contributes to motivate one's compliance), a shared dimension (there is a common awareness of each one's intentions to comply, or a common intention to comply), a cooperative dimension (the intention to comply concerns participation in a common project), a hierarchical-authoritative dimension (there an individual or collective agent who issues and implements the norms), a believed moral dimension (compliance appears to the concerned agent as the content of a moral obligation), a claimed moral dimension (those producing and enforcing the system claim that there is a moral obligation to comply with it), a moral dimension tout court (there is a moral obligation to apply the system or comply with it), etc.

I think however that such aspects, are complementary but independent of the model developed here, which only assumes that the addressees of a normative system adopt a policy-based intention to comply with it, regardless of the reasons supporting this intention and the ways in which the system's content is identified.

References

Alchourrón, C.E.: Logic of norms and logic of normative propositions. Logique et Analyse 12, 242–268 (1969)

Alchourrón, C.E., Bulygin, E.: Normative Systems. Springer, Vienna (1971)

Andrighetto, G., Campenni, M., Conte, R., Paolucci, M.: On the immergence of norms: A normative agent architecture. In: Proceedings of AAAI Symposium, Social and Organizational Aspects of Intelligence (2007)

Baron, M., Pettit, P., Slote, M.: Three Methods of Ethics: A Debate. Blackwell, London (1997)

Bénabou, R., Tirole, J.: Incentives and prosocial behavior. American Economic Review 96, 1652–1678 (2006)

Bicchieri, C.: Social norms. In: Stanford Encyclopedia of Philosophy. Stanford University (2011)

Boella, G., van der Torre, L.: A Logical Architecture of a Normative System. In: Goble, L., Meyer, J.-J.C. (eds.) DEON 2006. LNCS (LNAI), vol. 4048, pp. 24–35. Springer, Heidelberg (2006)

Bratman, M.E.: Intentions, Plans and Practical Reasoning. Harvard University Press, Cambridge (1987)

Bratman, M.E.: Intention and personal policies. Philosophyical Perspectives 3, 443–469 (1989)

Bratman, M.E.: Shared cooperative activity. Philosophical Review 101, 327–341 (1992)

Castelfranchi, C.: The micro-macro constitution of power. ProtoSociology 18-19, 208–265 (2003)

Castelfranchi, C., Paglieri, F.: The role of beliefs in goal dynamics: Prolegomena to a constructive theory of intention. Synthese 155, 237–263 (2007)

Conte, R., Castelfranchi, C.: Cognitive and Social Action. University College of London Press, London (1995)

Conte, R., Castelfranchi, C.: From conventions to prescriptions. towards a unified theory of norms. Artificial intelligence and Law 7, 323–340 (1999)

Conte, R., Castelfranchi, C.: The mental path of norms. Ratio Juris 19, 501–517 (2006)

Føllesdal, D., Hilpinen, R.: Deontic logic: An introduction. In: Hilpinen, R. (ed.) Deontic Logic: Introductory and Systematic Reading. Reidel, Dordrecht (1971)

Gelati, J., Governatori, G., Rotolo, A., Sartor, G.: Actions, institutions, powers: Preliminary notes. In: Lindemann, G., Moldt, D., Paolucci, M., Yu, B. (eds.) International Workshop on Regulated Agent-Based Social Systems: Theories and Applications (RASTA 2002), pp. 131–147. Fachbereich Informatik, Universität Hamburg, Hamburg (2002a)

Gelati, J., Governatori, G., Rotolo, A., Sartor, G.: Declarative power, representation, and mandate: A formal analysis. In: Proceedings of the Fifteenth Annual Conference on Legal Knowledge and Information Systems (JURIX), pp. 41–52. IOS, Amsterdam (2002b)

Governatori, G., Padmanabhan, V., Rotolo, A., Sattar, A.: A defeasible logic for modelling policy-based intentions and motivational attitudes. Logic Journal of IGPL 17, 36–69 (2009)

Governatori, G., Palmirani, M., Rotolo, A., Riveret, R., Sartor, G.: Norm modifications in defeasible logic. In: Proceedings of Jurix 2006, pp. 13–22. IOS, Amsterdam (2006)

Governatori, G., Rotolo, A., Riveret, R., Palmirani, M., Sartor, G.: Variants of temporal defeasible logics for modelling norm modifications. In: Proceedings of Eleventh International Conference on Artificial Intelligence and Law, pp. 155–159. ACM, New York (2007)

Grossi, D., Meyer, J.-J.C., Dignum, F.: The many faces of counts-as: A formal analysis of constitutive rules. Journal of Applied Logic 6, 192–217 (2008)

Hage, J.C.: A model of juridical acts: Part 1: The world of law. Artificial Intelligence and Law 19, 23–48 (2011a)

Hage, J.C.: A model of juridical acts: Part 2: The operation of juridical acts. Artificial Intelligence and Law 19, 49–73 (2011b)

Hart, H.L.A.: The Concept of Law, 2nd edn. Oxford University Press, Oxford (1994)

Hernandez Marín, R., Sartor, G.: Time and norms: A formalisation in the event-calculus. In: Proceedings of the Seventh International Conference on Artificial Intelligence and Law (ICAIL), pp. 90–100. ACM, New York (1999)

Holmes, O.W.: The path of the law. Harvard Law Review 10, 457–478 (1897)

Horty, J.F.: Agency and Deontic Logic. Oxford University Press (2001)

Jones, A.J., Sergot, M.J.: A formal characterisation of institutionalised power. Journal of the IGPL 4, 429–445 (1996)

Jori, M.: Del diritto inesistente. ETS, Pisa (2011)

Kelsen, H.: The Pure Theory of Law. University of California Press, Berkeley (1967)

Mill, J.S.: Utilitarianism. In: Gray, J. (ed.) On Liberty and Other Essays, pp. 131–201. Oxford University Press, Oxford (1991); (1st edn. 1861)

Olivecrona, K.: Law as Fact, 2nd edn. Stevens, London (1971)

Pattaro, E.: The Law and the Right, a Reappraisal of the Reality that Ought to be. Treatise of legal Philosophy and General Jurisprudence, vol. 1. Springer, Berlin (2005)

Pattaro, E.: From hägerstom to ross and hart. Ratio Juris 22, 532–548 (2009)

Pettit, P.: The consequentialist perspective. In: Three Methods of Ethics: A Debate, pp. 92–174. Blackwell, London (1997)

Pollock, J.L.: Cognitive Carpentry: A Blueprint for How to Build a Person. MIT, New York (1995)

Pörn, I.: Action Theory and Social Science: Some Formal Models. Reidel, Dordrecht (1977)

Prakken, H.: An abstract framework for argumentation with structured arguments. Argument and Computation 1, 93–124 (2010)

Prakken, H., Sartor, G.: Argument-based extended logic programming with defeasible priorities. Journal of Applied Non-classical Logics 7, 25–75 (1997)

Prakken, H., Sartor, G.: The three faces of defeasibility in the law. Ratio Juris 17, 118–139 (2003)

Sartor, G.: Legal Reasoning: A Cognitive Approach to the Law. Treatise on Legal Philosophy and General Jurisprudence, vol. 5. Springer, Berlin (2005)

Sartor, G.: Fundamental legal concepts: A formal and teleological characterisation. Artificial Intelligence and Law 21, 101–142 (2006)

Sartor, G.: Legality policies and theories of legality: From Bananas to Radbruch's formula. Ratio Juris 22, 218–243 (2009)

Sartor, G.: Defeasibility in legal reasoning. In: Ferrer, J. (ed.) Essays in Legal Defeasibility. Oxford University Press, Oxford (2011) (forthcoming)

Searle, J.R.: The Construction of Social Reality. Free., New York (1995)

Sergot, M.J.: A computational theory of normative positions. ACM Transactions on Computational Logic 2, 581–662 (2001)

Shapiro, S.J.: Law, plans and practical reasoning. Legal Theory 8, 387–441 (2002)

Siena, A., Mylopoulos, J., Perini, A., Susi, A.: Designing Law-Compliant Software Requirements. In: Laender, A.H.F., Castano, S., Dayal, U., Casati, F., de Oliveira, J.P.M. (eds.) ER 2009. LNCS, vol. 5829, pp. 472–486. Springer, Heidelberg (2009)

Simon, H.A.: Reason in Human Affairs. Stanford University Press, Stanford (1983)

Spaak, T.: Naturalism in scandinavian and american realism: Similarities and differences. In: Dahlberg (ed.) De Lege, Uppsala-Minnesota Colloquium: Law, Culture and Values, pp. 33–83. Iustus, Uppsala (2009)

Tummolini, L., Castelfranchi, C.: The cognitive and behavioral mediation of institutions: Towards an account of institutional actions. Cognitive Systems Research 7, 307–332 (2006)

Yoshino, H.: The systematization of legal metainference. In: Proceedings of the Fifth International Conference of Artificial Intelligence and Law (ICAIL). ACM, New York (1995)

Yoshino, H.: On the logical foundations of compound predicate formulae for legal knowledge representation. Artificial Intelligence and Law 5, 77–96 (1997)

Coherence-Based Account of the Doctrine of Consistent Interpretation

Michał Araszkiewicz

Jagiellonian University, Faculty of Law and Administration, Department of Legal Theory,
Bracka 12, 31-005 Kraków, Poland
michal.araszkiewicz@uj.edu.pl

Abstract. The aim of this paper is to provide a constraint satisfaction account of the doctrine of consistent interpretation developed by the European Court of Justice (now the Court of Justice of the EU) to protect effective and harmonious realization of the Communities' aims. The doctrine can be naturally seen as pursuit for establishing coherence in initially incoherent set of propositions. I represent the doctrine in the framework of coherence-based model of legal argumentation (CMLA). An attempt to represent *Marleasing* case in this framework is discussed.

Keywords: consistent interpretation, coherence, constraint satisfaction, European Court of Justice, legal argumentation.

1 Introduction

The relation between EU law on the one hand and Member State law on the other hand is one of the most discussed contexts regarding complexity of contemporary legal systems. In this paper I focus on a fragmentary, though very important, aspect of this relation, namely, on the doctrine of consistent interpretation as developed in a series of decisions of the European Court of Justice (now the Court of Justice of the EU, hereafter: the ECJ). It is natural to state that EU-friendly interpretation of Member State law by national courts is an instantiation of solving of a coherence problem. This is particularly visible when a domestic judge is faced with the following choice: either to declare the Member State norm inconsistent with the EU law and therefore inapplicable (which can create incoherencies in national legal system) or to find a reconciling interpretation of the norm in question (which can violate accepted canons of legal interpretation). I attempt to represent the process of legal argumentation relevant for EU-friendly interpretation within the framework of Coherence Model of Legal Argumentation (CMLA), outlined in [1]. However, the version presented below is much more developed. CMLA offers a possibility to highlight peculiarities of the doctrine of consistent interpretation. The existence of connectionist algorithms makes it possible to implement the framework presented here in working computer programs based on ECHO architecture [2, with many references to earlier work]. The order of investigations is as follows. In Section 2 I present the most important features of the doctrine of consistent interpretation as

M. Palmirani et al. (Eds.): AICOL Workshops 2011, LNAI 7639, pp. 33–47, 2012.

stated by the ECJ in its leading decision. I also attempt to emphasize the most problematic features of this doctrine discussed in legal literature. In Section 3 the most basic characteristics of CMLA are presented. In Section 4 I outline a CMLA-based account of the process of consistent interpretation. In particular, I present a CMLA-based representation of ECJ's reasoning in a well-known *Marleasing* case (Judgment of the Court of 13 November 1990. –C-106/89). On the basis of this example, in the last Section I try to justify the view that CMLA makes it possible to account for all the important peculiarities of the doctrine of consistent interpretation; *inter alia*, it is possible to show the degree of legal uncertainty as regards the application of this doctrine. Our analysis enables us to support the view, expressed by many scholars, that it would be better to accept the possibility of direct application of directives in also horizontal contexts instead of resorting to the doctrine of consistent interpretation.

2 The Doctrine of Consistent Interpretation

The doctrine of consistent interpretation is one of the so-called principles governing the relation between the legal order of the EU and legal systems of Member States. The other principles belonging to this group are: the principle of supremacy of EU law and the principle of direct effect of EU law. It is sufficient to define them on the basis of the representative textbook [3]. The principle of supremacy of EU law obligates the national judge to set aside any provision of national law inconsistent with the EU law. According to the principle of direct effect it is possible to invoke Community provisions in the proceedings before the national courts and these provisions may provide bases for decisions of the national courts, given the provisions satisfy certain conditions, which will not be discussed here [4]. These principles create a context of application of the doctrine of consistent interpretation, which reveals its peculiarities especially when application of EU directives is concerned. Let us recall that the directives are generally binding on Member States and not on their citizens and that the directives have to be implemented (transposed) into the national legal system [5]. In consequence, a Member State may fail to properly transpose the directive. As regards another distinction, the dispute before the national court may have vertical (individual-state) or horizontal (individual v individual) character. It is established in the jurisprudence of the ECJ that there can be no direct effect of the directives in the latter situation [3, 5]. Here's where the doctrine of consistent interpretation plays its important role of bringing about the Communities' aims. Although it is not the sole context of application of this doctrine, it is considered particularly important and gave rise to the most important ECJ decisions related to it. In consequence, we will focus mainly on the interesting context of directive-friendly interpretation.

2.1 The Nature of Consistent Interpretation in ECJ Jurisprudence

The doctrine of consistent interpretation is not a completely novel legal institution. It is often compared to the traditional canons of interpretation obligating the courts to interpret the national law in accordance with the Constitution and with the international law [5, 6].

However, the ECJ doctrine discussed here has its peculiarities, resulting mainly from characteristic features of the directives (the fact that the Member States and not private individuals are primarily bounded by them, and the abovementioned prohibition of direct effect of directives in horizontal disputes).

The first very influential formulation of the doctrine took place in the famous *Von Colson* judgment (Judgment of the Court of 10 April 1984. – 14/83), but in the later *Marleasing* decision, the ECJ formulated what became the standard reference for the doctrine of consistent interpretation:

[Marleasing Formula] *"(…) in applying national law, whether the provisions in question were adopted before or after the directive, the national court called upon to interpret it is required to do so, as far as possible, in the light of the wording and the purpose of the directive in order to achieve the result pursued by it (…)"*.

Although the doctrine was signalized in decisions earlier than *Von Colson* and reiterated many times after *Marleasing*, the two cases mentioned here remain the basis for academic and judicial understanding of the principle of consistent interpretation [3, 5, 6]. It is worth noting that both of the judgments were enacted in the context of troubles concerning implementation of directives.

2.2   Some Problems Concerning the Doctrine of Consistent Interpretation

The ECJ decisions on the principle in question, although formulated concisely, give rise to multifarious problems in legal research and in judicial practice. The first (and often overlooked) problem pertains to the very concept of 'interpretation' used here. Lawyers often refer to as 'interpretation' to any type of argumentative (heuristic or justificatory) move in legal reasoning. The question is, what is the structure of argumentative schemes which can be possibly used in the context of the principle discussed here. Second, as it is often emphasized, the obligation to interpret national law in accordance with the EU law is not absolute [3, 5, 6]. There are limits to this obligation, to mention the established rules of construction, the role of the judiciary in the legal system and general principles of (constitutional and international) law [5]. Third, it is often indicated that consistent interpretation may increase the level of legal uncertainty ([6] with many references to literature corroborating this point and in particular to the opinions of Advocate General Jacobs). These problems are accounted for in the following analysis.

3   Legal Argumentation as Constraint Satisfaction

3.1   Theory of Coherence as Constraint Satisfaction

CMLA (outlined in [1]) is based on Paul Thagard's theory of coherence as constraint satisfaction [2]. Let us recall the most basic features of the theory. It concentrates on the issue of solving the so-called coherence problems. A coherence problem may be defined as a process of finding the most acceptable subset among the initial set of data, which is inacceptable as a whole. A solution to a coherence problem can be found according to the following procedure. Let E be a finite set of elements $(e_1,….e_n)$. The (in)coherence relations between pairs of elements are referred to as positive constraints (C+) and negative constraints (C-). We need to divide the set E into two disjoint subsets, a subset of accepted elements (A) and a subset of rejected elements (R) so that the following two conditions are maximized (but not necessarily strictly satisfied):

1. If $<e_i, e_j> \in$ C+ then $e_i \in$ A if and only if $e_j \in$ A.
2. If $<e_i, e_j> \in$ C- then $e_i \in$ A if and only if $e_j \in$ R.

Each constraint is assigned with a number w – the weight of this constraint. The sum of weights of all satisfied constraints is symbolized by W and is equivalent to the degree of coherence of the result set. The solution to a coherence problem amounts to finding such division of the set E into subsets A and R which maximizes W. It is possible to compute (approximate) coherence of a given set of elements with the use of effective connectionist algorithms [2].

The proposed framework has several important advantages. First, it is rooted in tradition of coherentism, one of the most important and the most plausible theories of justification developed in the history of philosophy. The idea that the justification of a propositions stems from the degree of its coherence with other propositions has long history, but Thagard's account of the theory is one of the most precise ones to date. Moreover, the theory presented above is immunized against the most important objections typically raised against coherentist theories of justification [2, pp. 69-80]. Second, it is domain neutral and can be applied to different types of reasoning, including legal reasoning. Some papers concerning the application of constraint satisfaction theory of coherence to legal argumentation have been already published [1, 7, 8, 9]. Of course, it does not mean that the possibility of this application is not subjected to criticism [10, Chapter 2]. However, it seems possible to gradually overcome different limitations of Thagard's account. It should be noted here that domain neutrality of the framework presented above offers many possibilities of its application to different domains and the proposal outlined below is just one of these options. Third, Thagard's conception is relatively simple, but can be amended in many ways to attain representative power. This framework has practically no limitations as regards the set of relations between the elements which the piece of reasoning in question consists of. Fourth, it offers impressive computational possibilities due to the presence of connectionist algorithms. However, in the present paper we will not focus upon the last mentioned issue extensively, for the example discussed here below is relatively small.

3.2   Coherence Model of Legal Argumentation

Let us outline a model of legal reasoning based on Thagard's framework (CMLA). The version of CMLA presented here is a generalized and revised version of the model outlined in [1]. This concretization of Thagard's framework consists of the following types of elements.

Def. 1. Rule. Rule R is a conditional proposition of the following form: $\varphi_1, \varphi_2, \dots \varphi_n \Rightarrow \psi$, where $\varphi_1, \varphi_2, \dots \varphi_n$ is the antecedent of R, ψ is the consequent of R, and $\Rightarrow$ denotes defeasible implication.

Commentary. Several qualifications should be made as regards the relation of the presented framework and the existing systems of defeasible reasoning (for a general exposition of the subject cf. [12] - Chapter 2, pp. 55 ff. with literature referred there). CMLA encompasses the main feature of defeasible rules which is that although all conditions of a rule are satisfied, its consequent may not follow. This will become apparent from the foregoing discussion concerning constraints based on rules.

A few examples of rules may be presented as follows.

IF [causes damage] (x,y) AND [at fault] (x) THEN [liable] (x,y)

IF [obligated to pay tax] (x) AND [delays in payment] (x) THEN [obligated to pay fine] (x).

Let us note several important observations about these rules. First, the notation used here resembles standard first-order notation. We do not focus here on syntactical properties of this notation, though, for it is not our intention here to develop a logical framework, but a coherence-based model inspired *inter alia* by defeasible logic systems. Second, in consequence, it is not necessary to deal with metalogical properties of the elements used in the model (let is note that we do not discuss the truth values of the propositions). Third, the exact formulation of rules in a given case is shaped by the aims of the parties to the dispute. It is particularly advantageous for a party to find such rules which prescribe for the desired outcome in the case, that is, when the consequent of the invoked rule is the legal conclusion. Fourth, it is possible to have rules in the forms of defeasible bi-implications ($\Leftrightarrow$).

Def. 2. Legal Conclusion. Legal conclusion LC is a proposition of the following form ψ_C, where C represents the case which is to be decided.

Commentary. An example of a LC is for instance the following expression [liable] (Tom, John) which reads as "Tom is liable to John".

The problem how to arrive from abstract consequents of rules to concrete legal conclusions will be explained below while commenting on the definition of rule-based constraints. Note that in the representation of a case there will be at least two mutually inconsistent LCs.

The LCs need not be final answers to legal questions, but they may be intermediary solutions.

Def. 3. Optimization Command. Optimization Command OC is a proposition of the following form: V $_{<realized>}$ where V denotes a legally relevant value and the operator $_{<realized>}$ prescribes this value be realized to some acceptable level.

Commentary. Let us note that OCs enable us to model teleological reasoning in the law ([11] and the literature quoted there). In consequence the basic distinction of legal norms into rules and principles is represented in the CMLA [1].

The elements presented above are basic ones and it is possible to introduce many other types of them, for instance factor-outcome links (referred to as Subsidiary Rules in [1]). However, the three types of elements defined here are sufficient for the purposes of this paper.

Now it is necessary to introduce the types of constraints between the elements. The presentation here is an abstraction from the model presented in [1] and it is sufficient for the purposes of modeling the most basic features of consistent interpretation as presented in the jurisprudence of the ECJ.

Def. 4. Rule-Based Constraint. Rule-Based Constraint RBC has the following form: RBC_x <R, ψ_C>, where x: <+, 0, ->. RBCs represent the subsumption relation. The subsumption relation holds when the facts of the case can be classified as instantiations of predicates specified in the antecedent of the rule which is the first element of the constraint. If this is the case, then the rule supports the LC and the constraint is positive. Let us consider the following example. Assume that Tom destroyed John's car and that this is the only important element of the factual description of the case in question. Then we will have the following elements:

Rule: IF [causes damage] (x,y) AND [at fault] x THEN [liable] (x,y).
LC: [liable] (Tom, John)
~LC: [not liable] (Tom, John)

It is not difficult to state that there is a positive RBC between the Rule and LC if and only if Tom caused damage to John and Tom was at fault in it. There can be also a negative RBC between such elements if the factual description of the case changes. For instance, assume that John sues Tom on the basis of the Rule mentioned above but the destroyed car belongs to Bill; in such cases, in absence of any additional circumstances, Tom will not be successful in suing John and ~LC will be chosen instead of LC. Sometimes it is possible that we have doubts whether a rule is applicable to the facts in a given case and then it is convenient to state that the constraint between this rule and LC is neutral (this is represented by subscript "0"); yet, the LC still can be accepted as right conclusion to the case if it is supported by some other elements. It should be also clear from this exposition that CMLA is able to represent defeasible reasoning well. This is due to the fact that even of there is a positive RBC supporting LC, ~LC may still be the ultimately chosen conclusion because of other elements and constraints supporting it. Conversely, it is possible that although LC is not supported or even demoted by an existing RBC, it will be ultimately accepted due to the existence of positive constraints between it and other elements of the set.

Linguistic constraints as defined in [1] are specific kinds of RBCs.

Def. 5. Value-Based Constraint. Value-Based Constraint VBC has the following form: VBC_x < V <realized>, ψ_C>, where x: <+, ->.

Commentary. VBCs represent the assessment of a given LC from the point of view of realization of a given legally relevant value. CMLA is particularly fit for representing this kind of relations because the constraints possess assigned weight. This fact represents the "weighing of values" procedure in legal reasoning which was discussed in [1]. Let us note that it would make no sense to talk about neutral VBCs, because the values protected by OCs in these constraints would be irrelevant for LCs in a given case.

Def. 6. Inconsistency Constraint. It has the following form $\perp$<ψ, ψ'> and holds between each two elements of the set which are their mutual negations.

Commentary. This type of constraint is of course always negative, although it is possible for CMLA to account for such subset of accepted elements which encompasses inconsistency, unless mutually inconsistent propositions are LCs.

It is possible to add new types of constraints, for instance the ones based on multifarious canons of legal interpretation and argumentation moves.

Def. 7. Division Result Set. Division Result Set DRS is a structure $<D, \leq>$, where D is a finite set of all divisions $(d_1, d_2, \ldots d_n)$ of the initial set E into the subset of accepted (A) and rejected (R) elements, and $\leq$ is a weakly antisimmetric, transitive and reflexive relation representing the ordering of W numbers assigned to particular divisions, that is, the total weights of all satisfied constraints in these divisions.

Ideally, this relation should be total, although in actual legal argumentation it is often very difficult to order all possible divisions according to the relation $\leq$. If in a concrete DRS this relation is total, then we are able to indicate the maximal element of D with regards to $\leq$ and legal conclusion included in the subset A in this division would be the right legal answer. However, it is often possible to indicate such set of non-identical elements of D according to which we are not sure how they could be ordered according to $\leq$.

Def. 8. Problematic Division. A division d_i is a Problematic Division if and only if it belongs to the set PDS (Problematic Division Set) of such non-identical divisions belonging to DRS that: (1) are not comparable according to $\leq$ and (2) d_i is a maximal division in the subset of divisions with which it is comparable.

Def. 9. Certainty Ratio. Certainty Ratio CR is an expression of the following form: 1 / (1 + |PDS|), where |PDS| is the number of elements of PDS in a given case. The greater the number of doubtful divisions, the lower the degree of certainty of legal reasoning pertaining to a given set of initial elements.

Commentary. The definition of CR leads to the following consequences. If there is no PD, CR of the case is 1 and this result is highly intuitive. However, the existence of one PD leads to the decrease of PD to ½, which is problematic from the point of view of the rule of law. The task of the judge, then, is to make the PDS as small as possible. In the foregoing exposition we will attempt to show how the context of improper transposition of a directive increase legal uncertainty and how this problem can be dealt with by means of consistent interpretation.

The definitions presented above enable us to state how the cases can be represented in the CMLA. First, it is necessary to establish the set E – the set of concrete elements in a given case (instantiations of element types defined above). Second, the constraints between these elements should be defined. Third, a chosen algorithm should be chosen to compute the degree of satisfaction of satisfied constraints in different divisions of the set E into subsets of accepted and rejected elements. As simple as it may seem, the procedure may lead to complicated problems even when applied to relatively small sets (smaller than 10 elements). One of the most important problems is the possibility of existence of a PD, which may be a result of problems with defining constraints on elements or simply a result of a stalemate situation as regards assignment of weights to constraints in two different divisions.

It ought to be noted that in some cases the presented framework will be too weak to represent the judge's reasoning in the case realistically. However, this deficiency can be overcome either by adding new elements and constraints to the network or by imposing some external limitations on the set and divisions in questions.

4   CMLA Account of Consistent Interpretation

The following section is devoted to modeling an instance of legal argumentation concerning consistent interpretation. My aim is to represent the following features of the doctrine in question: (1) an exemplary explication of the term 'interpretation' as used in this context; (2) the problem of limits to consistent interpretation and (3) the problem of legal uncertainty arising from the application of this doctrine.

The problem of interrelations between different legal orders has not been much focused on in AI and Law literature. However, the work of Sartor on legal pluralism [12, p. 659 ff.], and recent paper of Dung and Sartor on private international law [13] should be mentioned here as pioneering logical accounts of this phenomenon. The idea discussed by Sartor in [12] that in pluralist legal world it is convenient to accompany representations of normative propositions with symbols denoting legal systems in which these propositions hold will be employed in the foregoing discussion. It is not my aim to exhaust the problem of relation between Member State legal orders and EU legal order here, for it would exceed the scope of this paper considerably; let us emphasize that legal pluralism is perceived as a "tremendous task for legal logic" [12, p. 660]. Bengoetxea's earlier paper [14] refers directly to the AI and EC law, but it has a more general character. The coherence-based framework seems to be particularly fit for representation of the problems of reasoning within the EU law, which is emphasized in jurisprudential accounts of the problem (e.g. [15]).

The representation of consistent interpretation in CMLA framework involves introducing a new aspect of the model, namely, the pedigree indexes. The pedigree index is attached to an element in the initial set E and it refers to the source of this element (the legal order from which it comes from). For instance, a rule R extracted from the EU legal order can be indexed as follows: $R_{<EU>}$ and a rule taken from the national law, say Spanish law, can be indexed as $R_{<ESP>}$.

I present only one CMLA representation of a consistent interpretation case, namely, the *Marleasing* case. It can be questioned if the following discussion of only one case can give rise to any interesting generalizations. My claim here is that some interesting generalizations are possible, and the lack of possibility of other generalizations is a symptom of important features of the doctrine in question. I will return to this issue after the discussion of the case's representation.

The case was heard before the ECJ due to the question asked by the Spanish court. The facts of the case can be summarized as follows. The Applicant in the dispute, the company Marleasing SA brought an application against the opposing parties, including La Comercial Internacional de Alimentación SA. One of the persons which established La Comercial was another company, Barviesa SA. According to Marleasing SA, La Cormercial was established to the detriment of the creditors of Barviesa SA. The Applicant invoked the Articles 1261 and 1275 of the Spanish Civil Code, according to which a contract which lacks a proper ground (cause) is invalid. In the opinion of the Applicant, the contract "was a sham transaction and was carried out in order to defraud the creditors of Barviesa". Simultaneously, the facts of the case were regulated by the Council Directive 68/151/EEC of 9 March 1968 on coordination of safeguards which, in its Article 11, provided for an exhaustive list of six cases in which the nullity of a company may be ordered. However, in 1990 the directive was not implemented in the Spanish legal order. The directive could not be directly applied to the case due to the prohibition of horizontal direct effect.

The elements of the initial set E were, then, as follows:

1. Two mutually inconsistent LCs: ψ ([nullified](LA COMERCIAL CONTRACT)) and ψ' ([not nullified] (LA COMERCIAL CONTRACT)). For the sake of readability we write ψ instead of ψ_C.
2. Rule $R_{<ESP>}$ of the form $\varphi_c \Rightarrow \psi$ (IF contract [lacks a proper cause] (x), THEN [nullified] (x)), based on the Spanish Civil Code.
3. Rule $R_{<EU>}$ of the form $\varphi_1, \varphi_2, ... \varphi_6 \Leftrightarrow \psi$ (IF AND ONLY IF [satisfies any of six premises] (x)), THEN [nullified] (x)), based on the Directive.
4. Optimization Command $OC_{<EU>}$ of the form TRUST $_{<realized>}$, where the protected value is trust of private individuals in the exact wording of the Directive.
5. Optimization Command $OC_{<ESP>}$ of the form PROTECTION $_{<realized>}$, where the protected value is protection of creditors.

The list of constraints present here is as follows:

1. $RBC_+ <R_{<EU>}, \psi' >$;
2. $RBC_- <R_{<EU>}, \psi >$;
3. $RBC_+ <R_{<ESP>}, \psi >$;
4. $RBC_- <R_{<ESP>}, \psi' >$;
5. $VBC_+ <TRUST_{<realized>}, \psi' >$;
6. $VBC_- <TRUST_{<realized>}, \psi>$;
7. $VBC_+ <PROTECTION_{<realized>}, \psi>$;
8. $VBC- <PROTECTION_{<realized>}, \psi' >$;
9. $\perp <\psi, \psi'>$.

The network of elements related to each other with multiple constraints can be presented as follows, where solid lines represent positive constraints and dotted lines represent negative constraints.

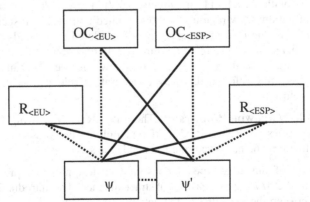

Fig. 1. A constraint network for the *Marleasing* case (consistent interpretation not included)

Ex post we know that ψ' ([not nullified](LA COMERCIAL CONTRACT)) was chosen as the proper answer to the legal question. How is this choice justifiable on the basis of information available in the network? Let us note that the network presented above is symmetrical and in absence of further information we are facing a stalemate situation. It seems at the first sight that we have a non-empty Problematic Division Set here, because the two plausible divisions are not comparable with regard to the relation of total weights of satisfied constraints. In consequence, we would have, presumably, Certainty Ratio at least as small as ½ , which is an unfortunate situation.

One could say that the choice of ψ' is grounded on the external assumption according to which EU-elements are weightier than ESP-elements here due to the principle of supremacy of EU law. This would lead to assignment of greater weight to the constraints involving these elements and to the ultimate choice of the following division as the most coherent one:

d_1: ACCEPTED: $< \psi'$, $R_{<EU>}$, $OC_{<EU>} >$; REJECTED: $< \psi$, $R_{<ESP>}$, $OC_{<ESP>} >$.

Unfortunately, the choice of this division can be interpreted only as direct application of $R_{<EU>}$, which is, in the present case, precluded by the prohibition of horizontal direct effect of (ill-transposed) directives. Due to this another external assumption we are forced to reject all divisions which assign the element $R_{<ESP>}$ to the subset R. $R_{<ESP>}$ ought to be assigned to the subset of accepted elements. However, the setting of the constraints between elements seems to make such solution impossible, because the following division:

d_2: ACCEPTED: $< \psi'$, $R_{<EU>}$, $R_{<ESP>}$, $OC_{<EU>} >$; REJECTED: $< \psi$, $OC_{<ESP>} >$

should be then seen as weightier than d_1, which counters the definitions of constraints specified above and the general coherence conditions.

How, then, the national court should behave in order to (1) see to it that EU-elements are accepted and (2) not choose an obviously suboptimal division as regards its level of coherence?

Ex post, on the basis of Marleasing decision, and in particular on the following passage: " (…) [I]t follows that the requirement that national law must be interpreted in conformity with Article 11 of Directive 68/151 precludes the interpretation of provisions of national law relating to public limited companies in such a manner that the nullity of a public limited company may be ordered on grounds other than those exhaustively listed in Article 11 of the directive in question.(…)", the following can be stated. There is a need for introducing of a new positive constraint between $R_{<EU>}$ and $R_{<ESP>}$. We may refer to this new constraint as rule narrowing-constraint and define it as follows.

Def. 10. Rule Narrowing Constraint. There is a Rule Narrowing Constraint RNC between two rules - $RNC_+<R_i$, $R_j>$ if and only if the presence of R_i makes it impossible to apply the narrowed rule R_j outside the scope of R_i.

The definition of this new type of constraint is designed to represent the quoted passage from the *Marleasing* case in abstract manner. The introduction of this new constraint between the two rules in the network changes the character of some of the already established constraints. As far as after introduction of RNC the scope of application of $R_{<ESP>}$ is no longer wider than the scope of application of $R_{<EU>}$, the former rule no longer supports ψ ([nullified] (LA COMERCIAL CONTRACT)), but it supports ψ' ([not nullified] (LA COMERCIAL CONTRACT)) instead. In consequence, we obtain the following, revised list of constraints:

1. $RBC_+ <R_{<EU>}, \psi' >$;
2. $RBC. <R_{<EU>}, \psi >$;
3. $RBC_+ <R_{<ESP>}, \psi' >$;
4. $RBC. <R_{<ESP>}, \psi >$;
5. $VBC_+ <TRUST_{<realized>}, \psi' >$;
6. $VBC. <TRUST_{<realized>}, \psi>$;
7. $VBC_+ <PROTECTION_{<realized>}, \psi>$;
8. $VBC- <PROTECTION_{<realized>}, \psi' >$;
9. $\perp_{<\psi, \psi'>}$;
10. $RNC_+ <R_{<EU>}, R_{<ESP>}>$.

Note that the value of TRUST is still promoted by ψ' and that the value of PROTECTION lacks now a rule which could count as a means of its realization.

The modified network for the case can be represented as follows:

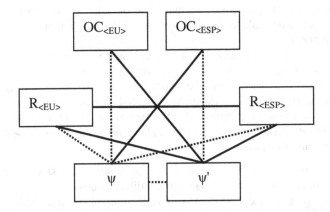

Fig. 2. A modified constraint network for the *Marleasing* case (consistent interpretation included)

The introduction of this new constraint type (and in consequence of a concrete RNC to the network) leads to serious doubts. Although it is not difficult to interpret the text of the decision in such manner *ex post*, it seems that this constraint type is added somehow *ad hoc* to explain the case and that it would be difficult for the Member State judge to figure it out *ex ante*. The main reason for this is that "rule narrowing" does not belong to the set of standard argumentative schemes employed in statutory interpretation in Western legal cultures [16]. This explains why the Spanish court was not eager to perform the consistent interpretation by itself but it was motivated to ask for a preliminary ruling from the ECJ. Let us add that the Spanish court's question referred to the possibility of direct application of the directive.

An important question is whether the analysis presented above leads to any interesting generalizations. On a very abstract level such generalization is possible. In the case of horizontal dispute involving EC legal rule, the latter being not directly applicable, the court should formulate its decision so as to assure that both rules: an applicable Member State rule and EC rule are in the set of accepted elements in the chosen division of the initial set of elements. As we could see from the discussion, this can lead to the choice of suboptimal division from the point of view of the degree of coherence. Therefore it is necessary to change the setting of constraints in the network, for instance by introducing new types of constraints as the RNC discussed above.

In consequence, the doctrine of consistent interpretation can be construed in the framework of CMLA as the process of modifying the setting of initial constraints in the network so as to assure that the desired division (containing in the set of accepted elements both rules: EU law rule and Member State rule together with LC supported by the EC rule from the beginning) is also the most coherent division.

The more interesting and less abstract generalizations are also more problematic. For instance, on the basis of the discussion above it could be stated that each time the relevant EU rule is a bi-implication, RNC should hold between this EU rule and the applicable Member State rule. But this is a hypothesis which should be verified on the basis of actual cases involving consistent interpretation. Moreover, it should be noted that consistent interpretation may make use of other types of constraints, too, and it can be a difficult task to account for the conditions of their application in a more general manner.

The explication of the doctrine of consistent interpretation also leads to the possibility of formulation of a general account of the limits to the consistent interpretation. Let C(CI) be the set of consistent interpretation constraints, *i.e.* such constraints which, after being introduced to the initial case networks lead *prima facie* to the situation that the desired division is also the most coherent division. The limits to the consistent interpretation could then be understood in the terms of CMLA as one of the following circumstances:

1. Showing in a concrete case that the proposed Consistent Interpretation Constraint cannot be construed in any intelligible manner.
2. Showing that, in fact, even after introduction of the new constraint, it is impossible to generate an optimal division of the initial set (this possibility is the most interesting one from the computational point of view).
3. Showing that introduction of a proposed Consistent Interpretation Constraint violates some external limitation we have agreed upon before engaging into computation of coherence in the network.

In my opinion, the representation of *Marleasing* case presented above supports the view that it would be more rational to accept the possibility of horizontal direct effect of the directives rather than to prohibit it. Let us recall that the chosen division for the *Marleasing* case was the following one:

$$d_2: \text{ACCEPTED: } < \psi', R_{<EU>}, R_{<ESP>}, OC_{<EU>} >; \text{ REJECTED: } < \psi, OC_{<ESP>} >,$$

which was possible after introducing RNC to the network. However, the structure of this division shows that the prohibition of horizontal direct effect of directives is illusory, for $R_{<EU>}$ is in the set of accepted elements and it is related to ψ' by a RBC. With respect to these elements, the situation would be identical if direct application of the directive was accepted, but the obtained solution would be simpler as regards the definitions of constraints. In consequence, CMLA offers for a precise formulation of the objection against the prohibition of direct effect of the directives in horizontal cases.

The representation presented here offers computational possibilities for measuring the degree of legal uncertainty as regards cases by means of Certainty Ratio value. However, our analyses clearly show that in the context of application of consistent interpretation doctrine the degree of legal uncertainty stems not only from the Problematic Divisions, but also from the unspecified character of the potential Consistent Interpretation Constraints.

Definitely, the results presented here are preliminary. In particular, they are too simplified with respect of the types of elements taken into account. However, they seem to offer interesting possibilities for further research. The results together with prospects for future analyses are summarized in the next Section.

5   Conclusions and Further Research

The investigations presented above enable us to formulate the following conclusions. They all need further elaboration, but I think that they are at least plausible in the light of the lightweight formalism presented above.

1. The CMLA based analysis of the *Marleasing* case showed that in the context of consistent interpretation, the word 'interpretation' can refer to determining of the scope of rule's application. This effect has been brought about here by Rule Narrowing Constraint, formulated from the *ex post* perspective. The general structure and of consistent interpretation was explicated. However, the possibilities of generalization of these results are limited and in order to overcome these limitations, at least to some extent, it is necessary to analyze a large database of ECJ cases involving consistent interpretation.

2. Although it is not difficult to determine the structure of consistent interpretation constraints *ex post*, it is difficult to establish it *ex ante*. This is why the national courts are motivated to ask for preliminary rulings from the Court. It seems that without resorting to the preliminary ruling the Certainty Ratio is no bigger than ½, and it is plausible to state that it is often smaller (the latter hypothesis calls for verification). However, the like existence of Problematic Divisions in the cases involving consistent interpretation is not the only source of legal uncertainty here. Another one is the indeterminacy and the open character of the set of Consistent Interpretation Constraints. This observation leads to two important conclusions. First, the possibility of generalization of the results on the basis of analysis of previously decided cases is unfortunately limited due to considerably large scope of discretion of the ECJ as regards the development types of consistent interpretation constraint. Second, if the horizontal effect of directives was accepted, then, presumably, it would be possible for the national courts to refrain from asking for preliminary ruling from the ECJ more often. It was shown here that the prohibition of horizontal direct effect of the directives is somewhat illusory, contrary to the direct wording of the jurisprudence of the ECJ.

3. The so-called limits of consistent interpretation seem to provide boundaries for designing *ad hoc* consistent interpretation constraints. However, it is far from clear which factors are relevant for determining the possibility of introducing such a constraint.

In consequence, I am of the opinion that the research direction set forth in the present paper ought to concentrate on the problem of limits of consistent interpretation. The CMLA account presented here enables us to explicate these limits as second-order constraints limiting the possibilities of introducing *ad hoc* consistent interpretation constraints, like Rule Narrowing Constraint as defined above. Also, the model presented here ought to be refined to encompass more complicated issues regarding the application of EU law and its transposition into Member State law, like the historical development of both national legal systems and EU legal system.

Acknowledgements. I would like to thank the organizers and participants of the AICOL 2011 – AI Approaches to the Complexity of Legal Systems workshop which took place on August 15-16, 2011, in Frankfurt as one of the events at XXV World Congress of Philosophy of Law and Social Philosophy where I presented a talk from which this paper emerged. In particular, I am grateful to Pompeu Casanovas, Jaap Hage, Giovanni Sartor and Radboud Winkels for their insightful questions and criticism. Moreover, I would like to thank the three anonymous reviewers for their constructive and stimulating opinions.

References

1. Araszkiewicz, M.: Balancing of Legal Principles and Constraint Satisfaction. In: Winkels, R.C. (ed.) Proceedings of the Twenty-Third Annual Conference on Legal Knowledge and Information Systems (JURIX), pp. 7–16. IOS, Amsterdam (2010)
2. Thagard, P.: Coherence in Thought and Action. MIT Press, Cambridge (2000)
3. Craig, P., de Búrca, G.: EU Law. Text, Cases and Materials. Oxford University Press, Oxford (2008)
4. Prinssen, J.L., Schrauwen, A. (eds.): Direct Effect. Rethinking a Classic of EC Legal Doctrine. Europa Law Publishing, Groningen (2002)
5. Prechal, S.: Directives in EC Law. Oxford University Press, New York (2005)
6. Betlem, G.: The Doctrine of Consistent Interpretation. Managing Legal Uncertainty. Oxford Journal of Legal Studies 22, 397–418 (2002)
7. Amaya, A.: Formal Models of Coherence and Legal Epistemology. Artificial Intelligence and Law 15, 429–447 (2007)
8. Amaya, A.: Legal Justification by Optimal Coherence. Ratio Juris 24, 304–329 (2011)
9. Joseph, S., Prakken, H.: Coherence-driven argumentation to norm consensus. In: Proceedings of the 12th International Conference on Artificial Intelligence and Law, Barcelona, pp. 58–67. ACM Press, New York (2009)
10. Hage, J.C.: Studies in Legal Logic. Springer, Berlin (2005)
11. Sartor, G.: Doing justice to rights and values: teleological reasoning and proportionality. Artificial Intelligence and Law 18, 175–215 (2010)
12. Sartor, G.: Legal Reasoning. A Cognitive Approach to Law. A Treatise of Legal Philosophy and General Jurisprudence, vol. 5. Springer (2005)

13. Dung, P.M., Sartor, G.: A Logical Model of Private International Law. In: Governatori, G., Sartor, G. (eds.) DEON 2010. LNCS, vol. 6181, pp. 229–246. Springer, Heidelberg (2010)
14. Bengoetxea, J.: AI, Legal Theory and EC Law: A Mapping of the Main Problems. In: Bankowski, Z., White, I., Hahn, U. (eds.) Informatics and the Foundation of Legal Reasoning, pp. 291–310. Kluwer, Dordrecht (1995)
15. Moral, L.: A Modest Notion of Coherence in Legal Reasoning. A Model for the European Court of Justice. Ratio Juris 16, 296–323 (2005)
16. MacCormick, N., Summers, R. (eds.): Interpreting Statutes: A Comparative Study. Dartmouth Publishing (1991)

Three Roads to Complexity, AI and the Law of Robots: On Crimes, Contracts, and Torts

Ugo Pagallo

Law School, University of Torino, via s. Ottavio 54, 10124 Torino, Italy
ugo.pagallo@unito.it

Abstract. The paper examines the impact of robotics technology on contemporary legal systems and, more particularly, some of the legal challenges brought on by the information revolution in the fields of criminal law, contracts, and tort law. Whereas, in international humanitarian law, scholars and lawmakers debate on whether autonomous lethal weapons should be banned, robots are reshaping notions of agency and human responsibility in civil (as opposed to criminal) law. Although time is not ripe for the "legal personification" of robots, we should admit new forms of both contractual and tort liability for the behaviour of these "intelligent machines." After all, this is the first time ever legal systems will hold people responsible for what an artificial state-transition system "decides" to do.

Keywords: Accountability, Agency, AI & Law, Complexity, Contracts, Criminal Law, Liability, Responsibility, Robot, Tort Law.

1 Introduction

This article examines the impact of robotics technology on contemporary legal systems and, more particularly, how the production and use of robots affect some tenets of criminal law, contractual obligations, and tort law. First used by Isaac Asimov in the 1942 novel Runaround [1], the word "robotics" concerns a panoply of disciplines such as computer science and cybernetics, physics, mathematics and mechanics, electronics, neuroscience, biology and more. Some argue that robots are machines basically built upon today's "sense-think-act" paradigm in AI research [2]. Others, as the director of the AI Laboratory at Stanford, CA., Sebastian Thrun, reckon that robots have to do with the ability of a machine to "perceive something complex and make appropriate decisions" [4]. While some others stress that robots should be able to learn and adapt to the changes of the environment, it is crucial we distinguish a number of applications as different as humanoids, adaptive service robots, unmanned underwater vehicles, drones, and so forth. Such differentiations are critical when ascertaining the impact of robotics technology on contemporary legal systems, because it is likely that drones and other types of autonomous (lethal) weapons mainly affect legal fields such as international humanitarian and criminal law, whereas other applications such as, say, da Vinci robot-surgeons mostly raise matters of contractual obligations and strict liability rules.

M. Palmirani et al. (Eds.): AICOL Workshops 2011, LNAI 7639, pp. 48–60, 2012.

The exponential curve of advancement in the field of robotics technology, e.g., the current rates of doubling amounts of computation also known as the "Moore's law," have nevertheless induced some scholars to exaggerations which are also popular in the legal field. Some argue that the current information revolution inexorably shapes the destiny of human beings and their societies. This sort of determinism is illustrated by the thesis of a distinguished researcher from Carnegie Mellon, Hans Moravec, who announced some years ago that intelligent robots will succeed humans and that we, as a species, would face extinction [4]. In other words, technology would decree robots replacing humans as the next step in evolution [5], so that lawyers should be ready to discuss a new generation of cases such as challenges to national sovereignty and robot revolutions [6], robotic sex crimes [7], or robots that choose to commit and, ultimately, carry out the crime, e.g., the adventures of the "Robot Kleptomaniac" [8]. According to these perspectives, new types of crime would emerge with robots accountable for their regrettable actions: the self-consciousness of the robot would not only materialize Sci-Fi scenarios as imagining a robot revolution and, hence, a new cyber-Spartacus. Furthermore, the meaning of traditional notions such as stealing and killing would change, since the culpability of the agent, i.e., its mens rea would be rooted in the artificial mind of a machine "capable of a measure of empathy" and "a type of autonomy that affords intentional actions" [9].

In order to prevent misunderstandings, this paper suggests we should grasp the new frontiers of complexity, AI and legal robotics, by reverting to the terra cognita of jurisprudence and its civil law-counterpart in Europe, i.e., "general theory of law." Since the late 1800s, after all, the "Law of the Automaton" has been a popular topic among legal scholars: Günther's Das Automatenrecht was published in 1891, Ertel's Der Automatenmissbrauch und seine Charakterisierung als Delikt as well as Schiller's Rechtsverhältnisse des Automen were printed in 1898, down to Neumond's Der Automat in 1899. More than a century later, a relatively strong consensus still exists about key legal notions regarding the production and use of robots: the common viewpoint excludes the criminal accountability and the "legal personality" of robots. For the foreseeable future, in fact, these machines will be held legally and morally irresponsible because they lack the set of preconditions for attributing liability to someone in the realm of criminal law. Since consciousness is a conceptual prerequisite for both legal and "moral agency" [10], it follows that even when Robbie CX30 assassinated Bart Matthews in Richard Epstein's story on The Case of the Killer Robot, the homicide remains a matter of human responsibility, for robots are not aware of their own conduct like "wishing" to act in a certain way [11].

However, we need not evaluate robots with Turing tests in order to admit a new generation of legal cases involving human liability as well as robots accountability [12]. Regardless of Sci-Fi scenarios where robots are provided with consciousness, free will and emotions, it is likely that "in a few years we are going to cohabit with robots endowed with self knowledge and autonomy in the engineering meaning of these words" [13]. Although robots do not raise new legal issues per se, e.g., homicides and other cases of criminal law, some applications really challenge the "completeness" of legal systems, i.e., the Hilbert-like capacity to decide every legal problem through the use of analogy (and the general principles of the system, according to the European civil law-tradition). It is enough to mention military applications of robotics technology and the civilian use of unmanned vehicles (UVs) as drones, besides the forthcoming generation of robot-traders, robot-doctors, robot-nannies and, why not?, sex robots [14], to stress that they challenge "the boundaries and efficacy of existing legal frameworks and raise a range of social and ethical constraints" [15]. In order to tell hard cases from routine cases, the paper proposes to tackle the new frontiers of complexity, AI & law, by distinguishing matters of criminal law, contracts, and strict liability in the design, production and use of robots.

In section 2, I examine the production and use of robot soldiers, and how they affect current principles of criminal law and, moreover, of international humanitarian law. In section 3, I take into account the 2004 "World Robotics"-report of the UN Economic Commission mainly focused on robots of peace such as industrial robots, surgical robots, edutainment robots, etc., so as to shed light on matters of risk and predictability in contractual obligations. In section 4, I consider issues of extra-contractual responsibility and strict liability induced by such "robots of peace" [16]. The conclusion is that the increasing autonomy and even "intelligence" of robotic behaviour impact on the complexity of legal systems, by altering the basis on which the principles of human responsibility and accountability are traditionally grounded. Although time is not ripe for the "legal personification" of robots, new forms of liability for artificial agents in the field of contracts [17–20], as well as models for distributing risk in tort law [15, 21], should be mentioned. This is in fact the first time ever legal systems will hold people responsible for what a robot "chooses" to do. Since robots are here to stay, the aim of the law should be to wisely discipline our mutual relationships.

2   Crimes

I mentioned the common legal standpoint that hooks "autonomous" and "intelligent" robots off all claims of criminal responsibility. Robots indeed lack psychological components such as intentions or consciousness, i.e., the set of preconditions for attributing liability to someone in the case of violation of criminal law. However, it is highly debatable to claim that robots lack all types of agenthood: after all, we already have a generation of proper "artificial agents" (AAs) that respond to stimuli by changing the values of their properties or inner states and, furthermore, that improve the rules through which those properties change without external stimuli. These robots suggest we are dealing with a new source of agency, in that they properly are interactive, autonomous, and adaptable [22]. Like animals, children and, obviously, adult human beings, robots can cause morally qualifiable actions such as good and evil [23]. Moreover, robots may represent a meaningful target of censorship as in the case of "monitoring and modification, removal to a disconnected component of cyberspace" or "deletion without backup" [12]. If it seems appropriate to extend the class of morally accountable agents so as to include the artificial agency of robots, however, we should further distinguish between this form of agency and the criminal liability of robots, that is, between the source of relevant moral actions and the evaluation of agents as being morally responsible for a given behaviour. This is what typically occurs with children's actions and the behaviour of animals: even though we assume that some sort of moral accountability is a necessary requirement for legal responsibility, the former does not represent a sufficient condition of the latter, because respondents ought to be subject to the ordinary process of moral appreciation in order to determine whether or not they are guilty in the name of the law. This is why, according to the current state-of-the-art, it would be pointless to debate before a judge whether a robot should be considered a "killer," a "robber," and so on. Simply consider the reasons underpinning the legitimacy of inflicting punishment in modern criminal law such as the theory of retribution, of special and general prevention, that, in the case of robots, would be devoid of meaning: Can we reckon robots paying their debt to society? Can we correct their moral character so that such machines fully understand why they ought not to repeat an evil action? Should we punish robots so as to dissuade human beings (and other robots) from committing similar wrongs?

However, there is another way robots affect today's principles and provisions of criminal law. As the field of computer crimes has shown over the past two decades, robotics technology makes possible what simply was unthinkable few years ago. In the mid 1990s, for example, the Legal Tender project claimed that remote viewers could tele-operate a robotic system to physically alter "purportedly authentic US $ 1000 bills" [24]. Since the early 2000s, research and development of new "robotic systems" have been particularly massive in military robotics (more than 50% of the American R&D in AI is sponsored by the U.S. Army). Similarly to previous technological advancements in chemical, biological and nuclear weapons, robots are already affecting a number of legal fields such as the laws of war, rules of engagement and provisions of international humanitarian law. Military applications of robotics technology such as MQ-9 Reapers and C-3PO Terminators are in fact challenging principles of military conduct like proportionate use of force or discrimination between soldiers and civilians. According to several scholars [25–27], what makes the use of such robots critical depends on the technical difficulty to design them so as to let machines distinguish between friends and foes. In their 2010 reports to the UN General Assembly, Philip Alston and Christof Heynes have significantly stressed that legal provisions are silent on two key points. Not only is analogy inadequate to determine whether certain types of "autonomous weapons" should be considered unlawful as such, but it is also far from clear the set of parameters and conditions that should regulate the use of these machines in accordance with the principle of discrimination and immunity in ius in bello [28]. In the first hypothesis, i.e., the case of a ban, political and military authorities would be responsible for AI soldiers violating principles of ius belli: once established what types of robotic weapons should not be permitted, e.g., autonomous lethal machines with no human supervision, it follows that their design and construction should be interpreted as a crime. In the second hypothesis, i.e., a UN-sponsored agreement defining the set of parameters and conditions that discipline the use of robot soldiers, the aim is to determine when the design, production and employment of military robotics technology are lawful.

Still, the production and employment of robot soldiers are not only provoking a number of loopholes in crucial fields as the laws of war, rules of engagement and provisions of international humanitarian law. Besides the impact of robotics technology on national security, public order and what is necessary in a democratic society in the interests of public safety (in the phrasing of art. 8.2 of the 1950 European Convention on Human Rights), the increasing unpredictability and autonomy of robotic behaviour are affecting legal notions of "causation" and "reasonable foreseeability." Some argue that this capacity of robots to operate in the real world without human control concerns a very core principle of the law, because no human could ultimately be held responsible [25]. In the hypothesis of a robot that causes serious harm by taking its own decisions on the battlefield, "this would indeed be a very tricky case legally. The only solution would be to simply withdraw all of the AW [autonomous weapons] of this particular design" [29]. Moreover, by affecting the idea of individual fault and the doctrine of "proximate causes" in deciding where to cut the chain of responsibility, some have suggested a "failure of causation," since it would be hard to predict what types of harm may supervene [17].

Yet, legal notions of "causation," "reasonable foreseeability," "apportioned liability," and more, do not only concern possible illegal uses of robotic applications. Even in military robotics, we should pay attention to the design and construction of such machines: when humans use robots in order to apply force in disproportionate ways, no lawyer doubts that the fault has to be attributed to the user of such robot, notwithstanding "unforeseeable" or "unpredictable" behaviour of the machine. But, when machines do not properly work within the limits of a given set of parameters, the fault will be attributed to the manufacturers of such artefacts, e.g., the case of the unintended movements of the Sword unites employed by the U.S. Army and the producer's claims to avoid liability [30]. In this latter case, focus should be on determining fault in complex software and hardware applications for autonomous AAs, pursuant to conditions, terms, and clauses that depend both on the voluntary agreement between private individuals that a court will enforce, and the commercial or non-commercial nature of that agreement. Let me deepen this different look at robotics, in connection with a new generation of "civilian robots" such as business-makers, AI traders, and their contracts: rather than the legitimacy of the ends, what is at stake in the next section concerns the means of robotics technology.

3   Contracts

The design and production of robots are disciplined by conditions, terms, and clauses established by the parties to a contract. Here, legal issues have to do with "causation," "foreseeability," and "apportioned liability" that depend on the range of goals and set of parameters of a given artefact. Consider the very controlled setting of operatory rooms in the case of the da Vinci surgical robots, which raise engineering problems that scholars routinely address as part of their research. On the basis of the probability of events, their consequences and costs, lawyers examine matters of unpredictability and risk provoked by such robots, as they did with previous technological innovations. For example, work on the da Vinci robot shows that only 9 out of 350 interventions (2.6%) could not be completed due to device malfunctions of the artificial doctor [31]. Likewise, others claim that only 4,8% of the malfunctions occurred in a New York urology institute, from 2000 to 2007, would be related to a patient injury [32]. However, the more we widen settings and goals of robotics programs, the more it is likely we will be dealing with growing amounts of complexity; but, the more the amount of computation is exponentially increased, the more the risks that emerge as a consequence of robotic behaviour. In order to cast light on matters of contractual liability, foreseeability and causation for some riskier robotics applications than the da Vinci, some words on the Zero Intelligence (ZI)-agents are necessary.

Archetypes in "double action markets" [33, 34], ZI agents are programmed to "generate bids and offers selected randomly from a uniform distribution subject only to the constraint it cannot deliberately lose money" [35]. Such agents are rudimental in that they are oblivious of their environment and do not control the timing of their actions: ZI agents even lack the capability of taking action so as to compensate their inability to respond to the environment. Since the robot tournaments at the Santa Fe Institute in 1990, it turned out that markets populated only by ZI agents have nonetheless the tendency of human markets to generate average prices and quantities of what economists traditionally present as a "competitive equilibrium." These artificial agents seem to confirm Hayek's idea that, in some circumstances as with social (i.e., contractual) interaction, "intelligence" emerges from the "rules of the game" rather than individual choices [36]. The level of autonomy that is insufficient to bring robots before judges and have them declared guilty in criminal courts is enough to have relevant effects in the field of contracts, where "the intentional stance represents usually the only possible viewpoint to explain and foresee the behaviour of complex entities that can act teleologically" [20]. After all, ZI agents achieve sophisticated goals as outperforming untrained human traders in double-oral auctions [37], so that, in "shopping around" or "planning ahead," the performance of ZI agents has been improved and "the design of a special-purpose agent that can trade in the simple asset markets examined in this article as well as, if not better than, humans seems clearly within grasp" [35].

Yet, even ZI agents may be risky and dangerous: their eagerness to trade has suggested troubling similarities with the greediness of human speculators and "real life" bubbles in markets, in that agents are overwhelmed by the complexity of the environment and appear extremely "inexperienced." By considering that, in many other cases, robots are "good" in decreasing the informational entropy of the system or enriching its informational properties [38], e.g., the new generation of "robot traders" which the UN Economic Commission illustrated in the 2004 "World Robotics"-report [39], it is thus necessary to address people's claims not to be ruined by their own robot's activities and intentions, that is, business run by ZI agents. In the light of how robots can be extremely fruitful in making contracts, or establishing rights and obligations between humans, e.g., cognitive automata in the form of software agents [20], how could we forestall any legislation that might prevent the use, rather than the production, of "robots traders" due to their risks?

Some scholars have proposed to introduce forms of limited responsibility through "personal accountability of robots" so as to discipline transactions mediated by artificial agents and tomorrow's smart ZI agents [18]. The wisdom of ancient Roman law suggests a kind of "artificial accountability" with the mechanism of "peculium": in the phrasing of the Digest of Justinian, it was "the sum of money or property granted by the head of the household to a slave or son-in-power. Although considered for some purposes as a separate unit, and so allowing a business run by slaves to be used almost as a limited company, it remained technically the property of the head of the household" [40]. In the case of robots, the aim should indeed be the same lawyers pursued in Ancient Rome: whilst some Roman slaves were estate managers, bankers, or merchants, though not "humans," rights and obligations established by robots could be guaranteed by the robot's own portfolio. The parallelism between robots and slaves is hence attractive, because a "digital peculium" guarantees that people would not be ruined by the "decisions" of their robots and that robots' counterparties would be protected when making business with them [21]. Besides further mechanisms of distributing risk through insurance models [17], or authentication systems [18], new forms of accountability such as the digital peculium might avert any legislation that prevents the use of robots due to the excessive burden on the owners (rather than, say, on the producers and designers) of these machines.

Legal issues concerning the design, production and use of autonomous robots, however, not only regard clauses and pacts between humans and "robot traders." Further robotics applications suggest that lawyers will increasingly discuss problems of extra-contractual responsibility, e.g., robots damaging "third parties" rather than affecting their contractual "counter-parties." This scenario transcends the mechanism of peculium and involves what Roman jurists defined in terms of Aquilian protection, namely, the form of responsibility which stems from the general idea that people are held liable for unlawful or accidental damages caused to others due to personal fault [41]. In the first case of "damages," we are still dealing with the technical difficulties of the project and clauses of the agreement between the parties, i.e., the design and construction of Swords, Warriors, Da Vincis, ZI agents, etc. It is noteworthy that work on robot trading in auction markets, as the Penn-Lehman Automated Trading project, showed relevant failures in programming ZI agents capable to effectively speculate against smart humans (sponsored by the University of Pennsylvania and Lehman Brothers, the project was suspended in 2005, that is, 3 years before Lehman Brothers' own collapse). Vice versa, in the hypothesis of extra-contractual responsibility, lawyers discuss obligations between private persons imposed by the government so as to compensate "damage" done by wrongdoing. After contractual obligations, let me examine what common lawyers define as torts.

4 Torts

In the field of extra-contractual responsibility, lawyers traditionally distinguish between intentional torts, negligence-related tortuous liability, and faultless liability or strict liability. In the first case, there is liability for an intentional tort when a person has voluntarily performed the wrongful action. Next, liability is based on lack of due care when the "reasonable" person fails to guard against "foreseeable" harm. Finally, faultless liability or strict liability is established, e.g., liability for defective products, when there is no illicit or culpable behaviour but, say, a lack of information about certain features of the artefact. In the field of robotics technology, the tricky part of this framework depends on the fact that, for the first time ever, legal systems will hold people responsible for the behaviour of expert systems that gain knowledge and skills from their own "decisions." A forthcoming generation of robot toys, "robot nannies," and even intelligent cars or UGVs [15], will learn from the features of the environment and, as any other proper agent, tomorrow's robot toys, robot nannies and even robot chauffeurs will improve the rules through which the values of their own properties change without external stimuli. The result is that the same model of AI "toy," AI "nanny," or AI "car," we will be possibly buying next Christmas, is going to behave quite differently after few days or weeks: in the event the machine causes harm to someone in the roundabouts, who is liable?

Leaving aside the hypothesis of intentional torts, we have to determine how social risk should be distributed via (new) clauses of extra-contractual responsibility for the behaviour of our robots. In some cases, e.g., unmanned ground vehicles or UGVs, we might address today's loopholes of the law by establishing forms of strict liability as in the aforementioned case of product liability for the damages caused by people's own dangerous activities, that is, regardless of the intent of the subject or her use of ordinary care. Employers, for example, are often held liable for any illicit action the employees engage in under their working contract activities. Such a policy could obviously be mitigated in the case of robots, so as to avert the risk that people think twice, before producing and using robots at all. We could perhaps make insurance compulsory as we have done in most legal systems with traditional cars. We might also extend the mechanism of peculium by determining that human extra-contractual liability should be limited to the value of their own robots portfolio (plus, eventually, the compulsory insurance set above). Yet, there are some other types of artificial agents, e.g., "robot toys" and "robot nannies," that suggest a different approach to tort policies: some claim that lawyers should frame human relations with such robots, as we do with animals rather than tin machines or smart fridges [23]. In the event that an "intelligent nanny" causes harm to someone in the roundabouts, people's liability would ultimately depend on how we treated our machine, rather than the ways, say, that machine was designed and constructed. In order to illustrate the ways such a responsibility may be established, it is important to understand how the burden of proof is allocated in these cases.

In fact, legal systems provide for some limits to the aforementioned clauses of faultless liability, as it typically happens to parents who evade responsibility for their children's behaviour, when they prove they could not prevent their children's actions. Likewise, this is what occurs to the owners of animals when they prove that a fortuitous event happened. While regarding the set of dangerous activities, some legal systems exclude liability when it is proved you have taken all the "appropriate measures" in order to prevent any sort of damage, we may guess what sort of limited responsibility fits this type of robot. Once the main legal issue revolves around how we educate, treat, or manage our autonomous machines, rather than around who owns, built or sold them, people's extra-contractual responsibility for the behaviour of their robots depends on the typology of the human-robot relation and, of course, on the circumstances of the case. As lawyers discuss in most legal systems, should we deny liability when it is proved that a fortuitous event occurred, i.e., robots disciplined as "pets," or should we deem that individuals fairly evade responsibility only when they prove it was not possible to prevent a machine's action, i.e., robots considered as "children"?

Such issues are particularly complex, in that answers would require more information than that conveyed by the same question [36]: for instance, it is more than likely that robots will raise psychological problems related to the very interaction with humans as matters of attachment and feelings of subordination, trust, reliability, and deviations in individual emotions. Back to the field of military robotics technology and the use of artificial agents on the battlefield, it is telling what The Economist reported in October 2010, that is a Lebanese newspaper editor declaring that "Americans and Israelis are cowards to send machines to fight for them," although the article recalls "another story of an officer in Iraq, so moved by the sacrifice of a bomb-disposal robot that he wrote a letter of condolence to its manufacturer ... " [42]. Despite conspicuous work on how robotics technology affects human psychology, we have not enough data on the probability of events, their consequences and costs, so as to determine levels of risk on which insurance models may hinge for the use of new artificial companions and helpers at home, e.g., robot toys and robot nannies programmed to provide love and take care of children and the elderly. Contrary to some robotics applications like the aforementioned da Vinci surgeons and different models of unmanned vehicles (UVs) undertaking repairs to oil rigs in the Caribbean Sea, or inspecting atomic plants in Japan, it is an open question the kind of tort liability-policy we should endorse to tackle the unpredictability of our multiple artificial agents' behaviour.

Still, from a legal viewpoint, we should not miss the crucial point: whether under forms of negligence-related tortuous liability or strict liability-rules, humans are going to be held responsible for what robots autonomously do. This is not the first time legal systems provide for the responsibility and agency of some "artificial persons" like governments, organizations, companies or corporations; yet, this is going to be the first time such a liability is not reducible to an aggregation of human beings as the only relevant source of their action. As previously stressed, besides cases of responsibility for the behaviour of their children, pets, and even employees, a new generation of robots induce novel types of human responsibility for others' actions. Whereas this latter kind of responsibility suggests we have to take into account multiple types of robot interacting with humans for different purposes, some argue that such machines should be deemed as actors in current legal systems [43]. Leaving aside the debate on whether robots represent "the" new actors of todays complex networks[5], it is crucial to determine what type of legal agency robots might have.

5 Conclusions

Let me sum up the analysis on how robotics technology is affecting today's legal systems, with research in network theory and the complexity of the law. Besides regulatory frameworks for the design, production and use of robots, we should in fact understand how current rates of doubling amounts of computation and widening of operational goals are challenging key tenets of the law, as it is shown by the debate on the "legal personification" of artificial agents [20, 43–45]. In particular, focus should be on how information is created and distributed in a given network through the "nodes," so that a system is complex when collective behaviour emerges from large webs of individual components with no central control or simple rules of operation. As work in "evolutionary algorithms," "adaptive social networks," and the "normative emergence" from a multi-agent system perspective illustrate [46], it is not necessary to grasp the nodes of the network from an anthropological point of view, thereby reducing such nodes to an aggregation of human beings as the only relevant source of their action. What matters, here, is the sophisticated signalling and information processing of the nodes, whether "natural" or "artificial," so as to adapt to the environment through learning and evolutionary processes that lead to "decisions." Whilst the class of morally accountable agents may legitimately include the artificial agency of robots as a source of "good" and "evil" [12], the class of legal personhood may analogously be expanded through, for example, the "actants" or "hybrids" of Latour's network theory [47]. In the phrasing of Günther Teubner, "the result is that the law is opening itself for the entry of new juridical actors [such as the] electronic agents" [43].

However, dealing with the legal "agency" and "accountability" of robots, we should distinguish three levels of analysis, that is, whether robots have to be considered as new legal "persons," "actors" or, rather, "sources" of novel sorts of legal responsibility. Although such different levels of analysis are interconnected, we should keep crucial distinctions firm: indeed, regardless of the legal personhood of robots and whether they should properly have "rights" and "duties," it is a matter of fact that such machines are affecting basic assumptions of the law because, like slaves in Ancient Rome, robots are "things" that, nevertheless, can play a crucial role in fields as different as family contexts, edutainment environments, or trade, commerce, and business [39]. Whereas lawyers needed more than 2000 years to recognize the human personhood of slaves, vice versa, the legal personification of robots does not represent a necessary condition for the acknowledgment of new forms of accountability and contractual responsibility for (some types of) robots. Likewise, as the mounting autonomy of robots is defining new "nodes of the network" in the field of extra-contractual responsibility, lawyers will increasingly discuss further types of liability for others' behaviour, besides the traditional human responsibility for damages caused by their children, animals, or employees. At the end of the day, even by considering robots as simple means or "objects" as it occurs in criminal law, it is evident that new critical issues will emerge in human rights law, international humanitarian law, rules of engagement in laws of war, and so forth. Accordingly, the paper aimed to pinpoint how robotics technology challenges the complexity of the law at three different levels.

The first challenge regards whether robots should be deemed as "legal actors," according to the set of preconditions for attributing liability to someone in criminal law. By averting some popular Sci-Fi scenarios, the paper stressed that robots cannot certainly be deemed as "guilty," that is, criminally liable, although they can represent a meaningful target of human censorship.

The second challenge concerns whether robots should be welcomed as "legal persons," having the faculty to autonomously establish rights and obligations in the field of contracts: insurance models, authentication systems, and mechanisms of accountability such as the "digital peculium," showed ways of distributing risk by making robots liable for (some of) their actions.

The third challenge has finally to do with robots as the "source" of human liability in social interaction, so that we should discern multiple types of robots when determining different forms of extra-contractual responsibility, i.e., negligence-related tortuous liability or faultless responsibility for the behaviour of robots.

All in all, this threefold notion of robotic agency as a new "node of the network" represents one of the most relevant topics for further research in complexity, AI & the law. Since robots are here to stay, the aim should be to wisely discipline our mutual relationships in connection with the new frontiers of crimes, contracts, and torts, brought on by the information revolution.

References

1. Asimov, I.: Runaround. Doubleday, New York (1942)
2. Bekey, G.A.: Autonomous Robots: From Biological Inspiration to Implementation and Control. The MIT Press, Cambridge (2005)
3. Singer, P.: Wired for War: The Robotics Revolution and Conflict in the 21st Century, p. 77. Penguin, London (2009)
4. Moravec, H.: Robot: Mere Machine to Transcendent Mind. Oxford University Press, London (1999)
5. Kurzweil, R.: The Singularity is Near. Viking, New York (2005)
6. Asaro, P.: How Just Could a Robot War Be? Frontiers in Artificial Intelligence and Applications 75, 50–64 (2008)
7. Barrio, F.: Autonomous Robots and the Law. Society for Computers and Law (2008), http://www.scl.org/site.aspx?i=ho0
8. Reynolds, C., Ishikawa, M.: Robotic Thugs. In: 2007 Ethicomp Proceedings, pp. 487–492. Global e-SCM Research Center & Meiji University, Tokyo (2007)
9. Hildebrandt, M.: Criminal Liability and Smart Environments. In: Conference on the Philosophical Foundations of Criminal Law at Rutgers-Newark (August 2009)

10. Himma, K.E.: Artificial Agency, Consciousness, and the Criteria for Moral Agency: What Properties Must an Artificial Agent Have to Be a Moral Agent? In: 2007 Ethicomp Proceedings, pp. 236–245. Global e-SCM Research Center & Meiji University, Tokyo (2007)
11. Epstein, R.G.: The Case of the Killer Robot. Wiley, New York (1997)
12. Floridi, L., Sanders, J.: On the Morality of Artificial Agents. Minds and Machines 14(3), 349–379 (2004)
13. Veruggio, G.: Euron Roboethics Roadmap. In: Proceedings Euron Roboethics Atelier, Genoa, Italy, February 27-March 3 (2006)
14. Levy, D.: Love and Sex with Robots: the Evolution of Human-Robot Relationships. Harper, New York (2007)
15. Gogarty, B., Hagger, M.: The Laws of Man over Vehicle Unmanned: the Legal Response to Robotic Revolution on Sea, Land and Air. Journal of Law, Information and Science 19, 73–145 (2008)
16. Sullins, J.P.: Introduction: Open Questions in Roboethics. Philosophy & Technology 24(3), 233–238 (2011)
17. Karnow, C.E.A.: Liability for Distributed Artificial Intelligence. Berkeley Technology and Law Journal 11, 147–183 (1996)
18. Katz, A.: Intelligent Agents and Internet Commerce in Ancient Rome. Society for Computers and Law (2008), http://www.scl.org/site.aspx?i=ho0
19. Pagallo, U.: Robotrust and Legal Responsibility. Knowledge, Technology & Policy 23, 367–379 (2010)
20. Sartor, G.: Cognitive Automata and the Law: Electronic Contracting and the Intentionality of Software Agents. Artificial Intelligence and Law 17(4), 253–290 (2009)
21. Pagallo, U.: Killers, Fridges, and Slaves: A Legal Journey in Robotics. AI & Society, Springers online first (2011)
22. Allen, C., Varner, G., Zinser, J.: Prolegomena to Any Future Artificial Moral Agent. Journal of Experimental and Theoretical Artificial Intelligence 12, 251–261 (2000)
23. McFarland, D.: Guilty Robots, Happy Dogs: the Question of Alien Minds. Oxford University Press, New York (2008)
24. Goldberg, K., Paulos, E., Canny, J., Donath, J., Pauline, N.: Legal Tender. In: ACM SIGGRAPH 1996 Visual Proceedings, pp. 43–44. ACM Press, New York (1996)
25. Sparrow, R.: Killer Robots. Journal of Applied Philosophy 24(1), 62–77 (2007)
26. Canning, J.: Weaponized Unmanned Systems: a Transformational Warfighting Opportunity, Government Roles in Making It Happens. In: American Society of Naval Engineers (ASNE) Proceedings of Engineering the Total Ship (ETS) Symposium, Falls Church, VA (2008)
27. Sharkey, N.: Grounds for Discrimination: Autonomous Robot Weapons. RUSI Defence Systems 11(2), 86–89 (2008)
28. Pagallo, U.: Robots of Just War: A Legal Perspective. Philosophy and Technology 24(3), 307–323 (2011)
29. Krishnan, A.: Killer Robots: Legality and Ethicality of Autonomous Weapons. Ashgate, Burlington-Surrey (2009)
30. Foster-Miller Inc.: Products & Service: TALON Military Robots, EOD, Swords, and Hazmat Robots (2008), http://www.foster-miller.com/lemming.htm
31. Borden, L.S., Kozlowski, P.M., Porter, C.R., Corman, J.M.: Mechanical Failure Rate of Da Vinci Robot System. The Canadian Journal of Urology 14(2), 3499–3501 (2007)

32. Andonian, S., Okeke, Z., Rastinehad, A., Vanderkrink, B.A., Richstone, L.: Device Failures Associated with Patient Injuries During Robot-Assisted Laparoscopic Surgeries: a Comprehensive Review of FDA MAIUDE Database. The Canadian Journal of Urology 15(1), 3912–3916 (2008)
33. Cason, T.N., Friedman, D.: An Empirical Analysis of Price Formation in Double Actions Markets. In: Friedman, D., Rust, J. (eds.) The Double Auction Market: Institutions, Theories, and Evidence, pp. 252–283. Addison-Wesley, Reading (1993)
34. Rust, J., Miller, J., Palmer, R.: Behavior of Trading Automata in a Computerized Double Auction Market. In: Friedman, D., Rust, J. (eds.) The Double Auction Market: Institutions, Theories, and Evidence, pp. 155–198. Addison-Wesley, Reading (1993)
35. Miller, R.M.: Don't Let Your Robots Grow Up to Be Traders: Artificial Intelligence, Human Intelligence, and Asset-Market Bubbles. Journal of Economic Behavior and Organization 68(1), 153–166 (2008)
36. Hayek, F.A.: Law, Legislation and Liberty: A New Statement of the Liberal Principles of Justice and Political Economy. Chicago University Press, Chicago (1982)
37. Das, R., Hanson, J.E., Kephart, J.O., Tesauro, G.: Agent-Human Interactions in the Continuous Double Action. In: The 2001 Proceedings of the International Joint Conferences on Artificial Intelligence, pp. 1169–1187 (2001)
38. Floridi, L.: On the Intrinsic Value of Information, Objects and the Infosphere. Ethics and Information Technology 4, 287–304 (2003)
39. UN Word Robotics 2004: Statistics, Market Analysis, Forecasts, Case Studies and Profitability of Robot Investment. In: UN Economic Commission for Europe Staff and International Federation of Robotics Staff (ed.). UN Publications (2004)
40. Watson, A. (ed.): The Digest of Justinian. University of Pennsylvania Press, Philadelphia (1988)
41. Zimmermann, R.: The Law of Obligations: Roman Foundations of the Civilian Tradition. Clarendon, Oxford (1988)
42. The Economist: Drones and Democracy (October 1, 2010)
43. Teubner, G.: Rights of Non-humans? Electronic Agents and Animals as New Actors in Politics and Law. Max Weber Lecture at the European University Institute of Fiesole, Italy (2007)
44. Solum, L.B.: Legal Personhood for Artificial Intelligence. North Carolina Law Review 70, 1231–1287 (1992)
45. Chopra, S., White, L.: Artificial agents - Personhood in Law and Philosophy. In: Proceedings of 16th European Conference on Artificial Intelligence (ECAI), pp. 635–639. IOS Press (2004)
46. Casanovas, P., Pagallo, U., Sartor, G., Ajani, G. (eds.): AI Approaches to the Complexity of Legal Systems: Complex Systems, the Semantic Web, Ontologies, Argumentation, and Dialogue. Springer, Berlin (2010)
47. Latour, B.: Reassembling the Social: an Introduction to Actor-Network-Theory. Oxford University Press, Oxford (2005)

The Legal Challenges of Networked Robotics: From the Safety Intelligence Perspective

Yueh-Hsuan Weng and Sophie Ting Hong Zhao

Peking University Law School,
No.5 Yiheyuan Road, Haidian District, 100871 Beijing, China
{yhweng,sophiezhao}@pku.edu.cn

Abstract. One of the reasons that future robots will enhance their intelligence and actions in an unstructured environment is because of their "networked" feature. Current robot designs have difficulty in understanding unstructured environments due to the inherent diversity and unpredictability of phenomena in the real world. However, new developments such as ubiquitous computing, cloud computing, the Internet of things and next-generation internet technologies will make it easier for networked robots to obtain structured information about their physical environment. The formation of cloud-enabled robotics by advanced technology will be tightly integrated into the virtual and real world, and this will strengthen the impact of cyberspace to the real world. Although these developments may help reduce Open-Texture Risk from the networked robots, risk will be transferred from the physical world into the virtual world. In this paper, we will try to address some of the resulting legal implications. This paper is divided into four parts, the first part defines the meaning of cloud-enabled robotics; the second part analyzes how the Collective Dynamics derived from virtual and real world with autonomous behaviors by intelligent robots affect Open-Texture Risk to expand a Larger Range and bring a Deeper Impact; the third part explains the dispute of legal issues in future technology of cloud-enabled robotics; the final part analyzes the Safety Intelligence of cloud-enabled robotics in a long-term perspective, and the theoretical control framework that we propose in solving Open-Texture Risk.

Keywords: Networked Robotics, Liability, Robot Safety, Law & Robotics.

1 Introduction

The openness of cyberspace has led to innovation; and it has facilitated massive "linkage" and self-developed "intelligence". The intelligence in this space is currently emerging from the virtual world into the real world, one of the examples being the development of networked robotic technology.

Because of the uncertainty inherent in an unstructured environment, current autonomous robots have difficulty completing tasks in everyday human settings. For example, robots must be able to recognize many real world objects in order to build a structured model of the world around it. The current development of ubiquitous

M. Palmirani et al. (Eds.): AICOL Workshops 2011, LNAI 7639, pp. 61–72, 2012.

computing, which endows everyday objects with intelligence and access to the wider networked world, will make this task easier [1]. In other words, networked robots can reduce the uncertainty of their physical world with resources from the virtual world, or cyberspace.

Along these lines, researchers have been studying the potential for robots to serve as bridges between physical and virtual worlds. As Google's Vice President, Vinton Cerf, observes regarding the Internet:

> "Virtual and real worlds will merge. Virtual interactions will have real world consequences. Control of the electrical grid and power generation systems could be made to appear to be part of a virtual environment in which actions in the virtual space affect actions in the real space." [2]

Networked robotics allow for multi-robots to operate together in coordination or cooperatively with sensors, embedded computers, and human users [3]. The concept of networked robotics is a combination of robotic resources including sensors, actuators, and computational components. Under this definition one composes resources to create robotic agents to perform tasks. Therefore, in networked robotics, we associate tasks with agents and in turn associate agents with resources, rather than the robot platform. This gives networked robotics a wider reach to sensor networks and ambient intelligence [4].

Networked robotic systems are formed with many networked resources, whether it is many robots working together, or whether it is one single robot performing a task, they all need to cooperate to accomplish the goal. However, the diversity of networked robots' abilities will cause complication in the combination of software and hardware. This is why a middleware is needed to in developing and operating networked robots [5].

Google formally introduced the term "cloud computing" in 2007, and thereafter opened the era of cloud computing. The goal of cloud computing is for computing, service, and application to become easily obtainable. Resource users only need to focus on the usage, since all other problems including production difficulties and technical problems are left behind the "cloud". Characteristics of cloud computing include [6]:

Dynamic: The ability to handle load fluctuations to improve resource utilization, and a high degree of fault tolerance.
Scalable: The ability to meet different application scales, with little adjustment costs.
Virtual: The ability to perform in common network environment without the need to hold a variety of hardware and software.

The significant impact that cloud computing has on networked robotics is the strong middleware platform it provides. A recent European project called RoboEarth has begun to develop a globally accessible database designed for sharing robots information required in terms of object recognition, navigation and task completion in the real world. This framework helps robots to learn and adapt to unstructured environments [7]. Other potential benefits to robots from cloud computing include the outsourcing of computing power and the acquisition of new skills directly from the cloud. All together, these features should help to make human-robot interaction smoother. James Kuffner, the professor of Carnegie Mellon and Google robotics researcher, summed cloud computing's potential to improve robotics by stating that "embracing the cloud could make robots lighter, cheaper, and smarter." [8] Since the applicability of intelligence from the virtual world depends upon having accurate information about the real world, cloud-enabled robots still need more sophisticated sensors [9]. This sensing ability is a primary part of how robots will tap into ubiquitous computing in the physical world, as expressed in the concept, "Identification, Location, Sensing and Connected." [10]

The final result of these developments is the mergence of the virtual and real world. Autonomous, social robots will use cyberspace as a bridge into the unstructured physical world. Thus, robots will no longer be stand-alone entities, but instead "Tethered Appliances" [11] linked to cyberspace. However, we must highlight that the networked robots this paper focuses on belong to cloud-enabled robotics, they are very different from tele-presence robotics operated by signal transmissions. Networked robots are operated through transmissions of the Internet, which is not under the designer's control [12].

For tele-presence robots, networks are the only channels to transmit signals. Yet with the ubiquitous model, robots are more closely linked together with a network. More specifically, robotics will integrate into the network, and the impact of Internet on robotics will change from signal transmissions to agent channels. Especially for cloud-enabled robotics, this will greatly reduce their actions in unstructured environments and lead to blurring the boundary between cyberspace and the physical world, presenting many new legal challenges.

2   The Open-Texture Risk in Cyberspace

The relationship between law and robotics is dependent upon the nature of a robot's contact with society. A group of Italian researchers have investigated how current legal regulations, such as traffic code, civil law and criminal law would impact autonomous mobile robots active in urban areas [13]. However with factory robots, due to its low contact with society, the current legal regulation is only limited to address their performance of highly repetitive tasks in structured environments.

A characteristic of general industrial product mechanisms is predictability. Machines that are built according to specific standards cannot alter their mechanisms to match changing environments, making them predictable. Thus, their safety standard is relevantly easy to achieve. The commonly used way is using risk assessment to measure machine risk in order to design mechanisms for achieving approved safety levels. Unfortunately, this regulation model does not fit well with the safety requirements of autonomous robots according to the legal concepts of "core" meaning and "open-texture". Risk assessment associated with robots' autonomous behavior faces with the problem that since the range of core meaning varies according to specified domains, points of view, time periods, it is difficult to define. Thus this result to what we refer to risks from machines' autonomous behaviors as Open-Texture Risk [14].

Safety problems with autonomous robots can be divided into risks from machine standards and risks from their autonomous behavior. Machine standard risks can be regulated via a process of assessment and design, but the autonomous behavior of robots makes their risks complex, changeable, and unpredictable, and thus requires a different approach [15]. Uncertainty causes risk and contiguity to increase, which leads to further demand in law and safety. But with the application and development of networked robotics technology, uncertainty in automation will expand from the physical world to the interlacement of physical world and cyberspace. In other words, the law binding robotics will develop from foreseeable risk in the structured environment to Open-Texture Risk in the unstructured physical environment and further to the combined in one physical and virtual environment. The networked robots we discuss here are mainly the robots that bring Open-Texture Risk to cyberspace, striding over the real and virtual world. Although there has been autonomous agents running activities in cyberspace in the past (such as e-commerce agent), the risk of these types of agents are only limited in cyberspace and does not affect the result of robots in the physical world.

Comparing to the original Open-Texture Risk, our definition of "Open-Texture Risk in Cyberspace" is more focused on how the "networked" characteristic affects robots' autonomous behaviors, and this is based on two factual elements. First, the autonomy of the robot, since its actions taken are not dependant upon signal transmissions and second, the autonomous actions that stride in physical and virtual world, causing physical effect. Judging by the contact level with the physical world, robots in cyberspace can be divided into two types: Tele-bot and Ubi-bot. Tele-bot refers to the traditional use of a remote-controlled robot, while Ubi-bot refers to a networked and autonomous robot in one. Ubi-bots are much more accessible than Tele-bots, and Ubi-bots in autonomy interacting with the Internet will cause Open-Texture Risk in cyberspace. In the trends of network intelligence, many Open-Texture Risks deriving from automation will appear, which denotes that the solution of Open-Texture Risk will soon become an inevitable problem in cyberspace. In particular, using the Ubi-bots as an example, we identify the following two main issues:

(1) **Larger Range:** The approach of Ubi-bot is to establish a structured environment, eliminating or reducing the risk of stand-alone robot. However, the risk now transforms to potential risk in a large group of robots, it is derived from the group dynamics of cyberspace and makes it harder to define. The risk has changed from stand-alone robots facing uncertainty in the physical environment to group robots facing "more definite physical environment(structured)" but "uncertainty in cyberspace"

(2) **Deeper Impact:** Cloud computing established a "virtual world" which forms a "robotic cyberspace" for robots to operate, but mistakes in this virtual world will cause threats in the physical world. In addition from the physical world perspective, robots accept many structured signals and thus Open-Texture Risk is reduced. For example, the original risk from the calculation of the world model is reduced by directly downloading ready-made models through the server. But on the other hand, Open-Texture Risk becomes difficult to define as it has already been transferred to the server.

Open-Texture Risk will expand to a "Larger Range" and bring a "Deeper Impact" due to the "Collective Dynamics" caused by the interaction between physical and virtual world. The original networked robots are in a world with a high degree of autonomy by code, but with the above reasons stated, human law must integrate within. In the next two parts, we are going to discuss the short-term and long-term legal issues accompany with intelligent machine's Open-Texture Risk accordingly.

3   Ubiquitous Computing and Its Legal Implications

Recently, the governments of Japan [16] and Korea [17] have adopted policies to assist the formation of an intelligent environment that utilize networked robots in private residences, larger buildings, and cities on a large scale. However, the major difficulty for these networked robots with ubiquitous functions ("ubiquitous robots") [18] operating in human society is that robots' contact with society is expanded from the physical world to cyberspace, engendering several new legal issues. We will discuss below the main issues linked with cyberspace below.

First, ubiquitous robots, being fully embedded in the Internet, become a part of cyberspace. A middleware can help them to better interact with both worlds. They can easily access a model of the real world through a middleware server platform such as RoboEarth. Although this might be a proficient way of reducing the original Open-Texture Risk, it will result in the transfer of risk from the physical world into the virtual world. One example is a kind of sensing mechanism called "Moving Object Sensing Technology," which refers to the whole process of data collection and reuse from ubiquitous sensor networks via corresponding middleware [9]. Although some raw data gathered by robots may not at first glance, appear to contain any personal information, it may reveal personal details otherwise obscured when combined with data mining techniques. Networked robots, being able to adjust their behavior according to circumstances and gather information by themselves, may on their own combine private personal information with innocuous technical data [19]. Furthermore, cloud computing allows autonomous Ubi-bots' behaviors become more dependant upon Internet resources. For example, world models by previous robots in the download server will reduce exploration of current robots and robots may download and inherit previous experiences or skills from the Internet. As previously mentioned, the harm done to personal privacy by robots collecting information automatically in open environments will become more severe when robots are connected to the Internet. This is because Ubi-bots may bring environmental information containing disputable privacy protection issues to the cloud server in the process of establishing or editing world models and uploading previous experiences to the cloud server. Next, the other robots interacting with the cloud server will continuously replicate this disputable information, which will cause it to spread like an infectious disease. The original Open-Texture Risk in stand-alone robots will start to contain collective dynamics of Open-Texture Risk, making it harder to define. Current privacy protection law falls short in coping with this issue about how sensitive information may be collected and spread autonomously between real world and cyberspace. Therefore, it is necessary for comparative legal research to address how to cover this legal gap.

Open-Texture Risk versus liability will be another crucial issue that needs further investigation. There are different views regarding the liability of physical damages by robots. Salvini defines robots as a type of property and liability is divided in two levels. The first level is to compensate directly by the manufacturer, the second level is the manufacturer can plea against the user (based on third party misuse of robot). This model can usually solve the problem of the civil liability on intelligent robots [13]. However, for networked robots, it is much more complicated. Being an open platform itself, it involves many sides of robot hardware, manufacturers, software service providers and network service providers that are bound to increase difficulty in distribution of liability.

The "openness" that we believe networked robotics possess is similar to the general purposes variety usages of robotic platform which Calo describes. Calo believes the general purposes platform consists of three features [20]. The first is that they are multifunctional as they lack a set of functions or a specific task; the second is that they are nondiscriminatory since the openness of the robotics allows them to accept third party software; the third is that they are modular in hardware design. Considering with these features and the nature of networked robotics, we see that there are three main legal problems with the "openness" of networked robots: foreseeability, liability distribution, and time delay [21].

In the open world, liability of damages done by robots is very complicated. The manufacturer will not be able to foresee all the potential damages with the robot and this could be used as a defense for manufacturers. Calo regards without specific tasks for robots, the product misuse defense which could have been used by manufacturers against consumers in the closed world is not applicable in the open world. Robots in the open world are meant to be modified and there are no limits to proper usage. Therefore, the third party misuse defense will be invalid [20].

In addition to the unforeseen functions and difficulty in defining software operating systems and applications, the boundary between the virtual world and the real world is hard to define. For example, it is hard to determine whether networked robots' autonomous mistakes originated from the host body or network server. This will cause difficulty in distributing the liability of manufacturer and cloud service provider. As mentioned before, stand-alone robots may not only upload information harmful to personal privacy, but they may also upload world models or previous skills that effect robots' autonomous behaviors. The minor effects are the delay in actions of the physical world in the downloading process, but this situation may also cause severe results of physical harm, in which the liability comes in complicated at this point. The owner, software manufacturer, cloud service provider, hardware manufacturer will all be involved in for the allocation of responsibility of the risk source. Another problem is when Ubi-bots are striding between the real and virtual world (during the Ubi-bot's interaction with the real and virtual world), it makes it hard to calculate the proportion of their autonomous behaviors dependant upon their own to their behaviors reliant on the cloud computing server. This calculation of proportion is the fundamental basis for the distribution of liability. One possible solution for this situation is to enable precaution, for example, building a Reputation Ranking System which allows robot owners to do credit rating to models on the cloud computing server. However, this method faces ethical risk, as some people may use improper methods to create fake credits, making it easier for robots to download flawed models. Another solution is to embed software into stand-alone robots to monitor the uploading and downloading process of the risk source. But this method gives great power to the monitoring authority, which might make them the "Big Brother". Therefore, this method must be used with great caution and under strict supervision of the law.

The final issue regarding liability of networked robotics is the time delay. In order to ease highly computationally intensive tasks, ubiquitous robots may tap into the outsourced computing power of a cloud computing data center. Robots may outsource their computing power in many ways, such as from IaaS (Infrastructure as a service), PaaS (Platform as a Service) to SaaS (Software as a Service) service models [22]. Outsourcing computing power will be a feature for cloud-enabled robots, however, its deployment into the marketplace faces a major issue of time delay. Normally, time delay is a main technical problem for Tele-bots, but Ubi-bots also face the same problem. For example, if the time delay of Cloud Service Provider (CSP) causes wrong decisions with the robot's actions, what kind responsibility would CSP have to bear? In addition, as stated before, a feature of Open-Texture Risk in Cyberspace is its larger range, so once time is delayed, it will affect a large number of robots. The compensation will be a very considerable amount. Considering the stability of the server, RoboEarth does not consider outsourcing computing power at the moment. However, Google carries an optimistic view and strongly supports cloud computing robots develop towards this. With all the facts stated above, we are forced in a dilemma. If we drop outsourcing computing power, it will limit cloud computing's powerful features, but using it will face extreme risks of liability.

4   Networked Robots with Safety Intelligence

This paper focuses on "virtual and real world in one" and the Open-Texture Risk will be gradually proliferate due to the collective dynamics from the cyberspace, so we propose an alternative risk control mechanism for the change of Open-Texture Risk under a long term consideration.

The current robot safety system can be divided into the pre-safety stage and the post-safety stage [23]. In the pre-safety stage, there is a safety standard and safety strategies are used to reduce robots' risks in architectural and mechanical ways. As for the risks that cannot be solved in the pre-safety period, the post-safety stage offers liability and insurance system to handle the remaining risks. For industrial robots, the current pre-safety and post-safety system can manage the robots' safety issues very well. However, the pre-safety strategies cannot cover all the risks for autonomous robots. Although part of their risks might be solved by liability and insurance system, the risk gap between the pre-safety stage and post-safety stage still exists. Furthermore, this risk gap will increase continuously according to the development of robot intelligence in the following decade.

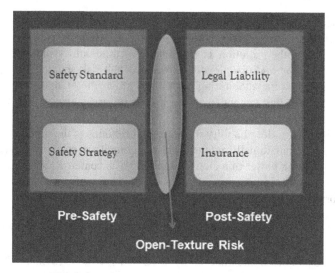

Fig. 1. Open-texture risk with robot safety system

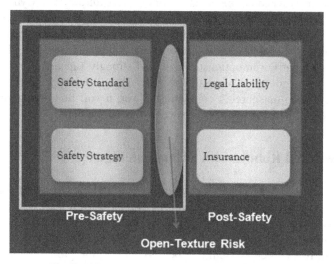

Fig. 2. Using pre-safety to solve open-texture risk

Facing the increase in Open-Texture Risk along with the development in robot intelligence, one solution is to keep the pre-safety system. These safety strategies reduce risks by machine architecture designs and mechanical skills. However, Open-Texture Risk is derived from the interaction between machine and unstructured environment. With the development of robot intelligence, the pre-safety system will meet new struggles in handling Open-Texture Risk. Another solution is to keep the post-safety system, which consists of liability and insurance. Yet liability itself cannot solve the Open-Texture Risk since liability is always distributed among people and robots themselves cannot be liable. Sartor claims that Software Agent (SA) can perform cognitive tasks and can be attributed intentional states. Even if Ubi-bot itself has a cognitive state as Sartor mentioned, it still should not be liable since SA are merely cognitive tools performing their tasks due to human users determinations. In addition, "obliged to compensation "is another standard to judge whether Software Agents are liable or not, therefore only human can be held liable [24].

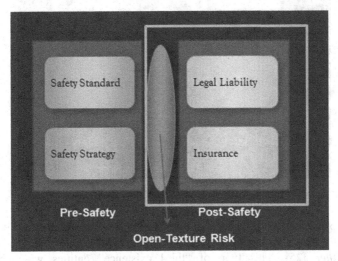

Fig. 3. Using post-safety to solve open-texture risk

Nevertheless, when robot intelligence develops to a certain degree, it will be unfair to distribute liability only among its user, manufacturer, seller etc. Insurance is only suitable as a supplement system and is usually used when liability of the manufacturer, seller cannot be held. Imagine if the risks that liability cannot solve are all passed over to the insurance system. It will face a problem involving the Open-Texture Risk on larger range and deeper impact. The insurance compensation will be very considerable, with most small to medium insurance companies not being able to provide robot insurance. When the insurance market is only left of a few big insurance companies with robot insurance, it will result in monopoly control of the market price. This way, it will be detrimental to the long-term development of the robot industry. Therefore, post-safety will not be able to solve Open-Texture Risk as well.

The alternative solution we propose to solve the risk gap between robot pre-safety and post-safety stages is called "Safety Intelligence". This framework attempts to control Open-Texture Risk through safety methods of using the robotic body to absorb risks, while maintaining the balance between the pre-safety and post-safety stage. However, this concept is still under the theoretical stage and has derived many Open Issues, such as "Robots as a Third Existence" and "Legal Machine Language"[15].

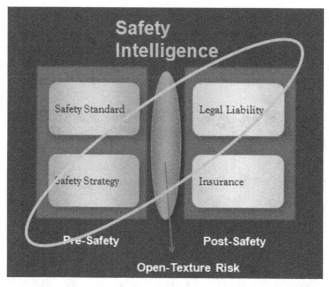

Fig. 4. Alternative solution: Safety Intelligence

Cynthia Breazeal has pointed out that there's a *"fuzzy boundary" that's very compelling for us, where we are willing to see robots not as human, but not exactly as machine either* [25]. This type of "third existence" entities which is neither living/biological (first existence) nor non-living/non-biological (second existence) will affect the accountability of the robot. For example, the traditional thought is worrying that robots may cause harm to human beings, but Salvini's empirical study has found that there are situations where humans bully robots as well [26]. On the other hand, robots' third existence features will also affect the behavior of robots for their own accountability. Dodig-Crnkovic and Curuklu emphasizes that *The intelligence must come in conjunction with ethics and through the concept of an artifact ethical by design* [27]. In addition to the moral ethics that engineers have to abide for Robot Ethics, we also need to take in consideration the artificial morality of the robots.

If the ethical by design mechanism could be adopted into the robot safety system, together with the advancement of mechanical to AI level in safety strategy of the pre-safety stage, such as the "Interaction Layer" that Haddadin pointed out [28], it will form a bridge across pre-safety and post-safety corresponding to liability. This may decrease the potential threat in Open-Texture Risk. Currently, Arkin is working on embedded code to enforce duties to autonomous military robots in order to solve the conflicts with international laws [29], however, Pagallo worries that this design approach will face a problem on ambiguity when the processed concepts are too abstract, such as "top normative concepts"[30]. This may be giving robots too much space for autonomy and increase the uncertainty of their behaviours. The ambiguity of this model will also become the greatest challenge in implementation of legalized machine language in the Safety Intelligence framework.

5   Conclusion

Virtual and real worlds will eventually merge. Ubiquitous robots will use cyberspace as a bridge to integrate into the unstructured physical world. Although this might be a useful way to reduce Open-Texture Risk in stand alone robots, risk is in fact transferred from the physical world into the virtual world. We have addressed several potential legal issues, including privacy protection and liability distribution. These issues also reflect that a critical challenge in regulating the future of cyberspace is to deal with the dynamics from users and the autonomy of web. Furthermore, from long term consideration, the Open-Texture Risk leaves a gap in the safety production of robots. Clearly delineating these issues now may give room for public debate and provide enough time to prepare legal regulation as the interaction between robotics and the Internet comes more and more close.

Acknowledgments. This work was supported in part by the Peking University Yahoo! Internet Law Center under 2010-2011 Yahoo! Grant "Cloud Computing and the Network Society: On a Legal Framework from Social System Design Perspective". Special thanks to Dr. Oliver Zweigle of the RoboEarth and Dr. Sami Haddadin of the German Aerospace Center (DLR) for their suggestions to this research.

References

 1. Kim, B.K., et al.: Networked Robots in the Informative Spaces. In: International Conference on Control, Automation and Systems (ICCAS), Busan, Korea, pp. 714–719 (2005)
 2. Cerf, V.G.: The Disruptive Power of Networks. Forbes Asia, 76–77 (May 7, 2007)
 3. Christensen, H.I.: EURON - the European Robotics Network. IEEE Robotics & Automation Magazine 12(2), 10–13 (2005)
 4. Mckee, G.: What is Networked Robots? In: Informatics in Control Automation and Robotics, Part I. LNEE, vol. 15, pp. 35–45 (2006)
 5. Mohamed, N., Al-Jaroodi, J., Jawhar, I.: A Review of Middleware for Networked Robots. IJCSNS International Journal of Computer Science and Network Security 9(5) (2009)
 6. Ahronovitz, M., et al.: Cloud Computing Use Cases White Paper - Version 4.0. Cloud Computing Use Case Discussion Group (2010)
 7. Zweigle, O., Molenraft, R., d'Andrea, R., Haussermann, K.: Roboearth – Connecting Robots Worldwide. In: International Conference on Information Systems (ICIS), Seoul, Korea, November 24-26 (2009)
 8. Guizzo, E.: Cloud Robotics: Connected to the Cloud, Robots Get Smarter. IEEE Spectrum (2011), http://spectrum.ieee.org/automaton/robotics/robotics-software/cloud-robotics
 9. Sato, T.: Moving Object Sensor Technology for Security and Safety. CREST Annual Research Report, Japan Science and Technology Agency, Tokyo (2007)
10. Weiser, M., Gold, R., Brown, J.S.: The Origins of Ubiquitous Computing Research at PARC in the late 1980s. IBM Systems Journal 38(4), 693 (1999)

11. Zittrian, J.L.: The Future of the Internet –and How to Stop It. Yale University Press, New Haven (2008)
12. Goldberg, K.: The Robot in the Garden. MIT Press, Cambridge (2000)
13. Salvini, P., et al.: An Investigation on Legal Regulations for Robot Deployment in Urban Areas: A Focus on Italian Law. Advanced Robotics 24(13), 1901–1917 (2010)
14. Weng, Y.H., Chen, C.H., Sun, C.T.: The Legal Crisis of Next Generation Robots: On Safety Intelligence. In: ACM 11th International Conference on Artificial Intelligence and Law (ICAIL), Palo Alto, CA, USA, June 4-8, pp. 205–209 (2007)
15. Weng, Y.H., Chen, C.H., Sun, C.T.: Toward the Human-Robot Co-Existence Society: On Safety Intelligence for Next Generation Robots. International Journal of Social Robotics 1(4), 267–282 (2009)
16. Japan Ministry of Internal Affairs and Communication: ICT Policy Outline, pp. 11–15 (2008)
17. Korea Ministry of Information and Communication: uIT 839 Policy Outline (2006)
18. Kim, J.H., et al.: Ubiquitous Robot: A New Paradigm for Integrated Services. In: IEEE International Conference on Robotics and Automation (IEEE ICRA), Roma, Italy, April 10-14 (2007)
19. Sanfeliu, A., et al.: Influence of the Privacy Issue in the Deployment and Design of Networking Robots in European Urban Areas. Advanced Robotics 24(13), 1873–1899 (2010)
20. Calo, M.R.: Open Robotics. Maryland Law Review 70(3) (2011)
21. Weng, Y.H.: Networked Robots: A Brief Look at Its Possible Legal Implications. In: 4th Workshop on Roboethics, IEEE International Conference on Robotics and Automation (IEEE ICRA), Shanghai, China, May 9-13 (2011)
22. Mell, P., Crance, T.: The NIST Definition of Cloud Computing - Version 15. National Institute of Standards and Technology, USA (2009)
23. Japan Ministry of Economy, Trade and Industry: Next-Generation Robot Safety Guideline (2007)
24. Sartor, G.: Cognitive Automata and the Law: Electronic Contracting and the Intentionality of Software Agents. Artificial Intelligence and Law 17(4), 253–290 (2009)
25. Lillington, K.: So Robots are Social Animals After All. Irish Times (2008), http://www.irishtimes.com/newspaper/finance/2008/1128/1227739081265.html
26. Salvini, P., et al.: How Safe are Service Robots in Urban Environments? Bullying a Robot. In: IEEE International Symposium on Robot and Human Interactive Interaction (RO-MAN), Viareggio, Italy (2010)
27. Dodig-Crnkovic, G., Curuklu, B.: Robots: Ethical by Design. Ethics and Information Technology (2011), doi:10.1007/s10676-011-9278-2
28. Haddadin, S., et al.: Towards the Robotic Co-Worker. Springer Tract in Advanced Robotics, vol. 70 (2011)
29. Arkin, R.C.: Governing Lethal Behavior: Embedding Ethics in a Hybrid Deliberative/Reactive Robot Architecture. In: HRI 2008: Proceedings of the 3rd ACM/IEEE International Conference on Human-Robot Interaction (2008)
30. Pagallo, U.: Robots of Justwar: A Legal Perspective. Philosophy and Technology 24(3) (2011)

Cloud Computing: New Research Perspectives for Computers and Law

Daniele Bourcier and Primavera De Filippi

CERSA - CNRS - Universite Paris II
{daniele.bourcier,primavera.de-filippi}@cersa.cnrs.fr

Abstract. Cloud computing represents a new business paradigm whereby a series of computing resources are offered as a service, available on-demand, on a pay-per-use basis, over the Internet. In this paper, we propose a hypothesis of how Cloud computing can be described as a complex system and we describe the various risks and opportunities connected with the current implementation Cloud computing. We then present a preliminary model for the implementation an automated system of certification based upon the formalization of contractual rules and consumers' preferences.

Keywords: Cloud computing, automated agents, contractual negotiations.

1 Cloud Computing: A Novel Paradigm of Complexity?

1.1 Definition

The Cloud consists of a distributed infrastructure that is made of a collection of interconnected computers, whose resources are pooled together into a virtual machine that maintains and manages itself. As opposed to other distributed architectures, the particularity of the Cloud is that its architecture is completely independent from the physical infrastructure it relies upon. This allows for extreme flexibility, as resources can be dynamically added or removed according to actual needs.

Although not a significant breakthrough in terms of technology (most of the technologies employed

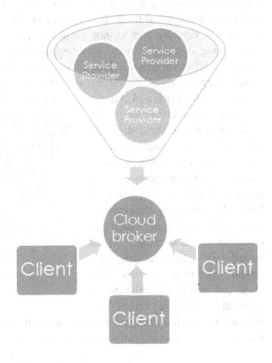

M. Palmirani et al. (Eds.): AICOL Workshops 2011, LNAI 7639, pp. 73–92, 2012.

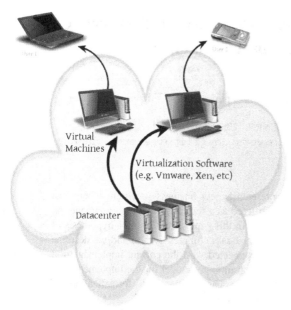

Virtual Machines

Virtualization Software (e.g. Vmware, Xen, etc)

Datacenter

in this model of computing were already available), Cloud computing has revolutionized the way in which technology is being employed. A new business paradigm has emerged where every application or resource is offered as a service, available on-demand, on a pay-per-use basis, over the Internet.

Virtualization, in particular, is a key technology for the implementation of Cloud computing environments. The idea is to pool together different physical resources into a single virtualized environment by means of specific virtualization software (such as Vmware, Xen, etc). The objective is to create a series of logical (virtual) machines with a dynamic set of resources. Virtualization permits a more efficient utilization of available resources. Indeed, thanks to this technology, the computing resources assigned to every virtual machine are not directly related to the underlying physical infrastructure, but are rather assigned dynamically according to the actual needs of the moment. Although a necessary attribute of Cloud computing, virtualization is not sufficient as such. It is the automated and self-provisioning aspect of Cloud computing that distinguishes it from former technologies in virtualized environments. Human intervention is no longer required in the case of Cloud computing, as resources are able to manage and re-organize themselves according to temporal and contextual contingencies.

Cloud computing is often broken down into three different categories: Infrastructure as a Service (IaaS), Platform as a Service (PaaS) and Software as a Service (SaaS). In the case of IaaS, the provider offers basic computing resources, such as computer networking, load balancing, data storage, and virtual operating systems. The client can benefit from the use of physical (hardware) resources,

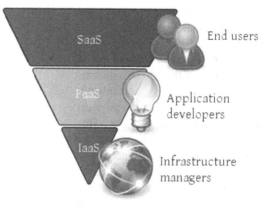

SaaS — End users

PaaS — Application developers

IaaS — Infrastructure managers

without the problems associated with the management thereof, and with the advantage that they can be dynamically resized according to the client's needs. PaaS provides a platform for users to develop software applications. It consists of a series of interactive tools, such as database management and application development, to support the making of powerful and flexible applications that will run in the underlying infrastructure of the Cloud. Finally, SaaS provides end-users with a software solution delivered over the Internet. It aggregates IaaS and PaaS together into a software application, which represents the service that end-users actually interact with.

Even though they qualify as different services, there is a definite overlap between IaaS, PaaS and Saas, as the latter cannot exist without the former two. It is, however, often the case that these different services are provided by different service providers.

From an operational perspective, the participation of separate but interconnected operators in the provision of one service is the main factor differentiating Cloud computing from former models of service delivery over the Internet. As opposed to previous business models, where one actor was responsible for the provision of one service, in its entirety, to a variety of clients, in the case of Cloud computing, the provision of a service requires the integration of a variety of services (infrastructure, platform, software, etc) provided by a variety of actors. This necessarily requires a complex network of contractual relationships amongst every actor involved in the provision of one integrated service to end-users.

Cloud computing is already widely deployed in the private sector and is, nowadays, also acquiring popularity in the public sector. Since government agencies must operate within a limited budget, Cloud computing can be used to decrease the costs and increase the efficiency of public administration, as well as to promote new services and initiatives that provide additional value to citizens. Given the central role it is starting to play in society, more and more lawyers and researchers are investigating the legal aspects of Cloud computing and debating strategies for the new challenges it engenders.

The Cloud's economic benefits are clear. Use of the Cloud enables both commercial actors and casual users to maintain as much or as little electronic data as they wish on a third party's mainframes without having to buy or maintain their own hardware systems. However, Clouds can be a legal minefield for companies and their lawyers. Data breaches, hosting of illegal content and inaccessibility of critical business information are just a few examples of difficult situations Cloud users can face.

There currently are a large number of initiatives, events, and international conferences on this topic taking place all over the world;[1] with this paper, we intend to launch the debate in the AI & Law community. We will describe, firstly, a hypothesis of how Cloud computing can be described as a complex system, and, secondly, the risks and opportunities connected with the current implementation Cloud computing. We will then present a preliminary model for the implementation an automated system of certification based upon the formalization of contractual rules and consumers' preferences.

---

[1] See, for instance, the State of the Art Analysis at http://crossroad.epu.ntua.gr/

1.1   Cloud Computing as a Complex System

Complex systems are characterized by (1) a large number of components, (2) partial knowledge of the relationships between them, and (3) limited predictability over the system evolution and dynamics due to the large number of actors involved. In the legal domain, the modeling of complex systems has been employed in a variety of fields (e.g. in France, in order to better understand and represent the network of articles referenced within and across legal codes[2]). In view of the large number of actors involved in the provision of Cloud applications, the complexity and lack of transparency that characterizes the network of contractual relationships, together with the ubiquitous and transient character of these relationships that must be dynamically updated or modified, the Cloud could ultimately be regarded as a complex system. As a consequence, the costs and complexity of managing and setting up the various aspects of an IT infrastructure (which can be significantly reduced thanks to the deployment of Cloud Computing), have been replaced by a new type of complexity, related to the management and coordination of the complex network of actors involved in the system.

Multiplicity of Actors
The network of contractual relationships in the context of Cloud Computing is becoming increasingly intricate and complex because a large variety of actors are generally involved in the provisions of one service. This creates a situation characterized by considerable **contractual complexity.**

Cloud computing modifies the relationship that subsists between users and service providers, but also amongst service providers themselves. While many operators are in charge of providing infrastructure and platform development, an increasing number of operators are offering services that rely on the infrastructure provided by others. Many Cloud applications involve a large number of actors that provide one or more services overlaid on top of the infrastructures tying them together (vertical integration) or based on the aggregation of services offered by others (horizontal aggregation). The higher is the number of services integrated together, the higher will the value of these services be to the users.

Even though, by exporting their resources in the Cloud, clients are moving away from the complexity of managing and coordinating the technological infrastructure, this complexity is being replaced by the necessity to coordinate the activities of different actors and to manage a complex network of contractual relationships.

Since they are purchasing a service rather than a product, it is important for clients to describe the service that they are willing to purchase. This is achieved by means of Service Level Agreements (SLA) – standard agreements intended to establish a common understanding between the clients and the Cloud provider with regard to the priorities and responsibilities of each party. Given that these agreements stipulate the minimum level of service that must be delivered by the Cloud provider, they

---

[2] MAZZEGA P., BOURCIER D., BOURGINE P., NADAH N., BOULET R., A Complex-System Approach : Legal Knowledge, Ontology, Information and Networks, in *Approaches to Legal Ontologies*, Theories, Domains, Methodologies Series : Law, Governance and Technology Series, Vol. 1 Sartor, G. ; Casanovas, P. ; Biasiotti, M. ; Fernández-Barrera, M. (Eds.) 1st Edition.,Springer, Heidelberg 2011, XIII, 279 p, Chap 7.

constitute the basis upon which clients can build up their expectations in terms of quality of service, infrastructure (uptime, response time, etc), security, privacy, responsibilities and potential liability of the service providers.

The problem is that clients generally enter into a direct contractual relationship only with one actor (i.e. the Cloud broker), but are generally affected by the choices of a large number of actors (i.e. different service providers) whose activities are critical to the provision of the service to which they have subscribed. While the Cloud broker might be aware of the client's expectations, there is no guarantee that the other actors involved will properly understand those expectations and actually fulfill them.

Opacity of the Network

One of the main advantages of Cloud computing is the reduction in costs resulting from an increased flexibility and scalability of resources. This has, however, to be counterbalanced with the higher costs that must be incurred to ensure the quality of the service. As the internal operation of the Cloud is inherently opaque, users inevitably lose control not only over the way in which they can access their own data, but also over the manner in which all data stored in the Cloud can eventually be exploited either by the Cloud provider or by third parties.

Nested Contractual Relationships

Before they enter into a contractual relationship with the Cloud provider, it is extremely important that users properly understand the terms of service. However, end-users are often reluctant to read the terms and conditions of the contract they agreed to because Service Level Agreements are often extremely complex and confusing. In addition, many end-user agreements are likely to change over time without any notice being given to end-users, who have already agreed to be automatically bound to the new terms and conditions.

The problem is further complicated by the fact that users usually enter into a contractual agreement only with the last actor in the supply chain (the Cloud broker) and are thus left without any recourse against the other actors involved in the actual provision of the service, who are not necessarily informed of the terms and conditions of the end-user agreement. Since the internal structure and operations of the Cloud provider or broker are generally not disclosed to the public, it becomes increasingly difficult for users and organizations to understand the actual scope of their contracts, and, in particular, to identify the terms and conditions that are not an integral part of the main contract.

Lack of Transparency

Whenever they move into the Cloud, users or institutions must export their data into the hand of a third party service provider. By doing so, they lose control over the way in which their data is being used, stored and processed by Cloud providers, as well as the manner in which the service will be delivered, as they have no knowledge nor control over the internal operations of the Cloud.

The terms of service can be defined by contractual means, by means of Service Level Agreements between end-users and providers, which have become a key aspect of Cloud computing. Due to the dynamic nature of the Cloud, ensuring that every provision of the SLA has been properly implemented and is still being respected requires however an active and continuous task of monitoring the Quality of Service (QoS) – and this is especially important in the case of enterprise customers that may outsource critical data. In particular, due to the raising concerns for privacy and data security, consumers may be hesitant to disclose certain details to Cloud providers.

Numerous other factors must be taken into consideration in order to assess the reliability and trustworthiness of a Cloud provider. These include, but are not limited to, the type of services provided, the overall accessibility and availability of these services, the formats, standards, and interoperability of the system, but also the respective roles and responsibilities of each party involved. Since different actors are likely to have different preferences and different adversities to risk, every contract must be carefully analyzed and assessed.

The higher is the number of parties, the harder it is to perform a proper assessment. The complex nature of the Cloud is therefore likely to introduce a series of challenges related to the **protection of privacy, the enforcement of intellectual property, the security and confidentiality of data, and, most importantly, the problem of liabilities and responsibilities** involved with the enforcement of various rights and obligations assigned to different actors, either corporate or consumers.

Unpredictability of Relationships

Cloud computing is used to provide flexible solutions that can automatically be adjusted to the changing needs of users. In a dynamic environment, relationships between actors need to be constantly changing or evolving. Clients are increasingly attracted to Cloud solutions because of the lower initial costs it entails, but mainly because of the possibility to pay only for the resources that they effectively use at any given period of time. This is the concept of utility computing, a new model of business whereby computing resources are no longer a product to purchase, but rather a service to subscribe to. This naturally requires a higher degree of elasticity with regard to the infrastructure, services and the different actors involved. Due to the dynamic character of the architecture of the Cloud, and to the temporary or transient character of every contractual relationship it made of, it becomes almost impossible to predict the way in which the Cloud will evolve over time.

Volatility of Actors

Cloud computing has disrupted the traditional value chain of service provision. A Cloud service is delivered by a variety of actors, whose identity can change over time without necessarily changing the nature or the quality of the service.

Even though they appear as infinite to end-users, the amount of resources available in the Cloud are of course limited to the resources provided by the various actors in the Cloud. Perfect elasticity requires the Cloud broker to be able to contract a new service provider whenever the need arises, and resource optimization requires that one service provider be replaced by another whenever the service of the latter is more valuable and/or less costly than that of the former.

As a result of virtualization, actors can come and go in and out of the Cloud in a completely transparent way. The identity of any actor whose role is to provide a particular kind of resources is ultimately irrelevant, provided that the resources it provides are actually interchangeable with each other. Users are not directly affected by the shift from one service provider to another, because most of the resources they provide are simple commodities, which have been gathered together into a virtual infrastructure that is completely independent from the underlying infrastructure.

The advantage is that, given that they are in a contractual relationship only with the last player in the supply chain, changing the identity of service providers does not have any impact on the usability of the system as a whole. Hence, users do not need to be informed of any change that is performed within the internal structure of the Cloud.

Dynamic Revision of Contractual Terms

A dynamic revision of contractual provisions is necessary in order to allow for a better re-organization of resources. Given that it has been designed to support unpredictable workloads, the architecture of the Cloud cannot itself be predicted. At any moment, clients' needs might either drop or drastically increase for a very short period of time. Clients might also decide to upgrade their subscription with a particular Cloud provider - in order to benefit from a broader range of services or resources – or even to subscribe to a completely new or different service, perhaps with a new service provider.

Because of the pace at which these revisions happen, terminating and re-creating a new contract each time would prove to be extremely tedious and inefficient. The solution is to integrate within the contract itself the possibility for the client (and sometimes even the service provider) to change the terms and conditions regulating the provision of the service. While this higher degree of flexibility significantly reduces the costs and complexity of contractual negotiations, it however considerably increases the level of complexity in the system, thus making it even harder to predict the manner in which the Cloud environment is likely to evolve in the future.

Transnationality

The widespread deployment of Cloud computing is likely to have a significant impact on the legal system as a whole, which traditionally relies upon the **concepts of jurisdiction**, national boundaries and territoriality.

Cloud computing services generally extend over several jurisdictions with a large number of data centers globally distributed around the world. In order to ensure a fast and reliable service at minimum costs, data is often replicated in several data centers and may end up distributed across multiple jurisdictions. Cloud computing technologies are designed for data to move around from one data center to the other according to the actual and expected utilization of available computing resources, but also depending upon the current level of congestion of the network. Minimum latency (i.e. the time required to access the data when requested) can be obtained by storing data simultaneously in multiple locations, whereas maximum storage and computing capacity requires a constant flow and transfer of data across different data centers. All these algorithms are unlikely to take national boundaries into account. Although certain service providers allow their clients to specify the country or region in which their data must be stored and/or processed, this is generally the exception rather than the rule, given that the geographical location of data is often difficult to determine ex-ante.

Overall, the issue can be traced to the fact that the fluidity and volatility of data stored in the Cloud is in conflict with the more static and deterministic character of the law. National boundaries are irrelevant in the context of Cloud computing, whose infrastructure exclusively depends on the architecture of the Internet and on the performance of the network. It becomes therefore very difficult to identify the applicable law, and, in the case of litigation, to determine who should ultimately be held liable for what.

2   Risks and Opportunities of Cloud Computing

2.1   Costs

The most obvious advantage of Cloud computing relates to its costs. Huge economies of scale make it possible for large service providers to offer their services at only a fraction of the costs that their clients would otherwise have to incur in order to set up an analogous infrastructure by themselves. Virtualization allows for a more efficient repartition of resources by separating the logical infrastructure from the technical and hardware architecture. A dynamic configuration of resources based on actual needs promotes a more efficient allocation of resources and reduces the risk of entering into a situation characterized by under/over utilization of resources. From the perspective of providers, this can be extremely valuable because it reduces the sunk costs that have been previously incurred in order to set up their underlying infrastructure. Virtually any resource that is not currently being used by the Cloud provider can be temporally assigned to one of its client. From the perspective of the clients, this can be very convenient because it completely eliminates the initial investment necessary to purchase hardware or software resources and properly setting them up. Most of the fixed costs (in terms facilities, hardware and software resources, technical management and engineering, etc) are fundamentally converted into variable costs.

Yet, the complexity of the Cloud introduces a variety of new costs related to complexity management. On the one hand, as the number of actors involved in the provision of a service increases, contractual negotiations becomes increasingly complex and costly. On the other hand, as the level of control over the infrastructure and the resources decreases, identifying a breach of the contract can become very difficult. The costs of monitoring the operations of the Cloud in order to ensure compliance with the agreed terms and conditions are likely to be very high whenever there is more than one actor involved in the provision of a service. This is even more critical when the identity of the actors responsible for providing the service has not been previously established or is likely to change over time. In an international context, the costs of enforcing contractual provisions can eventually overcome the benefits derived from an increased elasticity and scalability of resources.

2.2   Security

Cloud computing can either improve or reduce the security of a system. While most security mechanisms provided by Cloud providers are likely to be more robust and effective than those set up by end-users, the centralization of data into the hand of a few can make those players more prone to be attacked.[3] If Cloud Computing is characterized by the virtualization of a common pool of shared resources, every service provider must have a mechanism to control and access a variety of resources (e.g. storage, processing power, memory, bandwidth) from a centralized interface (the "hypervisor") in charge of re-organizing and re-allocating these resources according to the specific needs of the moment. To the extent that they are accessible through the Internet and that they provide access to a much larger quantity of data and resources, Cloud-based services constitute a more attractive target for attacks than more traditional servers.

The shift from traditional on-premise storage and operations to Cloud-based solutions can greatly reduce the costs for clients to set-up and secure their own infrastructure, which can generally be done in a more efficient and securely manner by a professional team of system administrators. However, this reduction in costs is to be compensated by the additional costs to be incurred to ensure that the security mechanisms adopted by every player in the Cloud are actually in line with the security requirements of each individual client. Clients often require their service providers to follow good security practices as an attempt to decrease the risks of attack and to diminish the consequences thereof. Yet, several actors are usually involved in the provision of a Cloud-based service. The greater is the number of actors involved, the higher are the risks that something will eventually go wrong. Besides, most of the service providers that clients communicate with are often unable (or unwilling) to provide everything on their own terms. They frequently aggregate a series of third-party services under a common framework, which - although presented as a single integrated service - is actually made up of a variety of services administered by a variety of actors with their own individual policies and security practices. Regardless of the degree of protection promised by the Cloud provider, the security of information is ultimately determined by the weakest link in the chain. Insofar as data is transferred through several intermediaries, only one of them needs to be violated for any malicious user to obtain the relevant information. Hence, the chances for inadvertent exposure increase substantially with every new intermediary and with every new layer of abstraction.

2.3   Privacy and Confidentiality of Information

There is an inherent security risk in the use of the Internet to transfer sensible information and personal data, but that risk has been considerably increased with the deployment of Cloud computing. The transfer and processing of personal information in the Cloud need to be carefully monitored in order to ensure that the privacy of end-users has not been infringed. The reason is that information stored in the

---

[3] See the ENISA report (2009) on Cloud Computing: Benefits, risks and recommendations for information security, which identifies the main risks of Cloud Computing in terms of information security as being due to loss of governance and user lock-in; isolation failure and compliance risks; management interface compromise; improper data protection; incomplete or insecure data deletion; and malicious insiders.

infrastructure of a third party has weaker protection than information that remains in possession of the data subject.

To begin with, the laws of certain countries oblige certain service providers to communicate to the authorities any information that constitutes evidence of criminal activities. This means that government agencies can, under certain circumstances, require the disclosure of personal or confidential information.[4] The international character of Cloud computing introduces an additional layer of complexity, given that information stored in the Cloud can be subject to a variety of different laws depending on the location where it is being stored or transmitted. Cloud providers might avail themselves of the services of other Cloud providers located in different jurisdictions, or they might distribute their data amongst multiple data-centers according to economic and/or legal incentives (i.e. forum-shopping). The difficulty for users to know with certainty which law applies to the information published into the Cloud raises a series of critical concerns in terms of privacy and confidentiality of information. Finally, while users generally disclose information voluntarily on the Internet (by means of e.g. through blogs, forums, newsgroups, mailing lists, search engines), problems would arise if the information given to separate (and apparently independent) services were actually aggregated together by one single entity (either because it is the common provider of said services, or because it has acquired the data from third parties). If one single entity were to provide a large variety of services and the data collected through all of these services were to be processed into an integrated framework of analysis, that entity would fundamentally be able to know much more about its user-base than what has been voluntarily disclosed by each individual user. This is problematic because, even though information had been voluntarily provided by users, aggregated data might provide further information about users, which they did not necessarily want to disclose.

2.4 Liability and Responsibilities

In front of such a large number of actors and such a diversity of regulations around the world, the traditional role of the law is getting less and less relevant and contractual relationships are assuming an increasingly important role.

Given the complexity of Cloud computing, particular attention should however be given to the specific rights and obligations assigned to each party to the contractual relationship. The dynamic character of the Cloud is such that any service provider could decide at any given time to out-source part of its infrastructure and operations to third-party providers, without ultimately informing the other parties to the contract. Although the operation is generally not visible to end-users, it might nonetheless affect the quality and reliability of the service as a whole. In order to preclude any

---

[4] For instance, in the USA, although the Electronic Communications Privacy Act (ECPA) provides a series of protections against the access by governmental agencies to personal information held by third parties (18 U.S.C. § 2510-2522 and § 2701-2712), these protections have been subsequently weakened by the USA PATRIOT Act, which entitles the FBI to compel, following a court order, the disclosure by Cloud providers of any record stored on their servers (50 U.S.C. § 1862).

responsibility in the eventuality of failure, most of the services provided to end-users (SaaS) are offered under specific Service Level Agreements that stipulate that the service provider cannot be held responsible or liable for the activities performed by third-party contractors.

This raises a series of legal challenges, which have still to be properly addressed. If service providers disclaim any form of liability towards end-users, what kind of recourse is available to users? Do they have a legitimate cause of action against the subcontractors who actually caused the damage, even though they are not in direct contractual relationship with them? If there is no recourse, who should be held responsible for a breach in the system? Who should be held liable for the improper transfer or illegitimate processing of data in the Cloud? Most importantly, if the players involved in the provision of a services have not been previously determined and are likely to change over time, how can users ensure that the level of service will remains the same? What are the legal consequences of any change in control? These questions have thus far not been addressed by the majority of Service Level Agreements. Given the strong asymmetries of information and the difference in bargaining power, not only is it very difficult for users to ensure that the service complies with the terms and conditions of the contractual agreement, but it is even harder to enforce these terms upon every actor involved in the provision of that service.

3   Formalization of Contractual Rules: Towards an Automated System of Certification?

As Cloud computing is being adopted by an increasingly larger number of businesses and individuals, the underlying technology and infrastructure is continuously evolving, but the law does not seem to follow the pace. Public regulation (such as intellectual property law, privacy law, and consumer protection law) is being superseded by **private regulation**.[5] Today, private parties - rather than legislators - are determining the rules of the game. What can or cannot be done is no longer a matter of law, but more a matter of what has been previously agreed upon between a variety of private entities. The problem is that if everything is to be regulated by contracts, the number and the complexity of contractual agreements will constantly keep increasing.

This complex and fluctuating system of contractual relationships requires more sophisticated means of management and enforcement, in order to embrace - rather than resist - the dynamic nature of the system. In this regard, we believe that semantic rules combined with Artificial Intelligence (AI) could reveal themselves useful, not only in order to simplify the work of lawyers in elaborating new contracts, but also in order to counter some of the concerns generated by use of these new technologies by way of technology itself.

---

[5] "As Facebook extracts commercially-valuable information from the aggregation and correlation of millions of users" in Gillian Hadfield, "Legal infrastructure and new economy" *USC CLEO Research paper* n° C10-7,
http://papers.ssrn.com/sol3/papers.cfm?abstract_id=1567712

3.1 Automated Contracts

The size and complexity of contractual relationships in a Cloud environment highlights the need for electronic support in every aspect of contractual activities. More precisely, the formalization of contractual rules can reduce the complexity associated with Service Level Agreements at the level of (1) the negotiation, by simplifying the procedure of identifying a common ground of agreement between each client and the different service providers involved in the provision of a Cloud service; (2) the formation of the contract, by allowing for the drafting of a contract to be performed automatically according to the specific criteria which have been previously agreed upon during the process of negotiation; (3) the performance, by providing a more efficient way of identifying the various rights and obligations assigned to the relevant actors; and (4) the enforcement, by providing a benchmark against which to compare the levels of performance of the services obtained from monitoring.

There is an unlimited range of possible tools that could be deployed to support the formation, performance and enforcement of contractual agreements in a Cloud environment, let us analyze a few.

Contractual Negotiation

The automatic negotiation of SLAs requires that every Cloud provider specifies in advance the terms of service that it will abide to - in terms of the resources provided (i.e. hardware infrastructure, software applications, network bandwidth, etc) and the manner in which these resources will be provided (i.e. costs, up-time, security, privacy, conditions, liabilities, responsibilities, etc); and that users expressly communicate the minimum level of service that they are willing to accept - in terms of the resources they want (e.g. storage, services and applications) and the way they want it (e.g. speed, up-time, security and privacy level, etc).

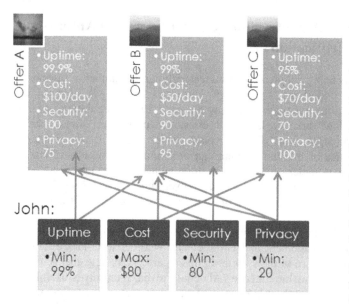

Through the formalization of the preferences of each party into a language that can be understood by a machine, it becomes possible to implement a mechanism whereby an automated system can autonomously determine whether the service offered by a provider actually complies with the individual needs and requirements each individual users (or other service providers) by merely comparing the formalized terms of the service provider with the formalized preferences of each client. The procedure can be repeated as many times as necessary, according to the number of actors involved in the provision of the Cloud service, and must be reiterated every time new service providers are incorporated into the Cloud, or whenever they change or update their terms of service.

For the purpose of clarification, we provide an illustration on how formalizing the preferences of the various actors involved can simplify the process of contractual negotiation. Let us take as an example the different services (A, B, and C) offered by different companies. Each offer is characterized by a series of attributes or guarantees that each service provider is willing to provide (e.g. computing resources, uptime, response time, degree of security and respect for privacy) and the various criteria or conditions at which it is willing to provide them (e.g. costs, liabilities, responsibilities). Offer A is very costly, but it is also extremely secure and is guaranteed to be working 99.9% of the time. It does not, however, guarantee a very high standard of privacy. On the contrary, Offer C is very concerned with the privacy of end-users, but does not however care too much about security. Finally, Offer B is much cheaper than the other two, but - although it is averagely good - the service does not actually excel at anything. New let us now consider the preferences of user John. John does not need a high level of privacy, but is rather concerned with the availability and security of the system. He is not willing to pay more than $80 for such a system. Provided that all those terms and conditions have been formalized into a machine-understandable language, John can rely on an automated system in order to identify the offer that best fits its criteria - according to the weight that has been assigned by John to every one of his preferences. In the case under analysis, B is the only offer that actually satisfies the four criteria stipulated by John, and is therefore the one that will be ultimately selected by the system.

The advantage of this approach is that every actor independently declares the minimum level of service that it is willing to provide or accept. The client enters into a contractual relationship only if the service as a whole (in aggregated form) fulfills all of the predetermined criteria. Not only can this significantly reduce the complexity involved in contractual negotiations, but this is also likely to increase the satisfaction of users who no longer have to commit to a standard-form agreement, but can actually obtain a service that specifically complies with the terms of the service to which they have subscribed.

Contract Formation

Once the best offer has been identified, it becomes possible to formulate a contract automatically without further negotiations, since all the relevant elements of the contracts have already been determined by the parties beforehand. A contract is an organized collection of concepts; a collection of rights, obligations, permissions, entitlements, and so on. It is also a collection of procedures that specify the operative aspects of the contract, e.g. how a particular exchange is to be conducted in practice, and a collection of parameters, such as the parties involved, the product of trade, the price of that product, and so on. [6]

---

[6] These notions have already been studied extensively in legal theory, namely, in the field of Artificial Intelligence, see : A. Daskalopulu & MJ Sergot, The representtaion of legal Contracts, AI & Society, 11, Nos 1/2, pp. 6-17.

Most importantly, a contract can be regarded as a collection of separate but interrelated sub-agreements. If contractual negotiations were to be guided by a formalized set of rules and constraints, contract formation could be supported by automated tools that understand the ways in which the contract is to be constructed in all of its components and sub-components (and where compliance with the rules and constraints of every part of the contract is a necessary requirement for the coherence of the contract as a whole). Provided that every user's criteria and every provider's condition can be linked to the corresponding, contractual provisions it refers to (in the form of a template), once negotiations are over, an automated system could subsequently proceed to the "composition" of the contractual agreement (as opposed to the drafting thereof). This is achieved by gathering together the relevant sections of the contract (i.e. a series of template provisions) and filling them up with the values that represent the common grounds of agreement between every service provider and client.

Performance

The formalization of contractual rules could strongly facilitate the exercise or the performance of contractual rights and obligations within a Cloud environment. Given that every individual user has entered into a different contractual agreement with different service providers, the proper execution of these contracts ultimately depends both on the identity of users and the distinctive characteristics of the service that every service provider has committed to give them. and could support the verification of the extent to which performance actually complies with the contractual provisions.

In a recent paper, Pankesh Patel, Ajith Ranabahu, and Amit Sheth[7] propose a mechanism for managing SLAs in a Cloud computing environment using the Web Service Level Agreement framework, developed for SLA monitoring and SLA enforcement in a Service Oriented Architecture (SOA). The authors suggest that all tasks performed within the Cloud can be defined by logical operators or functions. If this were to be the case, the contractual provisions of every SLA could be formally represented according to a series of logical standards in order to come up with custom logic-based tools capable of understanding and potentially even enforcing these contractual rules.

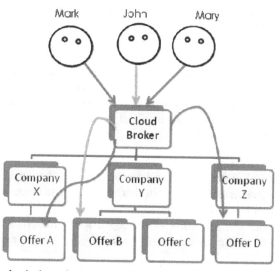

[7] See Pankesh Patel, Ajith Ranabahu, Amit Sheth, "Service Level Agreement in Cloud Computing" Cloud Workshops at OOPSLA09, 2009.

One particularity of the Cloud is that, in general, clients do not enter into a direct contractual relationship with every actor of the supply chain, but only with one particular agent that assumes the function of an intermediary between the clients and different actors involved in the provision of a service. This is the role of the Cloud broker, who is ultimately in charge of gathering together a large number of services offered by a variety providers and reorganizing them into a single integrated service that is offered to end-users.

The problem is that different clients might have different preferences, criteria, or expectations. Every user request needs therefore be processed by the Cloud broker before it can be forwarded to the actual service providers. Whether or not the request will be passed on to a particular service provider ultimately depends upon whether or not the service it provides is actually compliant with the terms and conditions incorporated within the SLA of the specific user in question. The same applies at deeper levels of analysis - e.g. if certain service providers decide to outsource part or all of their services to one or more third parties. Users' requests will only be forwarded to those service providers who can guarantee that the service provided by the external contractors is line with each and every user's preferences or requirements. The distinctive characteristics and attributes of each aggregated service (in terms of quality of service, security, privacy, etc) will therefore be ultimately determined by the least valuable or trustworthy of the services it aggregates.

In this respect, the role of the Cloud broker is to aggregate different service providers under a common framework, while ensuring compliance between each user's criteria and the terms of each service provided. Given that the internal operations of the Cloud are invisible to end-users, the Cloud appears to end-users as one comprehensive service, regardless of the number of actors involved in the actual provision thereof. Who is in charge of providing that service is ultimately irrelevant to end-users, who are only concerned with ensuring that they are actually getting a service that satisfy their criteria. This means that, provided that they all guarantee the minimum standard of service requested by a user, it is theoretically possible for the Cloud broker to shift from one service provider to the other without affecting the interests of end-users, nor infringing any contractual provision. Although this might be a very challenging task, the formalization of user preferences and service specifications into a formal language that can be understood by a machine could drastically reduce the complexity of identifying the routing assigned to each user requests, by allowing for every user's criteria to be assessed against the technical specification of alternative services.

The formalization of contractual rules and user preferences could even go further and extend to data itself. Indeed, it is often the time that one single user has different requirements for different types of data which has to be exported into the Cloud. For instance, while many users are likely to request that their personal data be subject to a higher standard of privacy, they might rather give more importance to speed, uptime and security when it comes to the storage or processing data that use on a daily basis. Temporary data of no or little importance could instead be assigned to a different service provider who does not guarantee much privacy or security, but whose cost is much lower than competing services.

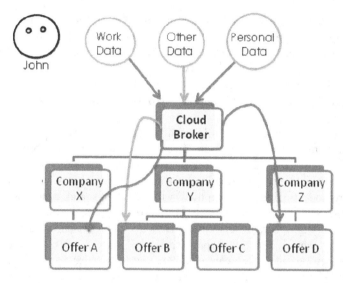

By means of metadata, data could be tagged in such a way as to automatically communicate to the system the various locations where it can be stored and the way in which it can be processed, e.g. 'this is personal data that must be treated according to UK law'. With such kind of information, the Cloud broker is able to determine, without human intervention, how to properly route the data in compliance with a series of criteria. All that the Cloud broker has to know is that the technical specification of the service provider comply the predefined requirements which have been contractually determined by the parties during contractual negotiations. The advantage of encode information directly into the data, rather than into the SLA, is that the conditions becomes inherently linked with the data itself, as opposed to the identity of the users. This allows for data to travel from one Cloud provider to the other without the necessity of entering into a new contract each time.

Contract Enforcement

Not only can the formalization of contractual provisions simplify and eventually enhance the performance of many SLAs, but it can also facilitate the procedure of enforcement. With the tools provided by recent developments in defeasible logic,[8] it is in fact possible to formalize the specific damages or reparation obligations that must be executed by one party whenever a right has been infringed or an obligation has not been properly fulfilled. This enables all parties to precisely understand the consequences of their acts and the compensation they can expect from the breach of any contractual provision. Whenever a particular event is triggered as a result of an action or non-action by one party, another party will be granted a new right, which generally constitutes an obligation to be fulfilled by the counterpart. With the representation of these rules into a formal and logical language, an automated system can communicate with the interested parties in order to inform them that a new obligation has emerged resulting from the breach or the improper performance of

---

[8] The activation of certain obligations in case of other obligations being violated is referred to as contrary-to-duty obligations (CTD) or reparation obligations (e.g. damages). These obligations are in force only when normative violations occur and are meant to 'repair' violations of primary obligations. See Governatori, Sadiq (2008), The Journey to Business Process Compliance.

another right or obligation, and, to the extent that it is practically possible, this new obligation can be automatically enforced from within the system.

The problem with said mechanism is that it fundamentally qualifies as a mere mechanism of **auto-certification,** based on the formalization of individual preferences or criteria, on the one hand, and the formalization of the service's provider terms of services on the other. The problem is that, while users have no incentive to lie about their own preferences, there is a strong incentive for service providers to commit to a much higher standard of service than what they are actually able or willing to provide, in particular because there is no way for users to actually find out whether or not their commitment has been properly or entirely fulfilled. Most Cloud services are offered as a black-box and provided to users without knowledge or visibility over the operational aspects thereof. Hence, any system of auto-certification ultimately depends upon the credibility and reputation of operators. Even if an operator is genuinely offering a service that purports to comply with certain standards or criteria, users can never be sure that it will actually succeed in fulfilling the prescribed standard of service. While it is always possible to introduce a system of liabilities and compensation in case of failure, there is no way for users to find out whether there has been a breach in certain provisions of the SLA (e.g. whether the proper level of security has been secured or whether the proper standard of privacy has been respected) before the situation gets out of hand.

In spite of the advantages provided by such a mechanism of auto-certification, the system is inherently flawed in that there is no guarantee that the terms of service stipulated by every Cloud provider will be respected, and, most importantly, there is no way to find out whether these providers are actually implementing the policies to which they promised to abide. The lack of transparency that is characteristic of every Cloud environment requires therefore the introduction of a new actor, whose function is to monitor and analyze the internal operations of the Cloud.

3.2   Third Party Certification

By and large, in Cloud computing contractual provisions are presented as pre-formulated standard contracts and the customers cannot audit the infrastructures of the brokers and providers: users can only ask to access to the audit report (ISO 27001 SAS 70).[9] If a system of auto-certification is not able to ensure accurate and transparent disclosure of information, the process of certification must be delegated to a trusted third party.

In this respect, the introduction of a new actor - the Cloud auditor - could further simplify the process of contractual negotiations by decreasing the costs of acquiring information and by reducing the risks of false or inaccurate declarations. Auditing Cloud-based services can however be quite challenging, not only due to the lack of transparency on the part of Cloud providers, but also because services are often deployed across different Cloud providers, each with their own distinctive attributes and characteristics.

In addition to the mechanism of auto-certification, a complementary mechanism of certification could therefore be adopted, whereby each Cloud provider whose services

---

[9] See I. Renard, J.M. Rietsch, (2012) Aide mémoire de droit à l'usage des responsables informatiques, Paris, Dunod, pp. 171-185.

actually satisfy a particular set of requirements would be granted a particular certificate by a third party certification authority. The duty of each certification authority is to investigate the internal operations of the Cloud and to issue a certificate whenever certain criteria are met. Certificates could theoretically refer to any aspect of the Cloud (e.g. the Certificate of Privacy, the Certificate of Security, etc) and could eventually be subdivided in different categories (e.g. level 1: minimum security, level 10: very high security) to precisely convey the range of minimum requirements that every service provider actually complies with.

In order to further facilitate contractual negotiations between service providers and end-users, these certificates could also be encoded into a format that can be understood by a machine, so as to make it possible for service providers to incorporate a certificate directly into their terms of services (i.e. to convey that the service complies with the requirements of that particular certificate), and for users to incorporate it into their own set of criteria (i.e. to convey that they are only willing to subscribe to a service to which that particular certificate has been granted).

To a certain degree, this mechanism of certification could be regarded as a preliminary system of standardization, given that each certificate can be regarded as a "tag" or "label" – acting as a guarantee that a service provider complies with a certain standard of service, regardless of the way in which the service is actually being implemented at the operative or technical level. For instance, to the extent that they all achieve a similar level of security, several service providers could be granted the same Certificate of Security regardless of the technology they use to actually secure their system. To the extent that users are only required to understand what does the certificate implies, rather than having to understand the pro and cons of the underlying technologies used by every service provider, each certificate can be regarded as a short-cut which has the potential of significantly reducing the costs for end-users to select the offer that best suit their needs.

One problem is due to the flexibility of Cloud computing and the inherent difficulty to predict the way in which the Cloud will evolve over time, since elasticity might requires new services or resources to be delivered in real time. Auditing the infrastructure of a Cloud is therefore a process that must be performed on an on-going basis - with the inevitable risk is that a certificate which has already been granted must subsequently be revoked. Certain service providers might no longer comply with the minimum set of requirements that had been previously satisfied, either because they have changed their policy over time, or because they have outsourced their services to other providers which are unable to guarantee the same standard of quality as before. Furthermore, Cloud computing can be an blockage for applying the procedure of e-discovery (legal obligations to retrieve, select and/or destroy data in a set of data).[10]

In that context, the transparency of the certification system and public disclosure of information by the service providers will be an important requirement for traceability. Before any certificate can be issued, the Cloud auditor must ensure that all relevant information necessary to assess the quality of a service has been disclosed and that this information is true. In the case of Cloud computing, given the inherent opacity of

---

[10] See I. Renard, J.M. Rietsch, (2012) Aide mémoire de droit à l'usage des responsables informatiques, Paris, Dunod, pp. 171-185.

the system, information can be obtained either by mandatory disclosure, i.e. by requiring that all relevant data logs be disclosed to the relevant certification authorities, or by internally monitoring the operations of the Cloud by means of automated software designed to assess compliance with every user's SLA.

In both scenarios, while the goal is to ensure a fair and transparent process of certification, the question is how to make sure that the certification authority will not be tempted to deceive the public in order to increase its profits. The issue arises from the fact that there is a conflict of interest given that the Cloud auditor is providing a service to the public at large, but is actually being remunerated by the service providers which it has been requested to certify.

Third party regulatory control might potentially help avoiding or reducing bias and corruption, although it does not really change the nature of the problem, but merely moves it at a different level. The fundamental question remains as to who is in charge of controlling the controlling authority.

We believe that this problem is however only a marginal one, given that natural market mechanisms might be able to resolve the issue without the need for any governmental intervention. Different certificates could be issued by different certification authorities according to different standards or criteria. Cloud auditors will not be tempted to deceive the public with a distorted system of certification, because their reputation is directly connected to the reliability of the certificates they have issued. The higher the trust of the public in a particular certification scheme, the greater the number of service providers who will request to be certified, and the higher the value of these certificates will be. Assuming that it is possible to preserve competition in the market for certifications, there would be no incentives for any Cloud auditor to provide false or erroneous information, because it would otherwise be immediately taken over by competition. Openness and transparency in the process of certification will instead be rewarded by a higher level of trust from the public. The result is likely to be an increased level of transparency in the private sector - in line with the various Open Data initiatives that are currently emerging in the public sector.

4   Conclusion: Legal and Technical Issues

Although still an evolving paradigm, Cloud computing has already been extensively deployed in the past few years and is already at the center of attention in many fields of business, industry and academia.

At the technical level, a large number of research projects are exploring and investigating the use of Cloud computing and ICT for Governance and policy modeling. RESERVOIR, for instance, is a EU FP7 funded project that purports to enable massive scale deployment and management of complex IT services across various administrative domains and governmental services;[11] DECISIA, is a new cloud based service for managing decision of Courts and other Tribunals.[12]

---

[11] For more details, see www.reservoir-fp7.eu

[12] For more details, see www.lexum.com

From a more socio-economic and legal perspective, some of the issues discussed are currently being explored in the French project ADAM on Distributed Architecture and Multiple Multimedia Applications (2010-2013), which is partially being undertaken at the CERSA (CNRS). In the next two years, our interdisciplinary team plans to investigate the specificities of Cloud computing, the social impact of this new paradigm of business, together with the new legal challenges it engenders. In this paper we launch the debate at the first step of this research. Given the current state of the art of Computers & Law in the context of Cloud computing, the objective of this paper is to propose a series of ideas that could eventually be implemented into practical solutions as an attempt to address the new legal challenges faced by different actors in the Cloud.

References

Daskalopulu, A., Sergot, M.J.: The representtaion of legal Contracts. AI & Society 11(1/2), 6–17

ENISA report, on Cloud Computing: Benefits, risks and recommendations for information security, which identifies the main risks of Cloud Computing in terms of information security as being due to loss of governance and user lock-in; isolation failure and compliance risks; management interface compromise; improper data protection; incomplete or insecure data deletion; and malicious insiders (2009)

Khan, K.M.: Establishing Trust in Cloud Computing. IT Professional 12(5), 20–27 (2010)

Macias, M.: Using resource-level information into nonadditive negotiation models for cloud Market environments. In: 2010 IEEE Network Operations and Management Symposium (NOMS), pp. 325–332 (2010)

Mazzega, P., Bourcier, D., Bourgine, P., Nadah, N., Boulet, R.: A Complex-System Approach: Legal Knowledge, Ontology, Information and Networks. In: Sartor, G., Casanovas, P., Biasiotti, M., Fernández-Barrera, M. (eds.) Approaches to Legal Ontologies, Theories, Domains, Methodologies. Law, Governance and Technology Series, 1st edn., vol. 1, ch. 7, p. XIII, 279 p. Springer, Heidelberg

Sim, K.M.: Towards Complex Negotiation for Cloud Economy. In: Bellavista, P., Chang, R.-S., Chao, H.-C., Lin, S.-F., Sloot, P.M.A. (eds.) GPC 2010. LNCS, vol. 6104, pp. 395–406. Springer, Heidelberg (2010)

Patel, P., Ranabahu, A., Sheth, A.: Service Level Agreement in Cloud Computing. In: Cloud Workshops at OOPSLA 2009 (2009)

Renard, I., Rietsch, J.M.: Aide mémoire de droit à l'usage des responsables informatiques, Paris, Dunod (2012)

Balancing Rights and Values in the Italian Courts: A Benchmark for a Quantitative Analysis[*]

Tommaso Agnoloni, Maria-Teresa Sagri, and Daniela Tiscornia

Institute of Legal Information Theory and Techniques (ITTIG)
CNR The Italian National Research Council
via de'Barucci 20, 50127 - Florence, Italy
{daniela.tiscornia,tommaso.agnoloni,m.t.sagri}@ittig.cnr.it

Abstract. In conformity with the global tendency, balancing is increasingly used in judicial practice as an argumentation technique for solving legal disputes; more and more, judges of all levels ground their decisions on the balancing of individual rights, interests, principles, needs, and values. Legal science has formulated theoretical and formal models to explain the argumentative structure of balancing and the criteria governing the argumentation process, but, in the absence of a conceptual model that encompasses all elements in play and enables a comparative mechanism to be abstracted, mapping instances of judicial practice to abstract theories is still difficult. In this context, the goal of the project here described is to allow the logic of judicial practice emerge from cases, verifying from the bottom up the assumptions of theoretical models. Starting off from a broad analysis of Italian cases, the paper aims at analysing the object of this operation, that is, what is 'balanced' and what is the nature of this process. The research was conducted by analysing the so-called 'massime' (case law abstracts) of the Italian High Courts (Constitutional Court, Supreme Court, Council of State), of the administrative courts (Regional Administrative Tribunals) and of a selection of lower court decisions. The methodology is divided into an initial phase of documentary collection and storage, a second phase of conceptual modelling and a third phase of data analysis.

Keywords: Reasonableness and proportionality in legal decisions, Forensic statistic, Legal conceptual modeling, Legal data management.

1 Introduction

The project originates from a cooperation between the University of Florence and the Institute of Theory and Techniques for Legal Information (Ittig-CNR).

Within a course on Legal Argumentation held at the Law Faculty of the University of Florence, a research group was set up for the purpose of conducting a bottom up

---

[*] This paper is a revised and extended version of : *Balancing rights and values in the Italian Courts: a statistical and conceptual analysis*, published in: Law, Probability & Risk, Special Issue: Proportionality and Quantitative Justice, 10(3): 265-275 (2011).

M. Palmirani et al. (Eds.): AICOL Workshops 2011, LNAI 7639, pp. 93–105, 2012.
© Springer-Verlag Berlin Heidelberg 2012

investigation into the nature and types of judicial argumentations based on the balancing of individual rights and constitutionally guaranteed values. The methodological approach was pragmatic, in that the research team proposed to build up a reference corpus of judicial decisions, on which to also carry out, apart from monitoring it on a statistical basis, a conceptual-type analysis. Judicial Decisions passed within the last decades (since 1970) were taken into consideration, selected from online legal information systems.

To enable us to make a comparative and integrated analysis of all the elements (means and goals, legislative instruments and constitutional principles, individual rights and values), the documents were organised and classified according to a conceptual model designed on the background of theoretical assumptions (see Sect. 2).

The role of the researchers at Ittig, which has considerable experience in modelling documents and legal knowledge, was, therefore, to build a conceptual model with which to organise the data, so that such a multi-faceted profile of analysis could be reached.

The methodology is divided into an initial phase of documentary research, a second phase of conceptual modelling and a third phase of data analysis. At the first level, lexical and statistical information (frequency of the use of the term 'balancing' combined with 'interests', 'needs', 'principles' 'values'; data intersection; statistical variants over time, etc.), provide a rich set of interesting indicators from the point of view of linguistic uses and forensic statistics. A deeper analysis conducted on these data, in relation to the legal concepts they express and their reciprocal interrelations, allows us to draw up a kind of semantic-conceptual map of the orientations of the courts. The third, more difficult, level of analysis is aimed at comparing the elements extracted from the data analysis with theories and models of legal theory, in order to formulate hypotheses for explaining and classifying jurisprudential trends.

2   Theoretical Background

Whilst resorting fairly frequently to argumentation techniques of balancing, Italian jurisprudence usually restricts itself to making an almost unthinking and 'intuitive' application of it; this is contrary to what happens in other modern legal cultures, such as, in particular, that in America and Germany, in which both the Courts (High Courts and the Constitutional Courts in particular) and the doctrine have made an effort to conceptualise and refine this form of legal reasoning.

Consequently, as a methodological pre-condition, we had to identify the conceptual premises that permit proper reasoning in relation to balancing. The first theoretical hypothesis was to place balancing among the methods for solving normative antinomies. In a technical sense, an antinomy outlines a logical situation of incompatibility. 'Antinomic' does not mean 'different' but 'incompatible' from the logic point of view. Incompatibility is also used in modern linguistics to define some semantic relations of language (opposition, complementarity, inversion).

Whilst antinomy has a binary nature, whereby, in logic as in linguistics, the assertion of one statements implies the negation of the other, the reasoning structure of balancing assumes the presence of two or more competing values coming into play, their scalability and the possibility to assign weights to them. [3]

It is indeed on the scalable nature of values that the differences between balancing and normative conflict resolution lays, the latter concerning pairs of normative premises from which antonymic outcomes can be inferred. "There exist no such analytically demonstrable, incompatibility *in abstracto* of principles. Collisions of principles occurs only in particular cases." [12, p. 81]

According to a widely agreed opinion among legal theorists, "balancing" can be classified within argumentative techniques, as an interpretative instrument aimed at providing parameters for semantic and non-logic conflict resolution. The interpretative process must be in accordance with an optimisation criterion in which, instance by instance, prevalence is given to the most effective and least harmful instrument towards the achievement of a normative goal. As a consequence, teleological interpretation is the basic reasoning step on which the judgement is grounded: "I shall argue that legislative decision making can indeed normatively be viewed as an exercise in balanced maximisation and that judicial review can be viewed as implementing constraints on this exercise. Some normative constraints are external to the teleological structure of the legislator's task...other constraints, on the contrary, address the legislator's teleology, ...to check whether these constraints have been violated we need to consider whether a certain value has been adequately taken into account."[19, p. 176].

Trying to give a formal specification of its argumentative nature, the models of *legal theory* locate balancing among the founding criteria for evaluating the practical rationality or the 'reasonableness' of decision making processes in the different fields of law, like public, private and international law [1], [2]. In public law and, in particular, in constitutional review, the reasonableness is understood as the intersecting point between moral and legal reasons that finds its essence in balancing [18].

In the evaluation of legislative decision-making the judgement of 'proportionality' measures the adequacy of the chosen means (that is, of the legislative instrument to be applied) regarding the general objectives in respect to the specific goal and the expected benefits the new norm intends to achieve. Balancing is the final decisive step in the assessment of the proportionality requirement, that implies: the preliminary assumption of alternative instruments having a different weight, the comparison and evaluation among values (and benefits) and the assessment in terms of maximisation and balancing[1].

---

[1] Several theories, either imported from economics or originated in legal theory have been proposed, aimed at setting objective parameters for defining the 'best choice'. Some rely on the economic concept of Pareto optimality: goods and resources are allocated in effective way when there is no alternative choice that creates advantages to at least one subject without carrying on disadvantages for others, In the legal domain the same concept, known as the "mildestes Mittel ", identifies the legislative means able to ensure the same level of fulfillment of public interest without constraining the individual freedom [17]. According to the [8], [20] theory, the best choices must balance the losses of one party by the introduction of proper remedies. On a principle of social justice is based the criterion of [23], that identify maximisation of advantages into the provision of equal opportunities.

The *formalization of balancing* as a reasoning process requires the use of argumentative logic, that in computational models have been studied and implemented within the framework of defeasible logic [15]. Defeasible logic, is a rule based non-monotonic formalism, that enables to derive "plausible" conclusions from partial and sometimes conflicting information. Inferred conclusions are defeasible, in the sense that a conclusion can be withdrawn when new pieces of information are introduced. Based on defeasible rules of inference, the argumentative reasoning patterns are formally represented, where arguments for and against a certain claim are produced and evaluated, to test the tenability of the claim.

Computational models including the representation of teleological aspects in argumentation have been proposed in [14] and in [5]. See also [4], [7].

2.1   Two Types of Balancing

From the viewpoint of judicial practice, balancing indeed can be reconstructed as an activity that, whilst containing assessment margins, does not necessarily translate into unbridled subjectivism, but is quite the opposite, namely, rationally controllable. Part of legal doctrine argues that a rational reconstruction of balancing is, in fact, possible by explaining the set of relevant properties in the light of which one of two competing principles prevails over the other. Furthermore, by isolating a set of relevant properties (or by designing a 'topography of conflict' [6]), it is possible to explain a rule that offers solutions that can be reproduced for all analogous cases that come up in the future, at least in 'central' cases. In this way, a co-ordination rule is produced between the two competing rights or principles that is subject to universalization (that is, is suitable for regulating future cases with analogous relevant characteristics) whilst nonetheless being a rule subject to revision where there are additional relevant properties. The identification of the rule, therefore, has the effect of transforming, at least apparently, the balancing judgement from a 'wisdom-based' judgement into 'procedural' reasoning [13].

Among the judicial practise, we need to distinguish between a strict notion of balancing, that properly refers to the process of constitutional review, and a wider notion of balancing applied by the ordinary courts. In the first case, the Constitutional Court judges on general and abstract issue addressed to evaluate the compliance of legislative norms with Constitutional provisions, while in the latter case ordinary courts judge on individual cases.

In constitutional review, the evaluation process follows a sequence of steps aimed at assessing: the fitness of the means for reaching the expected goal, the necessity of the rule (for the purpose of verifying whether the pre-established instrument, apart from being fit, is also the most mild, that is the least harmful of those fit for reaching the same result), and finally, its proportionality, that is, that in relation to the values to be protected leans towards a more equilibrated solution among the individual interests and the common good in play [20], [6].

The role of balancing before the courts is, to provide a *parameter*, that inside the specific pair of rights (and individual interests) coming in competition, can justify the decision [21]. Applying balancing implies, therefore, the absence of a pre-established rule that indicates a co-ordination or preference criterion, as happens in the case of conflicting rights that can be solved on the level of the hierarchy of sources. In the absence of a similar pre-established rule, each of the two rights has its own sphere of application that does not completely coincide with that of the other. Therefore, once ascertained that the competing rights   have the same rank because (*in se* or by means of teleological interpretation) constitutionally guaranteed, the judge has to assign weights to them in the light of the specific case, to choose which interest to hold as 'more' relevant, and therefore which prevails over the other.

What is interesting to note is that the application of balancing in specific cases can bring to different results, depending on the core components of fundamental rights that are opposed; one of the most frequent situation concerns balancing between privacy (including right to reputation, personal identity protection, etc) and freedom of expression (including freedom of speech, right to information, right to criticize, freedom of the press..). From our analysis emerges that the prevalence criterion is in some cases   grounded on the argument   that a fundamental right cannot be compressed beside a minimum threshold, whilst in other cases an effective proportionality between values can be recognized, (see Fig.3).

The following Figures (see Fig.1), show the statistic occurrences related to trends in Constitutional review and in High Courts.

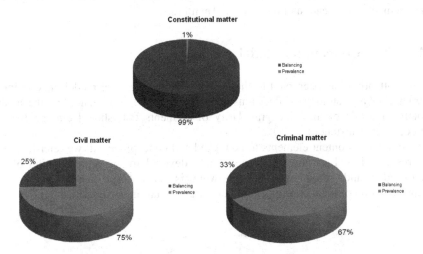

Fig. 1. Trends in Constitutional review and in High Courts

3   The Methodology

Based on these premises, the scope of the project is to create a data base of Italian case law, as a bench-mark for further analysis. Case selection was performed automatically, through lexical (full-text) search mechanisms or through keywords (*balancing* combined with *interests, rights, needs, principles, values*).

The corpus is composed of about 300 cases from 1990 to 2010.

According to the differences highlighted in sect. 2, two data bases have been created:

- Constitutional review: 178 Judgements of the Italian Constitutional Court
- High Courts:108 decisions of the Supreme Court (Corte di Cassazione) in Civil (58) and Criminal (42) Matters, and the Council of State (Consiglio di Stato) (8).

The first corpus analysis concerns the lexical level and terminological uses. In their judgements, Italian courts classify legal issues according to conceptual categories, to which they give different names. A legal factual element (*strike, health, economic initiative*) is sometimes called a 'diritto (right)', sometimes a 'principio (principle)'; other entities (*life, common good*) are sometimes a 'valore (value)', sometimes 'bene (asset)', while others (*financial stability of the State, good performance of the justice system*) are indifferently called 'interessi (interests)' or 'bisogni (needs)' and, finally, concepts like *reasonableness, proportionality, good performance of the public administration* are classified as 'bisogno (need)', 'interesse (interest)' and 'principio (principle)'. In the narrative structure of the text, the two entities that appear to be antinomic or competing are clearly identified and arguments pro and against their prevalence are discussed, but, the argumentation discourse is not supported by reference to a shared terminology nor to a systematic classification of the compared entities.

4   The Conceptual Model

Much attention has been paid to the design of the conceptual model underlining the structuring and storage of decisions; this is of crucial importance, as the analysis totally depends on the fine granularity of elements that should emerge from the processing of instances.

To identify content elements to be tagged and conceptualized, we refer to existing works in AI & Law literature, specifically devoted to the representation of legal knowledge and case law. The following is an excerpt from the ontological representation of legal concepts in judicial argumentation, as described in [22]:

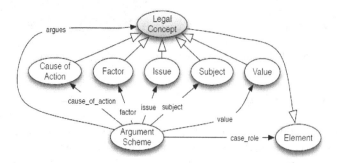

Fig. 2. Legal concepts in judicial argumentation

In the Figure (see Fig.2), judicial argumentation is framed in several components, including factual and legal elements; in our model, we do not consider factual elements, such as the facts of the case (the cause of actions) and the abstraction of facts (*factor*) as they lay outside the decisions of legality by the High Courts, but the legal items (Issue, Subject and Value) only. A similar model, based on the distinction among Object, Parameter(s) and Argument(s) is in [16].

Our conceptual scheme also refers to the Alexy's distinction, as reported in [19, p.177], between *action-norm* ("requiring the full accomplishment of a certain action or omission") and *goal-norm* ("requiring the appropriate pursuit of a certain objective").

In the schema below, (see Tab. 1) the object of the 'Object' field is to describe the *action norm* (and related *action duty*) introduced by the legislative measure subject to evaluation (in the case of Constitutional review) or backing up the low court decisions (submitted to decisions of legitimacy by a High Court).

The 'Goal' field represents the *goal-norm* (in terms of individual interest or constitutional values) inferred from the teleological interpretation of legislation, or explicitly claimed in the arguments; the 'Value' field intends to conceptualize goals at a higher level of abstraction, in order to enable a more sound conceptual analysis.

For the purpose of our analysis the abstraction of *Goals* into *Values* do not imply neither a conceptual distinction, nor a formal class/sub-class relationship, but just the need of clustering classification of rights into a more generalized model.

Our overall idea is to build a core reference system of a (flat) set of values, that should be considered common to several constitutional systems, and, therefore, re-used in further analysis. The *prevalence* of one of the two competing values is extracted from the motivation of the decisions of the High Courts.

Our aim is to build a basic model on the set of statistical data, that should in the future be enriched by a further layer of semantic information, such as the classification of values into classes organized according to (some) ordering criterion. This should enable interesting comparison between the practise in the courts and the theoretical arguments. In future developments, we intend to map the prevalence score arising from the cases analysis to the ordering criteria that, from a purely theoretical view, have been proposed, for instance the classification in *fundamental*, *general* and *supreme* principles. Fundamental principles are the *genus* (due to their vagueness and ambiguity), from which the generals are the *specie* (as they are specified by a direct or indirect constitutional connotation). The supreme principles are a sub-species, as they are both fundamental and general, but with a particular status that distinguishes them as they are indisposable and unchangeable. [11]

With reference to the Italian system [10], *supreme rights* are recognized by the first twelve articles of Italian Constitution, (for instance: art. 2, individual rights; art. 3, equality right; art.10 and art. 6 democratic participation, etc.); *general* principles are those inferred from the second part of the Constitutional text (art. 36, right to work; art. 41, economic freedom), and, finally, *fundamental* or *institutional* principles have a more generic scope (art. 97, reasonableness principle; efficiency and impartiality of P.A., art. 39, freedom of trade unions organization; art. 25, personal liability).

Table 1. The data model

Metadata	Content
Case id	Year, number
Object	Legislative measure (action norm and action duties)
Goal 1	Interests, social rights, individual rights (goal norm)
Value 1	Constitutional rights, fundamental principles (inferred from Constitutional rights)
Normative references	Italian Constitution
Goal 2	Interests, social rights, individual rights (goal norm)
Value 2	Constitutional rights, fundamental principles (inferred from Constitutional rights)
Normative references	Italian Constitution
Ratio decidendi	Text
Prevalence	1 < 2 or 2 >1

5   Data Analysis

The three following cases exemplify the behaviour of the Supreme Court in Civil Matters (Corte Cassazione Civile) in cases where individual right to reputation and freedom of speech are competing. The Court reached through a balancing process at opposite solutions, establishing, in 1996 (Tab.2), that the right to other people's reputation constraints the exercise of the freedom of speech, while in 2005 (Tab.3) stated that the right to information can be considered to comply with the right to privacy whenever it meets the following criteria: a) social benefits of information; b) objective truth, resulting from careful and serious research; c) a fair way of setting out the facts, requiring the exclusion of any intention to defame. This criterion is confirmed by the 2006 case outlined in Tab. 4. where the right to minor protection prevails over freedom of speech whenever a public interest to information cannot be justified.

Table 2. Example about case Civil Italian High Courts

Case id	Cass. Civ. Sez. III, n. 465/1996
Object	critics against person playing public roles, exposed to public opinion
Goal 1	individual interest to reputation
value 1	privacy protection
Normative references	Italian Constitution Art. 15 Cost. Art.10 c.c. ; Art. 2 Cost.
Goal 2	right to criticize
Value 2	freedom of speech
normative references	Italian Constitution Art. 21
ratio decidendi	right to criticize is subject to public and social interest
prevalence	1 >2

Table 3. Example about case Civil Italian High Courts

Case id	Cass. Civ. Sez. III, n. 379/2005
Object	subjective interpretation of facts
Goal 1	individual interest to reputation
value 1	privacy protection
Normative references	Italian Constitution Art.10 c.c. ; Art. 2 Cost
Goal 2	right to criticize
Value 2	freedom of speech
normative references	Italian Constitution Art. 21 Cost
ratio decidendi	public interest to information encompasses critical reformulation of facts
prevalence	2 >1

Table 4. Example about case Civil Italian High Courts

Case id	Cass. Civ. Sez. III, n. 1969/2006
Object	Disclosure of information and images concerning minors
Goal 1	minor protection
value 1	privacy protection
Normative references	Italian Constitution Art.10 c.c. ; Art. 2 Cost
Goal 2	right to information
Value 2	freedom of speech
normative references	Italian Constitution Art. 21 Cost
ratio decidendi	public right to information prevails over privacy protection only when justified by a social interest
prevalence	1 > 2

5.1   A First Data Exploration

In the following, some (statistical) items of the behaviour of the Italian Courts are presented.

Fig. 3 outlines the pairs of values in competition in civil law matters (High Court in Civil Matters (Corte Cassazione Civile) and Council of State (Consiglio di Stato) on Public Administration Matters ; it also sets out the prevalence score (1= deep grey, 2 = light grey) and the cases where the two values have been evaluated as well balanced (medium grey).

In criminal law (High Court on Criminal Matters), as shown in Fig.4, the most frequent conflict arises between rules governing the administration of justice (competence, rationalization of administrative services, etc..) and the individual right (to the impartiality of judge and to a proper duration of judicial trial), the latter quite always prevailing over the former); it is also interesting to note that balancing is frequently applied when human rights and equality right are competing.

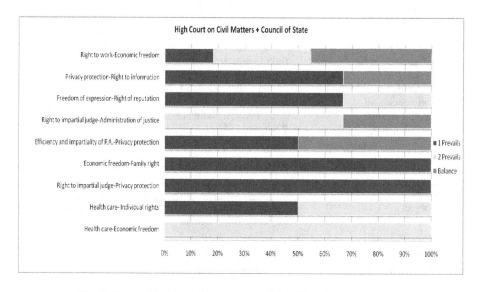

Fig. 3. Competition in civil law matters (High Court in Civil Matters)

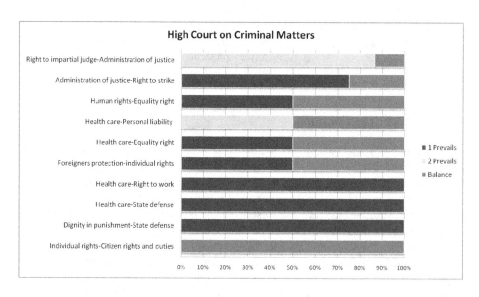

Fig. 4. Data relating to criminal law

Fig. 5 shows a detailed overview of the institutional values competing with the fundamental right to *health care*:

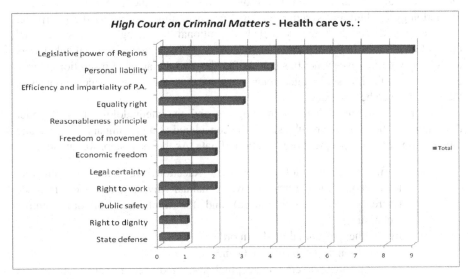

Fig. 5. Data relating to Health care vs. other values

Fig. 6 outlines the pairs of values in play in Constitutional review. It should be noted the statistical predominance of the *equality right*, balanced with further fundamental, institutional and inferred values:

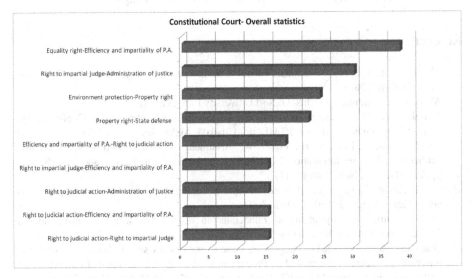

Fig. 6. The values in play in Constitutional review

6   Conclusion and Future Work

Our first goal, in carrying on the project was to provide a benchmark and a methodology for building systems for legal practice, aimed at giving the interpreter support in understanding the so-called 'definitional balancing'. The conceptual description of cases in the database enables to design a picture of judicial behaviours and of its evolution over time; this could offer, beyond the areas of discretion a judge is allowed, a guideline so the interpreter can properly argue the assessment choices of the facts that he/she has made.

The project presented here is in progress and the methodology is still under development; therefore the results should be considered as a preliminary outcome and a first testing of the conceptual organization of data. As the next step, we intend to:

- Refine the methodology for case selection: we intend to integrate lexical/conceptual retrieval with manual investigation (for instance references to legal literature) and the application of data mining techniques);
- Refine the conceptual model in order to:
  ○ perform a fine grained data analysis,
  ○ consolidate a core conceptual system of set of values, to enable comparison and mapping to trans-national case law;

This last step will be crucial in order to explain quantitative information in terms of qualitative data (formal models). This should enable the mapping of quantitative/ statistical outcomes to abstract models of 'proportionality' and the construction of a bottom up extensional definition of 'reasonable' decision making.

References

1. Alexy, R.: The theory of rational discourse as theory of legal justification. Clarendon Press, Oxford (1989)
2. Alexy, R.: Constitutional rights. Oxford University Press, Oxford (2003)
3. Araszkiewicz, M.: Analogy, Similarity and Factors. In: 13th International Conference on Artificial Intelligence and Law, ICAIL, pp. 101–105. ACM, New York (2011)
4. Bench-Capon, T., Prakken, H., Visser, W.: Argument schemes for two-phase democratic deliberation. In: 13th International Conference on Artificial Intelligence and Law, ICAIL, pp. 21–30. ACM, New York (2011)
5. Bench-Capon, T., Sartor, G.: A model of legal reasoning with cases incorporating theories and values. AI&Law J. 150, 97–142 (2003)
6. Bin, R.: Diritti ed Argomenti, Il bilanciamento degli interessi nella Giurisprudenza costituzionale. Giuffrè, Milano (1992)
7. Grabmair, M., Ashley, K.D.: Facilitating Case Comparison Using Value Judgements and Intermediate Legal Concepts. In: Proceedings of 13th International Conference on Artificial Intelligence and Law, ICAIL 2011, pp. 161–170. ACM, New York (2011)
8. Hicks, J.: The Foundations of Welfare Economics. The Economic Journal 49(196), 696–712 (1939), http://jstor.org/stable/2225023

9. Kaldor, N.: Welfare Propositions in Economics and Interpersonal Comparisons of Utility. The Economic Journal 49(195), 549–552 (1939), http://jstor.org/stable/2224835
10. Modugno, F.: Voce Principi generali del Bilanciamento. In: Enciclopedia del Diritto, vol. V, pp. 1–24. Giuffrè, Milano (2001)
11. Modugno, F.: voce Principi fondamentali, generali, supremi. In: Cassese, S. (ed.) Dizionario di Diritto Pubblico, vol. V, pp. 4490–4496. Giuffrè, Milano (2006)
12. Peczenik, A.: On law and Reason, vol. 81. Kluwer, Dordrecht (1989, 2008)
13. Pino, G.: Teoria e pratica del bilanciamento: tra libertà di manifestazione del pensiero e tutela dell'identità personale. Riv. Danno e Responsabilità, 577–584 (2003)
14. Prakken, H.: A exercise in formalising teleological case-based reasoning. In: Prakken, H., Winkels, R. (eds.) 13th Jurix Conference on Legal Knowledge and Information Systems, pp. 49–57. IOS Press, Amsterdam (2000)
15. Prakken, H., Vreeswijk, G.: Logics for Defeasible Argumentation. In: Gabbay, D., Guenthner, F. (eds.) Handbook of Philosophical Logic, 2nd edn., vol. 4, pp. 219–318 (2002)
16. Damele, G., Dogliani, M., Mastropaolo, A., Pallante, F., Radicioni, D.: On Legal Argumentation Techniques: Towards a systematic approach. In: Biasiotti, M., Faro, S. (eds.) From Information to Knowledge - Online Access to Legal Information: Methodologies, Trends and Perspectives. Frontiers in Artificial Intelligence and Applications, vol. 236, pp. 105–118. Springer, Heidelberg (2011)
17. Rawls, J.: A theory of justice (revised edn.). Oxford University Press, Oxford (1999)
18. Sartor, G.: A sufficientist Approach to Reasonableness in Legal Decision Making and Judicial Review. In: Bongiovanni, G., Sartor, G., Valentini, C. (eds.) Reasonableness and Law, pp. 17–68. Springer, Heidelberg (2009)
19. Sartor, G.: Doing justice to rights and value: teleological reasoning and proportionality. AI&Law J. 177, 175–215 (2010)
20. Scaccia, G.: Gli "strumenti" della ragionevolezza nel giudizio costituzionale. Giuffrè, Milano (2000)
21. Tarello, G.: L' interpretazione della legge. Giuffrè, Milano (1980)
22. Wyner, A., Hoekstra, R.: A Legal Case OWL Ontology with an Instantiation of Popov v. Hayashi. Knowledge Engineering Review (in press, 2012)
23. Zoonil, Y.: Das Gebot der Verhältnismäßigkeit in der grundrechtlichen Argumentation. Peter Lang, Frankfurt am Main (1998)

Survival of the Fittest: Network Analysis of Dutch Supreme Court Cases

Radboud Winkels and Jelle de Ruyter

Leibniz Center for Law, University of Amsterdam
winkels@uva.nl, jacobus.deRuijter@student.uva.nl

Abstract. In this paper we present the results of a study to see whether the number of citations to cases is an indication of the relevance and authority of these cases in the Dutch legal system. Fowler e.a. have shown such results for the US common law system, but given the different status of case law in continental tradition it is not clear whether this will hold in the Netherlands. Moreover, we introduce an alternative way to validate the results using selections made by human experts for legal education. We discuss the results and conclude that network analysis of cases is a useful tool for legal research.

Keywords: relevance, authority, case law, citations.

1   Introduction

"Law is a seamless system with its own autonomy. It provides one correct answer to any case, difficult or not, by application of its rules, precedents, principles and spirit." [1]. In Dworkin's theory, a judge finds *one* solution in each case, however difficult it is. He starts looking for the solution with the best 'fit', searching for rules, cases and other sources of law that are most 'on point'.

With the growing number of published sources of law, this task is becoming increasingly difficult for legal practitioners. What are the relevant sources of law for a particular case? And which ones are the most important ones?

In this paper we discuss the result of a small study we conducted on Dutch Supreme Court cases to see whether we can aid in determining the importance of a case by analysing its place in the network of sources of law. Network analysis has already been used to determine importance of scientific publications [2], patent requests [5] and judgements of the US Supreme Court [3]. In [8] network analysis is used to get an idea of the complexity of French legal code.

Not all cases are being published[1], courts make selections and the importance of the ruling will play a role in these decisions, but not all published cases are of equal importance. Our hypothesis is – not surprisingly – that people refer more to important cases than to other ones. In order to test this hypothesis we need some other way to

---

[1] In 2004 not even 1% of all cases were being published in the Netherlands on the Dutch official portal (rechtspraak.nl) [10].

M. Palmirani et al. (Eds.): AICOL Workshops 2011, LNAI 7639, pp. 106–115, 2012.
© Springer-Verlag Berlin Heidelberg 2012

decide upon the importance of cases. It seems plausible to ask legal experts what they consider to be important cases.

2   The Data

For this study we limit ourselves to one type of source of law: cases, and moreover, only cases from the Dutch Supreme Court ("Hoge Raad") as published in the Dutch periodical NJ ("Nederlandse Jurisprudentie" or Dutch Case law) between January 1st 1965 and December 31st 2008[2]. In total this collection contains 15,053 Supreme Court cases. References to other cases in these decisions are unfortunately not marked or otherwise machine readable. A first step is to detect these references automatically and build up the network of citations.

2.1   Building the Citation Network

There are several places in a Supreme Court decision where references to other cases can be found. First, the history of the case in earlier instances is summarized, the attorneys or the public prosecutor plead their case, then the court explains its decision, the Attorney General sometimes gives his opinion, the decision may be annotated and finally the editorial board of the publisher may add links to other cases deemed relevant. Most of these parts of a decision can be detected quite easily by their header, but surprisingly the actual opinion of the court is sometimes harder to detect. Often it starts with the text: "The Supreme Court, etc.", and ends where the opinion of the Attorney General starts, signalled by the text "[OPINION]", or the annotations or the editorial additions start, or where the complete text ends.

Table 1. Validation of the approach for finding references

Number of Cases	100		
Correct	862	F1	98,7 %
False Positives	19	Precision	97,84 %
False Negatives	4	Recall	99,54 %

In earlier research we have built a parser to detect references in legislation using a context free grammar approach [7]. The present corpus is simpler since the publisher uses a more or less standard way to refer to other cases in their own database by a so-called NJ-number[3]: The letters "NJ" followed by one or more spaces, the year of publication, followed most of the time by a comma, backslash or forward-slash and then a follow-up number. We used regular expressions to find these references.

---

[2] The official Dutch portal contains more, but not all, Supreme Court cases per year, but only since 1999.

[3] This excluded potential references to cases not published in NJ.

We found 106,559 references this way. To validate the approach we examined a random sample of 100 cases by hand. Table 1 presents the results; an F1 score of almost 99% is quite good, especially recall is very good. Since the database only contains Supreme Court decisions, only references to Supreme Court decisions can be resolved.

3   The Supreme Court of the Netherlands

According to the Dutch Constitution the Supreme Court has two main functions. Article 118 section 2 Dutch Constitution states that "The Supreme Court is, in the cases stated in and determined within the confines of the law, charged with the cassation of judicial rulings on violation of the law." The second function is stated in article 119 of the Dutch Constitution and is the trial of ministers and parliamentarians for abuse of the office; this article has never been used. Other functions are attributed to the Supreme Court by law, such as the trial of judges of all courts, for judicial misconduct.

As the name suggests the Supreme Court ("Hoge Raad") is the highest judicial instance in The Netherlands in matters of civil law, criminal law and tax law. The highest court in matters of administrative law is the Central Court of Appeal (Centrale Raad van Beroep). The Netherlands does not have a constitutional court.

The Supreme Court is organized in four chambers:

- First Chamber or Civil chamber
- Second Chamber or Criminal law chamber
- Third Chamber or Tax chamber
- Fourth Chamber or Ombuds chamber

Each chamber has two or three vice-presidents and nine to thirteen judges, the President of the Supreme Court is also the only permanent judge on the Ombuds-chamber, the highest instance of complaint for misconduct by judges, the other judges in this chamber are appointed on an ad-hoc basis. In a cassation case, one of the vice presidents and four judges of the Supreme Court give judgment on the case. The five judges deliver one judgment, without dissenting (or concurring) opinions. Whether or not the judges disagree is not shown in the judgment in any way.

As the Supreme Court is a cassation court it only gives a verdict on the question of whether the lower courts have correctly motivated their decisions and whether or not these lower courts have applied the law correctly. No verdict is given on the facts of the case.

Dutch law formally doesn't recognize *stare decisis*, art. 12 Law of general provisions ("Wet algemene bepalingen") states that: "No judge may do verdict by means of a general statute...". In other words, a verdict concerns only the parties to the case and may not be seen as a general rule applying to everyone. In practice however, lower courts, but also the Supreme Court itself, tend to adhere to previous rulings by the Supreme Court. Deviation from (or reconsideration of) a previous judgment is in practice the exception[4].

4   A First Analysis of the Network

A network of case law is different from most other ones like social or computer networks; it evolves rather simply. Prior cases do not disappear, though they may be cited less and less, links between cases remain and a new case can only cite older cases.[5] Figure 1 below shows the average number of in- and outgoing references used by the Supreme Court itself per year. The first observation is that this average is low, about 1 till the 1980s and increasing afterwards to about 2.[6] Not surprisingly the average number of incoming references decreases at the end, because there are fewer years from which these citations could come.

The Supreme Court also does not refer much compared to other parties. Table 2 shows these figures. The Advocate General uses the most citations (40%), followed by the editorial board of the publisher. This last category is added later on and may grow over time and *can* refer to future cases. This category will be ignored in the remainder of this paper, as will the Introduction and the 'Essence' or summary of the decision that is also written by the editorial board.

Table 2. Number of references per section of a decision

Position in the decision	References	%
Introduction	1.440	1,4%
Essence	2.528	2,4%
Prior instances	5.930	5,6%
Supreme Court	7.992	7,5%
Advocate General	42.858	40,2%
Annotations	10.246	9,6%
Editorial links	35.565	33,4%
Total	**106.559**	

---

[4] Over the period 1980-1993, Franx found 23 rulings where the Supreme Court had explicitly reconsidered a previous judgement. Franx,, 1994 [5].

[5] Not completely true, the publisher may later on add links to newer cases, but we will ignore these citations in this research.

[6] If we compare this to the Supreme Court of the US [3] we notice that they refer about ten times as much.

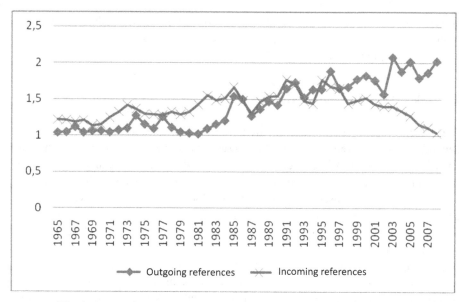

Fig. 1. Average in- and outgoing references of the Supreme Court per year

Figure 2 shows the total number of incoming references from the prior instances, the Supreme Court, the Advocate General and the Annotator for the first 500 most cited cases. It declines logarithmically.

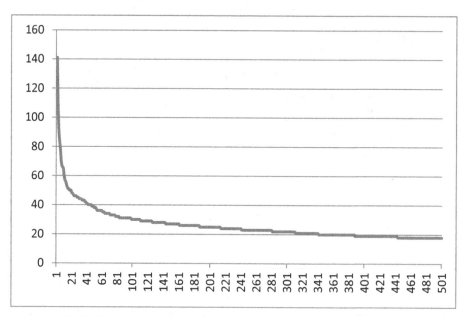

Fig. 2. Total number of incoming references of first 500 most cited cases

4.1   The Top-10 Cases in the Network

Based on the number of incoming references we can determine a top-10 of cases. Table 3 gives an overview of these cases. It does not display the number of times the case was cited, but the relative ranking for a particular actor or party. The overall ranking is based on the average ranking over the four actors: Supreme Court, Advocate General, Annotator and Other parties. So the first place is taken by the "Zwolsman" case, a decision concerning police investigation methods in criminal procedures. These methods led to the so-called "IRT affair", a parliamentary investigation and finally an addition to the Dutch penal code for special police investigation competences (title IV added in 1999). Since that code became effective, the number of citations of the Zwolsman case also drops as can be seen in Figure 3. The number of citations peaks in 1999 and drops after that, especially for the court itself and the annotators.

Table 3. Top-10 cases based on incoming references

Nr	NJ-number	Case Name	Other parties	Court	Advocate General	Annotator	Area
1	1996/249	Zwolsman	1	3	2	1	Criminal procedure
2	1981/635	Haviltex	5	5	1	4	Civil procedure
3	1986/723	Heesch/Van den Akker	10	10	3	2	Civil procedure
4	1982/503	Boon/Van Loon	33	10	4	3	Family law
5	2000/721	Reasonable period	10	1	8	21	Civil procedure
6	1986/242	Enka/Dupont	5	15	5	51	Criminal procedure
7	1994/427	Limits oral testimony	3	8	10	33	Criminal procedure
8	1993/659	Vredo/Veenhuis	33	23	14	6	Civil procedure
9	1989/4	HBM/Wielenga	3	33	22	79	Civil procedure
10	1986/3	- <no name>	7	8	143	51	Family law

In the Zwolsman case and the Heesch/Van den Akker case (nr 3), there was no applicable law and the Supreme Court forms law itself. In the Haviltex case (nr 2) there is an applicable article of law (art. 1378 of the old Dutch Cicil code), but the court interprets it differently from before.

Most of the cases in the top-10 are about procedural law, except for the cases Boon/Van Loon (nr 4) and case nr 10, which has no name. Both these cases deal with family law issues and in both cases the court breaches a line of thinking from the past. They are pointed out in Figure 4 by arrows in a network that shows two clusters around them. The cases that connect the two, that cite either Boon/Van Loon or case nr 10, are all cases in which the court also changes a line of reasoning. Especially case NJ 1980/353, the so-called bull-calf case in which a calf injured a farmer that crossed its meadow, in which the court moves from a 'suspicion of guilt' to a 'risk liability' line of reasoning. The Supreme Court itself only refers once to the bull-calf case (in this dataset), in another case about risk liability (NJ 1984/2), but the attorney general cites the case 19 times and the annotators 3 times, mostly to point to the breach in reasoning and not to the content of the case.

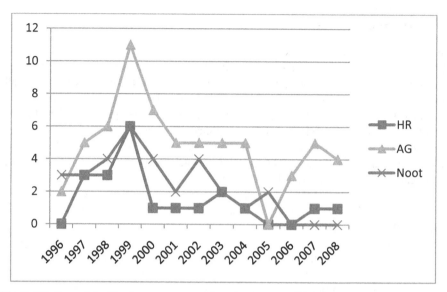

Fig. 3. Citations of the Zwolsman case over time for the court (HR), attorney general (AG) and annotators (Noot)

4.2   The Top-10 Cases According to Legal Experts

To get an idea which cases of the Supreme Court are considered important by legal experts, we decided to analyse collections that are being used in teaching law at Dutch universities. We analysed five of these collections from two different publishers and again only considered decisions by the Supreme Court between 1965 and 2008. In total we selected 376 cases and ranked them according to the number of citations to these 376 cases in the database of 15,053 cases we created as discussed above. Results are presented below in Table 4 the same way as in Table 3 above.

The first four cases are also in the top-10 of our network cases: Zwolsman, Haviltex, 'reasonable period' and 'limits oral testimony' (1, 2, 5 and 6 in Table 3), the

Table 4. Top-10 cases based on case law collections

Nr	NJ-number	Case Name	Other parties	Court	Advocate General	Annotator	Area
1	1996/249	Zwolsman	1	3	2	1	Criminal procedure
2	1981/635	Haviltex	5	5	1	4	Civil procedure
3	2000/721	Reasonable period	10	1	8	21	Civil procedure
4	1994/427	Limits oral testimony	3	8	10	33	Criminal procedure
5	2004/376	Waste pipe	33	2	16	198	Criminal procedure
6	1977/241	Bunde/Erckens	1231	764	7	115	Private law
7	1982/411	Roof tiler	19	369	14	581	Criminal procedure
8	1974/450	Meer en Vaart	217	51	16	581	Criminal procedure
9	2006/393	Pot nursery	1231	23	10	5370	Criminal procedure
10	1991/393	State/Windmill	217	15	328	8	Civil procedure

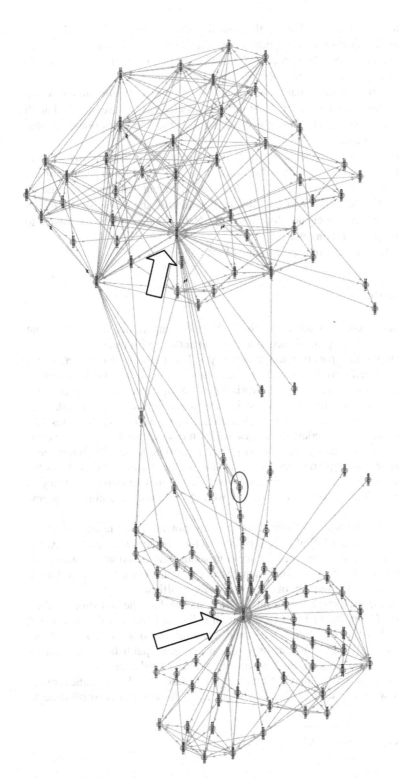

Fig. 4. Network of cases that cite Boon/Van Loon (left arrow) and case nr 10 (right arrow). The bull-calf case is circled.

other six are not. Cases 7, 8 and 9 are all about the 'duty to respond' for judges, i.e. duty to motivate a rejection of an appeal by a defendant. Case 5 is about rules for evidence in criminal cases. Cases 6 and 10 have in common that there is no article directly applicable and general principles of private law apply.

In general it is not the case that these collections of cases only contain cases that are cited a lot by the Supreme Court. 46 of the 376 selected cases were not cited at all in our database, but the overall citation distribution over the 376 cases resembles that of the complete database.

None of the top-10 network cases from Table 3 appear in collections of both publishers, but 4 of them appear in collections of the same publisher as that of the database (Kluwer). The other publisher also uses different sources and citation methods, i.e. not the NJ-number. If all publishers would use the recent standard way for citing cases that has been developed for the Netherlands,[7] it will be easier to compute and compare citation rates (cf. [10]).

The family law cases do not appear in any of the collections.

5   Conclusions

Does the number of citations tell us anything about the importance of case law and can it be used to help legal practitioners find relevant authoritative cases?

The answer to the first part of the question is yes, but it is not the complete story. The cases that are cited most seem to have certain things in common: They are mostly about procedural law, which is not surprising since the database only contains Supreme Court cases, the highest and last instance court whose primary task is to judge judicial procedures and not so much material facts. Secondly, the cases 'fill holes in legislation', i.e. the court forms law where the legislator has not (yet) done so. This was best illustrated by the citation pattern over time to the Zwolsman case; once the legislator had 'repaired' the hole, the number of citations dropped. So the number of citations is not the only criterion that determines relevance, recency is another one and probably the type of court as well (but we only examined Supreme Court cases).

Network analysis of sources of law seems to have potential as a research tool for legal scholars. Fowler e.a. [3,4] already showed this for US Supreme Court cases, but given the different status of case law in our continental law tradition, we wanted to see if this would also be the case for Dutch cases. Our court cites much less than in the US, but despite that the number of citations seems significant.

Fowler e.a. used prediction for future citation to validate the outcome of their network analysis. Given the fact that case citation is less important in the continental legal tradition and our courts cite much less and less long over time, we could not rely on future citation prediction. Comparison to the selection by publishers and teachers for case collections used in educating law students seems a good alternative.

A 'sudden' drop in the number of citations of a case may be an indication of codification, as we have seen in the Zwolsman case. It also seems worthwhile to

---

[7] Juriconnect, http://www.juriconnect.nl/

further investigate the network structures around cases where the court changes a line of reasoning. Perhaps the structure presented in Figure 4 has some features in common with similar cases that can be used to 'mine' for these breaches in case law.

Future research will extend the type of cases to other courts than the Supreme Court, and to cases published by other publishers or by the official Dutch portal.

Finally, the *(material) content* of a case is of course also an important criterion for its relevance in a particular case. Outcomes of network analysis need to be combined with more traditional or other ways to search and match cases on their content (cf. "reason for citing" as used in [11]).

References

1. Dworkin, D.: Taking rights seriously. Harvard University Press, USA (1978)
2. Egghe, L., Rousseau, R.: Introduction to Informetrics. Elsevier, Amsterdam (1990)
3. Fowler, J.H., Jeon, S.: The authority of Supreme Court precedent. Social Networks 30, 16–30 (2008)
4. Fowler, J.H., Johnson, T.R., Spriggs II, J.F., Jeon, S., Wahlbeck, P.J.: Network Analysis and the Law: Measuring the Legal Importance of Supreme Court Precedents. Political Analysis 15(3), 324–346 (2006)
5. Franx, J.K.: De Hoge Raad: voorgaan, doorgaan of omgaan. Kluwer, Deventer (1994)
6. Jaffe, A., Trajtenberg, M.: Patents, Citations and Innovations: A Window on the Knowledge Economy. MIT Press, Cambridge (2002)
7. de Maat, E., Winkels, R., Van Engers, T.: Automated Detection of Reference Structures in Law. In: van Engers, T. (ed.) The Nineteenth Annual Conference on Legal Knowledge and Information Systems, JURIX 2006, pp. 41–50. IOS Press, Amsterdam (2006)
8. Mazzega, P., Bourcier, D., Boulet, R.: The network of French legal codes. In: ICAIL 2009, pp. 236–237. ACM (2009)
9. de Mey, J.M.: Toegang tot rechterlijke uitspraken. Rapport van de VMC-studiecommissie Openbaarheid van rechtspraak, Mediaforum (April 2006) (in Dutch)
10. van Oppijnen, M.: A Public Index of Case Law References –the End of Multiple and Complex Citations Source. In: Proceeding of the 2006 Conference on Legal Knowledge and Information Systems, JURIX 2006. Frontiers in Artificial Intelligence and Applications, vol. 152. IOS Press, Amsterdam (2006)
11. Zhang, P., Koppaka, L.: Semantics-based legal citation network. In: ICAIL 2007, pp. 123–130. ACM (2007)

Ontology Framework for Judgment Modelling

Marcello Ceci and Monica Palmirani

CIRSFID, University of Bologna

Abstract. The paper shows how to model judgments starting from the text and capturing not only the structural parts, but also the basic arguments used by the judge to reach its conclusions. We have also included a qualification of citations following the Shepard's method. The goal of this approach is to build a complete ontology framework capable of detecting and modelling knowledge directly from the judgment's text, providing the basic metadata to the logic and reasoning layers.

1 Introducton

Precedent is a main element of legal knowledge worldwide: by settling conflicts and sanctioning illegal behaviours, judicial activity enforces law provisions within the national borders, therefore supporting the validity of laws as well as the sovereignty of the government that issued them. Moreover, precedents (or case-law) are a fundamental source for law interpretation and it paradoxically happens that the exercise of jurisdiction can influence the scope of the same norms it has to apply, both in common law and civil law legal systems – even if to different extents. The AI & LAW community has presented very significant research outcomes in this topic since the '80, with different approaches: legal case-base reasoning (HYPO, CATO, IBP, CABARET) and more recently also argumentation (Carneades [10]).

The goal of the present research is to define a complete framework for the precedent modelling following the Semantic Web cake, starting from the text and filling the gap with the rules annotation. Cornerstone of the framework is the ontology, intended in its computer science meaning: a shared vocabulary and taxonomy which models a domain of knowledge by defining objects, concepts, their properties and their relations. A formalization of the main structure of the case-law, the metadata connected with the judicial legal concepts, and the ontology constitute the basis of a semantic tool which enriches the XML mark-up of precedents and supports legal reasoning. We believe that the new features of OWL 2.0 could unlock potentialities for legal concept modelling and reasoning [8], to be combined with those of rule modelling. Our aim is hence to formalize the legal concepts and the argumentation patterns contained in the judgment in order to check, validate and fully reuse the discourse of the judge as expressed by the text and the argumentation he produces. Four different models are necessary:

a) a document metadata structure, capturing the main parts of the judgment to create a bridge between text and semantic annotation of legal concepts [17][2];

M. Palmirani et al. (Eds.): AICOL Workshops 2011, LNAI 7639, pp. 116–130, 2012.

b) a legal core ontology, modelling the abstract legal concepts and the institutions to capture the main parts of the rule of law [12];

c) a legal domain ontology, modelling the main legal concepts in the specific domain concerned by the case-law (e.g. contracts, e-commerce, etc.)[14];

d) an argumentation system, modelling the structure of argumentation (arguments, counterarguments, premises, conclusions, rebuttal, etc.) [1][6][18].

The paper, following this path, presents the XML syntax of the judgment metadata, describes the core and legal domain ontologies and finally introduces the argumentation system.

This approach allows, under the technical point of view, to reach results such as:

- IR and query: it becomes possible to perform some very complex querying, applying the Semantic Web techniques (SPARQL-DL) on the qualified parts of the judgment text. It is possible, for example, to make the following request: "give me all the judgments in the last year, with a dissenting opinion, in the e-commerce field and where the main argument of the decision is the application of Consumer Law, art. 122";

- NLP: the machine can detect relevant parts of the speech using the semantic annotation and the ontology;

- Rules: the ontologies provide information for the rule engine to perform the legal case-based reasoning.

2   Multi-layer Legal Document Modelling

In the last ten years, the e-law scientific community has made relevant efforts to model and represent the legal resources using the XML standards (Metalex/CEN [4][5] for legislative documents, NormeInRete for laws [7], Akoma Ntoso for parliamentary activity and judiciary documents [17], Australian Judgment XML standard [16], LegalXML OASIS initiative). Some projects even allowed the XML markup of the structure of precedents and of different interpretations [2]. But in order to capture and represent the legal knowledge embedded in the case-law – including the judge's reasoning – a further semantic layer is required following the Tim Berners-Lee semantic cake, and in this perspective ontologies can actually give a fundamental contribution. Although based on a shared ontology capturing the semantics of legal concepts [13][15], the actual project is more focused on the representation of the contents of judicial decisions as they are expressed in the text.

The approach adopted is based on a multi-layer paradigm, where the legal resource is managed in separated levels which are linked to each other but organized in order to allow multi-annotation, multi-interpretation, and multi-ontology with redundancy of representation [14]. The syntactical approach was based on the following schema:

- Text annotation in XML: the Akoma Ntoso standard grants proper mark-up of the structure of the judgement and of citations;

- Metadata annotation: the Akoma Ntoso metadata block captures not only the metadata concerning the lifecycle of the document (e.g. workflow of the trial, formal steps, jurisdiction, level of judgments), but also the legal qualification of relevant parts of the decision, such as the minority report or the dissenting opinion;

- Ontology annotation: using external OWL definitions and linked through special mechanism to the XML document;
- Rules: unfortunately OWL, even with the functionalities of version 2.0, is unable to represent complex and defeasible legal arguments. It is therefore necessary to extend the model with rule modelling, using the argumentation theory [11] [9].

3   Judgement Structure

The judgment in Akoma Ntoso is a particular type of document modelled for detecting the main significant parts of the precedent document: header for capturing the main information concerning the parties, coram, neutral citation, document numbers and identification information; body for representing the main part of the judgment including the decision; conclusion for detecting the signatures.

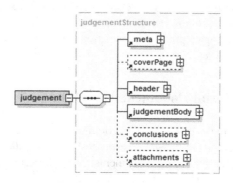

Fig. 1. Judgment main structure in Akoma Ntoso

Structure of the main part of the judgment
The body is divided into four main blocks:

- the `introduction`, where usually (especially in common law decisions) the story of the trial is introduced;
- the `background`, dedicated to the description of the facts;
- the `motivation`, where the judge introduces the arguments supporting his decision;
- the `decision`, where the final outcome is given by the judge.

This division is fundamental for detecting facts and factors from the background: in the `motivation` we detect arguments and counterarguments and in the `decision` the final conclusion of the legal argumentation process. Those qualified fragments of text should be annotated by legal experts with the help of a special editor tool (e.g. Norma-Editor) that allows an easily linking between text, metadata and ontology classes.

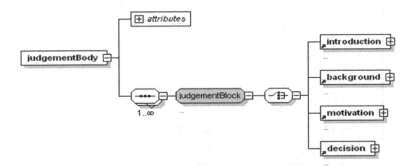

Fig. 2. Judgment Body sub-elements in Akoma Ntoso

Metadata of judgments

The metadata of the judgment are divided in four main blocks:

- **Descriptive** metadata objectively tracking judgment data such as the date of publication, the number of the case, the natural citation, the names of the judges, the jurisdiction, the level of the judgment, the nature of the case, the type of court, the parties, the lawyers, and so on.
- **Classification** metadata concerning the matter of the case (thesaurus), together with the reportable or not reportable case-base. These metadata represent a filtering station of reportable judgments, following the common law tradition that underlines the cases producing a new *rule of law*.
- **Lifecycle:** the history of the document, useful for versioning.
- **Workflow** metadata, tracking each step of the document production process. Since multi-annotation of the same fragment of text is allowed, each actor in the workflow chain can annotate the document with his/her specific metadata.
- **References**: metadata remarking all documents citing/cited by the judgment or links all documents which are logically connected to the judgement.
- **Semantic annotation**: the classification of the text under the legal point of view, especially in the motivation part.
- **Ontology**: a definition of mechanisms for linking the fragment of text to macro classes such as People, Organization, Role, Actions, Event, Terms, Location, etc.

Using these metadata it is possible to annotate very specific knowledge. In the following fragment of text, we have to capture the role of each person involved in the trial: Mr. Du Plessis is a lawyer, with the role of advocate of the appellant, instructed by the Kruger Inc. I can annotate these information with XML to allow complex queries such as: "give me all the judgments where Du Plessis is playing the role of instructor of the appellant on behalf of a third Inc. company".

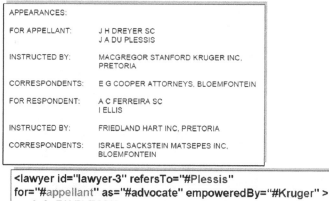

Fig. 3. Actor qualification

Qualification of the citations in the judgment

Each judgment citation could be qualified using Shepard's method that permits to understand which references are in favour of the current judgment argumentation. The list of qualifications include: support, meaning that the cited judgment supports the current decision; isAnalogTo, meaning that the current case-law is analogue to a cited precedent; distinguished, meaning that the current precedent is distinguished from the cited case-law. Particular importance is played by overrules, detecting the case-law cited by the judgment's motivation whose rule of law the judge intends to overrule. This qualification mechanism helps to reinforce the main arguments used by the judge to provide evidences and parameters (e.g. list of the cited case-law with the role played in the argumentation).

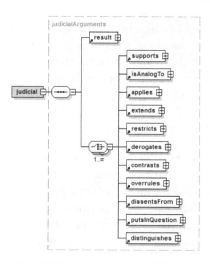

Fig. 4. Shepard's method qualification

Rule of law, stare decisis and ratio decidendi, obiter dicta
One of the main principles in common law judgments is to define a *rule of law* fixing the pattern for the similar future cases. This monotonic mechanism is called *stare decisis* and guarantees the equal application of justice to comparable cases. The *stare decisis* is applied only to a particular and relevant part of the decision called *ratio decidendi*, excluding accessories arguments called *obiter dicta*. Now the research conducted on the ontology framework (see section 4) reveals the importance of marking up those relevant and meaningful parts of the text. By marking up the *ratio decidendi* text and using this information in combination with the Shepard's qualification method for cited cases-law, it is possible to provide richer information to the argumentation engine devoted to the legal reasoning. For these reasons we will define new metadata in the analysis block of Akoma Ntoso in order to qualify also the *ratio decidendi* and the *obiter dicta*.

4 Legal Ontology Methodology

This research is based on a middle-out methodology: bottom-up for capturing and modelling the core and legal domain ontology classes and top-down for modelling the argumentation theory components and their relationships.

The research starts from the analysis of a sample of Italian case law constituted by a set of 27 decisions of different grade (tribunal, court of appeal, Cassation Court) taken from a precedents database and concerning the same law subject, consumer law.

This domain is particularly fit for this research because it includes situations where the strong rules and the strict deductive logic are not sufficient to cope with the legal reasoning of the judge. We need to evoke the defeasible logic for the representation of the "defeasible rules" concerning the subject. In fact, many norms concerning contracts are not absolutely mandatory: they can be overlapped by different discipline through specific agreements between the parties. The problem of representing "defeasible" rules, in fact, is a core problem in legal knowledge representation. Exploring how the OWL 2.0 could prepare the background for the application of defeseable logic is therefore a main goal of the present research: in fact, the OWL language (even in its 2.0 version) is not fitted for managing the defeasibility, being only able to capture the static factual and legal knowledge to be reused in the rule layer. Nevertheless the gap between ontology and rules is often underestimated, and the benefits coming from the OWL 2.0 computation are neglected. For this reason, well aware of the limitations of the OWL 2.0 in representing defeasibile logic, we have the intention to stress the axiom definitions as much as possible to improve performances, computability, and management of the classes over the time, and to foster the Semantic Web tools and applications which are already available in this sector.

Under a different perspective, the law of contracts is an interesting field because the (either automatic or manual) markup of contract parts allows the highlight of single clauses and their comparison to general rules as well as to case law concerning the matter. These possibilities can be used to introduce a semi-automatic check compliance of a contract draft.

The legal field taken as a first sample is the discipline concerning oppressive clauses in Consumer Contracts. The matter is specifically disciplined in the Italian "Codice del Consumo" (Consumer Law) as well as in most foreign legal systems, which will allow an extension of the research to foreign decisions (and laws).

5   Ontology Framework

The knowledge will be modelled using a set of three ontologies:

- a *Core Ontology* describing the domain's main elements in terms of general legal concepts through an LKIF-Core extension;
- a *Domain Ontology* containing the modelling of both the concepts and the rules expressed by (and used in) the Italian "Codice del Consumo" (Consumer Law) and in artt. 1241-1242 Civil Code, as well as all relevant knowledge extracted from Italian sentences containing interpretation of private agreements in the light of those laws.;
- an *Argumentation System* for the modelling of argumentation patterns followed by the judge during the interpretation process.

Following these principles, an ontology was built starting from a sample of 33 Judicial decisions issued by Italian 1st grade tribunals and courts of appeal and concerning the subject of consumer contracts.

5.1   Core Ontology

The core ontology introduces the main concepts and interactions in the legal domain, defining the classes which will be later filled with information taken from the judicial decisions. Even though the core ontology should be domain-generic and not modeled upon a specific legal subject, the model presented here was conceived to successfully represent the interaction in the civil law subject, when contracts, laws and judicial decisions come into play. Obviously, it will be necessary to add further classification prior to successfully insert knowledge about a different domain (es. Public contracts, administrative law, tort law).

The backbone of the Core Ontology is represented by three classes already existing in LKIF-Core: `Qualificatory_Expression`, `Qualification` and `Qualified`.

- `Qualificatory_Expression` (subclass of `Mental_Entity>` `Mental_Object>Proposition>Expression>Legal_Expression`) represents a legal expression which ascribes a legal status to a person or an object (for example, "x is a citizen, "x is an intellectual work", "x is a technical invention).

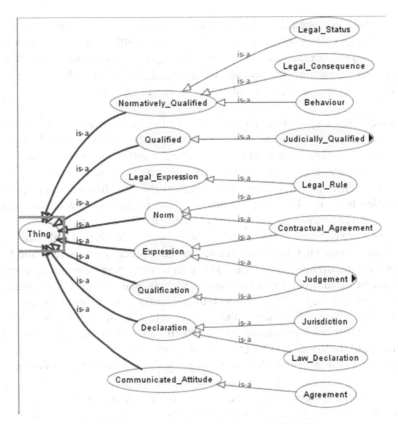

Fig. 5. Core Ontology's specification of LKIF-Core

- Qualification
  (Mental_Entity>Mental_Object>Proposition) expresses e.g. a judgement: the thing qualified by the qualification is comparable to something else.
- Qualified represents anything that is qualified by some qualification.

Since the main object to be represented in the present set of ontologies is the normative/judicial qualification brought forward by contractual agreements, legal rules and judicial interpretations, the classes presented above constitute the nucleus of the Core Ontologies. Unfortunately, because of the limits of OWL language, the LKIF-Core Qualification and Qualified classes are linked only by a single property (qualifies/qualified_by), but what we want to model is an n-ary relation between (1) a qualifying expression, (2) the kind of qualification 1 gives and (3) the object qualified by 1. In order to represent this, the property "qualifies" has been forked into two new properties: "considers" and "applies". The first one, "considers" (modelled as superclass of the LKIF-Core properties "evaluates", "allows", "disallows") represents the object of the qualification. The second property, "applies", shows *towards which concept* the qualification is made. For example, a

Contractual_Agreement considers a Material_Circumstance and applies a Legal_Status; a Legal_Rule considers a Legal_Status and applies a Legal_Consequence; a Judicial_Interpretation considers a Material_Circumstance and applies a Legal_Status; an Adjudication considers a Judicial_Claim and applies a Judicial_Outcome.

Qualifying Legal Expression. To overcome the limited expressivity of the original LKIF-Core classes a new conceptual category called "Qualifying_ Legal_Expression" has been created, putting together the characteristics of the Qualificatory_Expression and Qualification classes, enhanced by the fork of the qualifies property. This class represents the formalization of a disposition, and includes the three legal expressions involved in contract law-related judicial decisions: Contractual_Agreement, Legal_Rule and Judgement.

As Qualificatory_Expression sub-classes, the three Qualifying Legal Expressions contain all information related to the "speech act"; its semantic bonds with the externalization, the legal power and the agents ensure a complete representation of all aspects that can come into play when facing a legal issue (the legitimacy of the legislative body/court/legal party, the characteristics of the corresponding legal document, the identity/characteristics of people/bodies involved...). Their main properties are "medium" and "attitude" (see below for a specification of the Medium, Attitude and Agent classes).

As Qualification subclasses, the three Qualifying Legal Expressions contain all information related to the effects they have in the legal world: the legal categories/obligations/legal effects they create, modify or repeal. A subdivision can be made between one direct subclass (Judgement, which in this perspective is furtherly divided into the Judicial_Interpretation and Adjudication subclasses) and two subclasses of Norm (Legal_Rule and Contractual_Agreement). As explained before, the property "qualifies" - linking the qualifying expression to the Qualified expression - has been forked into two new properties: "considers" and "applies", representing respectively the direct object and the "destination" of the qualification.

Qualified Expressions. All the ranges of the "considers" and "applies" properties presented above are subclasses of the Qualified class. Its subclasses are Normatively_Qualified (a class already present in LKIF-Core) and Judicially_Qualified (created anew).

Normatively_Qualified expressions include Material_Circumstance, Legal_Status and Legal_Consequence. They represent the expressions that can be directly taken into consideration by a Norm: while Material_Circumstance represents any factual circumstance that can be taken into consideration by a Norm, Legal_Status represent an institutional fact (i.e. fulfillment of contract, oppressive clause, contract breach) that is normally considered_by a Legal_Rule and applied_by a Contractual_ Agreement or a Judgement. As we will see, the link between a Contractual_Agreement and the Legal_Rule it applies is a "weak" link until a Judicial_Interpretation has confirmed (or denied) it. Finally, Legal_Consequence represents the sanction provided by the law in the presence of some Legal_Status or Material_Circumstance. It covers all cases when the Legal_Rule considers some Normatively_Qualified expression, but does not simply allows, disallows or evaluates it.

Judicially_Qualified expressions include Judicial_Claim, Judicial_Outcome and all elements susceptible of being taken into consideration during a legal proceeding (i.e. Contractual_Agreeement, but also Legal_Rule, expecially in Cassation Court and Costitutional Court).

Judicial_Claim is the claim which is object of the legal proceeding. It is considered_by an Adjudication, the answer of the judge to the legal claim (subclass of Qualification>Judgement). The content of the answer (rebuttal/acceptance of the claim or any other possible outcome foreseen by the law) is represented by the Judicial_Outcome class, applied_by the Adjudication. So the representation is the following: a Judicial_Claim is considered_by an Adjudication that applies a Judicial_Outcome.

The miscellaneous elements that can be taken into consideration during a legal proceeding are included in the Judicially_Qualified class as long as they are actually considered_by some Judicial_Interpretation. So, for example, a Contractual_Agreement can be considered_by some Judicial_Interpretation who applies some Legal_Status to it (i.e. the agreement is oppressive, is inefficacious, represents an arbitration clause, is specifically signed by both parties). In these cases, a OWL 2.0 property chain directly links the Contractual_Agreement to the Legal_Status judicially applied to it. This "strong" link, represented by the property "interpreted_as", is the the fundamental information that we want to represent – and manage – through this set of ontologies.

Mediums, Propositional Attitudes and Agents. The Medium class identifies the support through which the proposition is expressed. It does not represent the material support of the Expression instance but rather its *genus* (Contract, Precedent, Code).

The latter two classes, ranges of the Expression properties of the Qualifying Legal Expression, describe its background. The LKIF-Core class Propositional_Attitude was specified with the Jurisdiction, Law_Declaration and Agreement subclasses, representing the enabling powers that stand behind a Judgement, a Legal_Rule and a Contractual_ Agreement. On the contrary, to represent the possible "authors" of a Qualifying Legal Expression there was no need to specify the subclasses of Agent already present in LKIF-Core (Person and Organization). The knowledge that agents and attitudes represent can be important in some cases: i.e. if a claim is based on the lack of contractual power by one of the parties, or on the identity/characteristics of a part, or on the lack of force by some law or other regulation (which can in turn depend by the lack of legitimacy of one of its authors).

Modularity of the Core Ontology. The expansion brought by the Core Ontology to the LKIF-Core concepts is currently oriented to the representation of the elements involved in civil-law cases regarding contract law. Nevertheless, the Core Ontology provides general – and relatively open - categories for this kind of judicial activity to be represented, and can therefore be considered as a core to be "expanded" with categorization from other branches of law, but not to be "substituted", since the basic concepts introduced here can come into play also in judgements concerning different subjects.

5.2   Domain Ontology

Following the structure outlined in the Core Ontology, the knowledge taken from judicial decisions is represented in the Domain Ontology under the perspective of the contents of the documents involved. The modelling should be carried out manually by experts in the legal subject, as automatic information retrieval and machine learning techniques do not yet ensure a sufficient level of accuracy: the activity of building a domain ontology is very similar to that of writing a piece of legal doctrine, thus it should be manually achieved in such a way as to maintain a reference from the model to the author, while at the same time keeping an open approach (i.e. allowing different modelling of the same concept by different authors).

The modelling of the ontology is explained here through a simple example of data insertion and knowledge management by the Domain Ontology:

Example. *In the decision given by the 1ˢᵗ section of the Court of Piacenza on July 9ᵗʰ, 2009[1], concerning contractual obligations between two small enterprises (α and β), the judge had to decide whether a contract clause "14" of α/β contract, concerning the competent judge could be applied. The judge cites art. 1341 comma 2 of Italian Civil Code who says "a contract clause determining the competent judge is invalid if not specifically signed". In the contract signed by the parties there is a distinct box for a "specific signing" where all the clauses of the contract are recalled (by their number). The judge, with the support of precedents (he cites "among others" 9 Cassation Court sentences) interprets the "specific signing" as not being fulfilled through a generic recall of all the clauses, and therefore declares clause "14" of α/β contract invalid and inefficacious.The claim of inefficacy of clause 14, brought forward by α, is thus accepted.*

In order to represent the knowledge contained in that sentence, we have to represent three documents: Art. 1341 of Italian Civil Code, the contract between the two enterprises α and β, and the decision by the Court of Piacenza.

---

[1] Sent. N. 507 del 9 Luglio 2009, Tribunale di Piacenza, giudice dott. Morlini.

Modelling of the Law. The law involved in the judicial decision (Art. 1341comma 2 of Civil Code) is represented as a Qualifying Legal Expression (Legal_Rule) called "art1341Co2" (with a Code medium, a Law_Declaration attitude and a Parliament as agent). The Legal_Rule considers any individual which has the characteristics required by the law, and either allows/disallows/evaluates or applies some Legal_Consequence to it. In the example, each Contractual_Agreement which applies both a "CompetentJudge" and a "NotSpecificallySigned" statuses will be considered_by "art1341Co2", which in turn applies the Legal_Consequence of "invalidityExArt1341co2". The individuals "competentJudge" and "notSpecificallySigned" are thus created as Legal_Statuses that can be considered_by a Legal_Rule and applied_by a Contractual_Agreement, and the individual "invalidityExArt1341co2" is created as a Legal_Consequence applied_by the Legal_Rule "art1341Co2".

Modelling of the Contract. The α/β contract is a composition of one or more Contractual_Agreements (Contract, Contract_Clause), each of which represents an obligation arising from the contract sharing the same attitude, the "meeting of minds" between the Agents. A Contractual_Agreement normally considers some Material_Circumstance and applies some Legal_Status to it. Contract_Clause "α/βClause14" is created and linked to a Contractual_Agreement which applies the Legal_Status of "competentJudge"[2].

Modelling of the Decision. The "tribPcI09/07/2009" Judgement is created, composed of different instances: an Adjudication and at least one Judicial_Interpretation. They share a common attitude (a Jurisdiction power) a Precedent medium and some agents (claimant, defendant, and court). The Adjudication contains the Judicial_Outcome of the Judicial_Claim: in the example, the Court is incompetent because the contractual clause concerning the competent judge is invalid. The Judicial_Interpretation considers the Contractual_Agreement contained in "α/βClause14" and applies the "notSpecificallySigned" Legal_Status.

Reasoning on the Knowledge Base - To check the consistency of this knowledge we will use Pallet queries. This tool was built to extract data from the OWL ontology, but could also be used to check if the ontology gives a unique and correct answer to some formalized question (i.e. asking about the validity of some proof, or about the

---

[2] This is done because there is no argue between the parties about whether clause 14 concerns the competent judge. However, as explained before, this kind of link is a "weak" one, considering that the contractual parties have no power to "impose" a legal status to a contract, and that reconducting a contractual agreement to the legal figure it evokes is the main activity brought forward by judicial interpretation in the contracts field. For this reason, the property "applies" related to a Legal_Status is very weak when its domain is a Contractual_Agreement, and likely to be overridden by a contrasting application performed by a Judicial_Interpretation.

qualification of factual events under legal principles). When a `Contractual_Agreement` (the expression brought by a `Contract_Clause`) is considered_by some `Judicial_Interpretation`, the ontology gathers all relevant information related to the three documents presented above: contract parties, judicial actors, legal status applied to the agreement (eventually in comparison to the one suggested by the contract/judicial parties), the law rules which are relevant to the legal status, the final adjudication of the claim and the part played in it by the interpreted agreement, and so on. In this perspective, the citations to case law constitute the first element to be represented using Akoma Ntoso metadata (Shepardizing) to classify interpretations and argumentations.

The first objective is to gather all this semantically-rich information for advanced querying on precedents, but more can be achieved by combining different `Judicial_Interpretation` with knowledge coming from the contract and the applicable law: the ontology reasoner is in fact capable of predicting – to some extents – the outcome of the judge (i.e. predicting that a clause will be judged as valid/invalid) and to run inferences about the agreement (i.e. as interpreted, the clause is irrelevant for the whole Italian Consumer Law/for the legal rule contained in article 1342 comma 2 of Italian Civil Code).

In the example, when all the relevant knowledge is represented into the ontology, the reasoner is capable of inferring that "The agreement contained in clause 14 of the α/β contract is invalid ex article 1341 comma 2". At the actual stage, this result is reached through a sublcass of the `Contractual_Agreement` and `Qualified` class, defined by an axiom representing the rule of law: clauses that fulfill the axiom are automatically classified in that class, and thus considered_by the proper law. At this point, a simple property chain gives the clause its final (validity/invalidity) status under that law (see section 6 for further exploitations of OWL 2.0).

This inferred knowledge is important for two reasons: *a.* by "predicting" the judge's final statement on the clause (even if not that on the claim), this knowledge represents a logic and deontic check on the legal consequences the judge takes from its interpretation; *b.* it gives a fundamental element for the argumentation system to support the explanation of the adjudication of the claim. The argumentation system, in fact, will be able to use the (inserted and inferred) elements of the decision's groundings to support and explain the `Adjudication` contained in the last part of the judgment.

6   OWL 2.0

OWL 2 introduces several features to the original Web Ontology Language, some of which allow a richer representation of knowledge, mostly when dealing with properties and datatypes. Some of these would be useful, but also lead to a great increase of complexity in the models: for example, disjointness between properties has been introduced, but in order to exploit this feature it would be necessary to create as many properties as possible statuses, which in turn would greatly affect computability.

At least two of these new constructs concerning properties deserve attention because they could enhance expressivity without affecting (or even reducing) the complexity of the model built so far:

Keys. An **HasKey** axiom states that each *named* instance of a class is uniquely identified by a (data or object) property or a set of properties - that is, if two named instances of the class coincide on values for each of key properties, then these two individuals are the same. This feature can be useful for identifying the unique "actors" of the judicial claim, such as the parties, the contract, the norm, and the decision itself.

Property Chains. The OWL 2 construct ObjectPropertyChain in a SubObjectPropertyOf axiom allows a property to be defined as the composition of several properties. Such axioms are known as *complex role inclusions* in SROIQ. In the previous section the Expression having all the properties required by the relevant law is inferred as being an instance of an "anonymous qualified class" which is, in turn, linked to the applicable law through a property. Here, a property chain unifies the two properties (from the qualified expression to the law, and from the law to the legal consequence) and brings their semantics to the surface by creating a direct property linking the contract clause to its status (judged_as Invalid). A better exploitation of the OWL 2.0 property chains could lead to an ever more direct and complete solution, mainly by removing the need for the anonymous subclass in order to identify the clause instances considered_by the relevant law.

7   Future Work

Next step is to develop an arugmentation system in support of the knowledge base already built. The argumentation system will delve deeper into the interpretation process to capture relevant argumentation schemes and other informations that should enhance the inferential capabilities of the set of ontologies. The domain-specific ontology, in fact, does not go beyond stating that a judicial interpretation was made towards some normative classification, but cannot describe how the judge came to that conclusion. With the argumentation system, the set should be able to highlight similarities between different judicial decisions not only comparing their normative anchors or factual/processual circumstances, but also the argumentation schemes followed (and the abstract legal figures recalled) by the judge in the decision's text. The study of argumentation technology and argumentation theory will be fundamental in this perspective.

8   Conclusions

This paper presents an innovative approach to manage knowledge contained in the case-law filling the gap between text, metadata, ontology representation and rules modelling, with the goal of detecting all the information available in the text to favour the legal reasoning through the argumentation theory. This approach allows to directly annotate the text with peculiar metadata representing the hook for the core, domain and argument ontologies. On the other hand, the envisaged ontology framework brings to the surface some weak points in the Akoma Ntoso structure that need to be reinforced, such as the metadata detecting and qualifying the *ratio decidendi* and the *obiter dicta* in the text. Finally, OWL 2.0 is used to get as close as possible to the rules, in order to exploit the computational characteristic of description logics.

References

[1] Ashley, K.D.: Ontological requirements for analogical, teleological, and hypothetical legal reasoning. In: ICAIL 2009, pp. 1–10 (2009)

[2] Barabucci, G., Cervone, L., Palmirani, M., Peroni, S., Vitali, F.: Multi-layer Markup and Ontological Structures in Akoma Ntoso. In: Casanovas, P., Pagallo, U., Sartor, G., Ajani, G. (eds.) AICOL-II/JURIX 2009. LNCS, vol. 6237, pp. 133–149. Springer, Heidelberg (2010)

[3] Bench-Capon, T.J.M., Prakken, H.: Using argument schemes for hypothetical reasoning in law. Artif. Intell. Law 18(2), 153–174 (2010)

[4] Boer, A., Winkels, R., Vitali, F.: MetaLex XML and the Legal Knowledge Interchange Format. In: Casanovas, P., Sartor, G., Casellas, N., Rubino, R. (eds.) Computable Models of the Law. LNCS (LNAI), vol. 4884, pp. 21–41. Springer, Heidelberg (2008)

[5] Boer, A., Hoekstra, R., de Maat, E., Hupkes, E., Vitali, F., Palmirani, M., Rátai, B.: CEN Metalex Workshop Agreement (August 28, 2009) (proposal), http://www.metalex.eu/WA/proposal

[6] Brüninghaus, S., Ashley, K.D.: Generating legal arguments and predictions from case texts. In: ICAIL 2005, pp. 65–74. ACM Press, New York (2005)

[7] Circolare AIPA/CR/40, Formato per la rappresentazione elettronica dei provvedimenti normativi tramite il linguaggio di marcatura XML, GU n. 102 del (May 3, 2002)

[8] Gangemi, A.: Design Patterns for Legal Ontology Construction, in Trends in Legal Knowledge. In: The Semantic Web and the Regulation of Electronic Social Systems, pp. 171-191. European Press Academic Publishing (2007)

[9] Gordon, T.F., Governatori, G., Rotolo, A.: Rules and Norms: Requirements for Rule Interchange Languages in the Legal Domain. In: Governatori, G., Hall, J., Paschke, A. (eds.) RuleML 2009. LNCS, vol. 5858, pp. 282–296. Springer, Heidelberg (2009)

[10] Gordon, T.F., Walton, D.: Legal reasoning with argumentation schemes. In: ICAIL 2009, pp. 137–146 (2009)

[11] Gordon, T.F.: Constructing Legal Arguments with Rules in the Legal Knowledge Interchange Format (LKIF). In: Computable Models of the Law, Languages, Dialogues, Games, Ontologies, pp. 162–184 (2008)

[12] Hoekstra, R., Breuker, J., Di Bello, M., Boer, A.: The LKIF Core Ontology of Basic Legal Concepts. In: Casanovas, P., Biasiotti, M.A., Francesconi, E., Sagri, M.T. (eds.) Proceedings of LOAIT 2007 (2007)

[13] Mommers, L.: Ontologies in the Legal Domain. In: Poli, R., Seibt, J. (eds.) Theory and Applications of Ontology: Philosophical Perspectives, pp. 265–276. Springer (2010)

[14] Palmirani, M., Contissa, G., Rubino, R.: Fill the Gap in the Legal Knowledge Modelling. In: Governatori, G., Hall, J., Paschke, A. (eds.) RuleML 2009. LNCS, vol. 5858, pp. 305–314. Springer, Heidelberg (2009)

[15] Sartor, G.: Legal Concepts as Inferential Nodes and Ontological Categories. Artif. Intell. Law 17(3), 217–251 (2009)

[16] Supreme Court of Western Australia, in partnership with the Department of Justice. Proposed XML Schema Definition of Supreme Court Judgements (June 15, 2011), http://www.aija.org.au/info/techn/JudgmentsVersion1.9Web.doc

[17] Vitali, F.: Akoma Ntoso Release Notes (1997), http://www.akomantoso.org

[18] Wyner, A., Bench-Capon, T., Atkison, K.M.: Towards Formalising Argumentation about Legal Cases. In: ICAIL 2011, Pittsburg, June 5-10. ACM (2011)

Eunomos, a Legal Document and Knowledge Management System to Build Legal Services

Guido Boella[1,2], Llio Humphreys[1], Marco Martin[1,2],
Piercarlo Rossi[3], and Leendert van der Torre[4]

[1] Università di Torino, Italy
{guido,humphreys,notmart}@di.unito.it
[2] Nomotika s.r.l., Italy
[3] Università del Piemonte Orientale, Italy
piercarlo.rossi@unipmn.it
[4] University of Luxembourg, Luxembourg
leon.vandertorre@uni.lu

Abstract. We introduce the Eunomos software, an advanced legal document management system with terminology management. We describe the challenges of legal research in an increasingly complex, multi-level and multi-lingual world and how the Eunomos software helps expert users keep track of the state of the relevant law on any given topic. We will describe in particular the editorial process for building legal knowledge.

1 Introduction

To operate efficiently, law firms and financial institutions need reliable and up-to-date information on the state of the law on the relevant topics. It is now more difficult than ever to know what the law is on a particular topic. Italy, for example, produces thousands of laws every year, with many pieces of legislation containing a number of norms on a range of different topics. While Italy is well known for legislative over-production, any one or any business that needs to operate in an international context will have to deal with multiple legislations. In Europe, legal research is today more complicated than ever due to subsidiarity and the number of laws that have to be considered from sources at different levels - international, European, national, regional and municipal. For large institutions, internal regulations may be drafted to ensure a standard way of complying with legal obligations. European, national and regional governments have a public duty to make laws available to all. In turn, citizens and organisations have a duty to ensure they are informed of the law and comply with laws and regulations. This duty is very strict for banks and insurance companies. Not only must they respond promptly to changes in their legal obligations, they must demonstrate that they have systems and procedures for searching for legal changes. So they employ specialist lawyers who trawl through various sources to find relevant legislation and influential cases.

M. Palmirani et al. (Eds.): AICOL Workshops 2011, LNAI 7639, pp. 131–146, 2012.

The world wide web is a wonderful resource to find information about almost anything. But in using this resource, lawyers have to ask whether the information is authoritative, relevant, complete and up to date. It is no longer difficult to find authoritative sources online because in many regions in Europe and beyond, there are now official online portals making laws and decrees available to citizens and organisations. These portals are updated on a regular, often daily basis. It is not so easy to ensure that the information is relevant, complete and up to date, especially since the information is presented differently on each portal.

The Normattiva website contains Italian national legislation in HTML format. For each legislation, there is another page listing modifications from other legislation. An useful feature of their website is that it is possible to view a modified legislation as revised text, with modifications from other legislation inserted in the text. Indeed, users can view not only the original and current version, but all intermediate versions as well. This is useful for finding which modifications are relevant for a particular date in the past, although care must be taken as the modification may come into force at a later date. At the moment, the regulations are not classified in many cases. In the future, Normattiva will use Eurovoc, a controversial EU classification system which is inspired by the bureaucratic organizational structure of the EU commission and is not commonly accepted by practitioners. So the usefulness for public offices, businesses and citizens is limited. Moreover, there is no way of knowing whether new interesting legislation has been issued on a particular topic, or a new piece of legislation overrides existing legislation of interest.

For this reason, besides institutional repositories, there are websites maintained by public offices responsible for different areas of governance that list the particular regulations relevant for their daily work. For example, the Piedmont regional tax office website lists among other things national and regional legislation and ministerial decrees about car tax. The website also serves businesses and citizens who require legal information about about particular domains, and its usefulness is attested to in surveys and logfile analysis carried out by the Regione Piemonte government. But the regulations are in heterogeneous formats (HTML or PDF) and do not have in-text hyperlink references. Another problem is that the user cannot be entirely certain that all the regulations are up to date, since the site is managed by hand and there are no automated processes to assist officers with revised legislation.

There are professional services delivering up to date legislation, but they too present several problems: often they are costly; the updates are not always communicated fast (sometimes the services are distributed on paper or on CD-ROM); the user does not know which updates are relevant; the classifications - if any - are not necessarily relevant for users and are not adaptable by users; finally, their data are copyrighted and thus it is not obvious how to build further services on top.

The UK legislation.gov.uk website publishes UK legislation, regional government statutory instruments, county council and church acts. It is possible to search for modifications made by and to any legislation. They also provide a co-operative editorial tool to enable other stakeholders to work with them to create and maintain, open, free to use, up to date revised legislation. Recognising variety of uses and needs, every document published on their website is available not only in HTML and PDF but in machine readable XML format in accordance with the Metalex standard. Their aim is to give others a stake and an incentive to work with them to resuse their documents and provide further services. Our work presupposes this philosophy, building on top of existing legislative portals to offer a customizable service. For us, the future is customised comprehensive information management that processes information from official sources, presenting only legislation that is relevant to clients' particular areas of interests, and updating clients on legislative changes that are relevant for their work.

Official government portals address the need for authoritative sources of information on legislation. But the research process continues to be manual and ad-hoc, and for law firms and financial institutions especially, there is a real risk that legal researchers might miss important information and misinterpret the law, resulting in significant costs in terms of reputation as well as legal payments. In summary, key problems are:

Viewing Laws from Different Sources: legal norms can come from regional, national or European authorities;

Classifying Laws into Different Topics: many pieces of legislation contain norms on a range of different subjects;

Explicit References: some portals do not contain clickable links to other referenced legislation, which makes navigating laws more difficult;

Implicit References: some legislation modify or override existing norms but do not explicitly say so;

Knowledge: legal terms acquire different meanings within different concepts, jurisdictions and over time and are sensitive to interpretation in judgements and doctrinal work;

We believe that there is a need for innovative products that addresses the increasing complexity of the legal domain and economic incentives towards greater efficiency. The Eunomos software described in this paper was developed in the context of the ICT4LAW[1] project with the following requirements in mind:

- the ability to view legislation at regional, national and European level displayed in the same manner from the same web interface;
- a mechanism for supporting classification of norms within legislation in user defined categories such as taxation, immigration etc.;
- hypertext links between legislation that contain references to other legislation;
- a list of similar legislation to help expert users identify legislation that may have been implicitly modified or overridden;

---

[1] ICT4LAW: "ICT Converging on Law: Next Generation Services for Citizens, Enterprises, Public Administration and Policymakers" funded by Regione Piemonte 2008-2013, call Converging Technologies 2007.

– multilevel, updatable ontologies that allow multiple definitions of terms to allow for different contexts, and which enables users to track the evolution of terms over time;

In this paper we show how the Eunomos software meets the above requirements, providing a description of the user workflow and some details on technical implementation.

2   General Overview

Eunomos is a knowledge management system that enables users to research laws and legal concepts, and make sure they comply with their legal obligations. It is much more than a website of legislation. It is a web-based interface for legal researchers and practitioners to manage knowledge about laws and legal concepts in different sectors and different jurisdictions. By offering a highly structured framework with legislative XML, norm classification and ontologies, Eunomos can be used as an in-house software that enables expert users to search, classify, annotate and build legal knowledge and keep up to date with legislative changes. Alternatively, Eunomos can be offered as a combined software and services package so that legislation monitoring is effectively outsourced. Eunomos knowledge engineers would be responsible for maintaining the data, while practitioners would be able to search for information and receive updates on legislative changes.

The Eunomos software is based on a legal inventory database of norms downloaded automatically from legislative portals (about 70,000 Italian national laws in the current demo) converted into legislative XML format and semi-automatically classified, with links between related legislation by analysis of in-text references. The legal terminology is represented using a legal ontology tool called Legal Taxonomy Syllabus which connects to relevant norms. Figure 1 shows the components of the system and the flow of documents into the system. More technical details are discussed in Section 4.

The legal document management part of the system is composed of a database of laws with the relationships among norms expressed by references and the classification of laws, articles or even single paragraphs. The laws can be inserted in the database via a web interface or collected by means of web spiders from portals like Normattiva and converted into XML. Then, references are extracted to build a network of links between the norms citing each other. The editorial process of the norm proceeds in manual manner with a classification phase which is supported by tools suggesting categories on the basis of different clues as discussed below. Finally, relevant concepts can be extracted and modelled using the ontology. The same holds for specialized concepts like roles and prescriptions (obligations, permissions) which are beyond the scope of this paper. An alert message is generated by the system to users if a newly downloaded legislation seems to be relevant to the user's domain of interest.

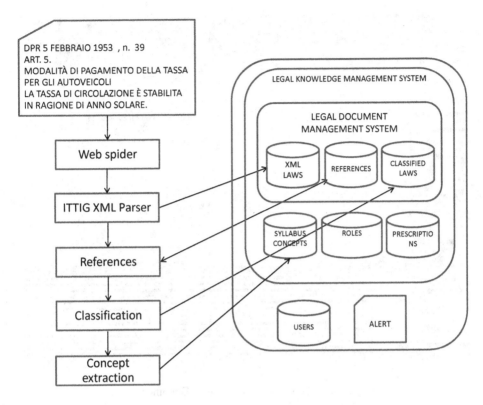

Fig. 1. Key components of the Eunomos system

3   User Workflow

3.1   Viewing Laws from Different Sources

Eunomos provides a web-based interface for lawyers and Eunomos knowledge engineers to find information about laws and legal concepts in different sectors and different jurisdictions. The legislation and user navigation are the same, whatever the source of the legislation. Users can not only search for legislation by name, which is translated into a Uniform Resource Name, but also by year, number, quoted text from the legislation or user comments associated with elements in the legislation.

For each piece of legislation, users can click on different options to view useful information about the legislation:

- the *Testo* (Text) option shows the full text of the legislation. The legislation is available in HTML, PDF or XML. Users can then choose whether to view the legislation in its original form or as revised text i.e. modifications via subsequent legislation to norms in the legislation in question are inserted

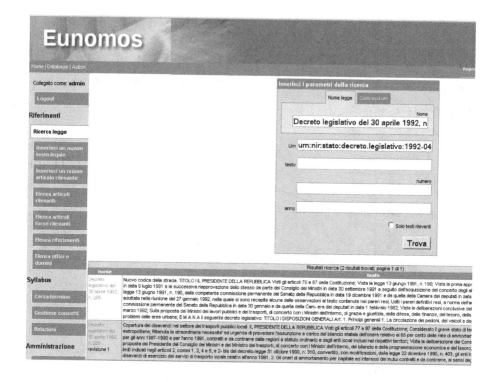

Fig. 2. The search interface of Eunomos

into the text of the modified legislation. Each article in the legislation is listed separately. References in the text to other articles or other legislation are linked to the relevant articles or legislation using URNs that conform to the NormaInRete standard. Users can click on the link to view the referenced legislation. Alternatively, they can hover their mouse over the link, and the relevant article appears in a text box above.

– the *Leggi o articoli rilevanti* (Relevant laws or articles) option provides a list of articles relevant to the domain selected by the user. The articles have been classified by a knowledge engineer. Users can click on relevant articles to view the text or hover their mouse to see the article in a text box.

– the *Candidati articoli rilevanti* (Candidate relevant articles) option provides a list of articles that may be relevant to the domain on the basis of links to classified legislation. If the reference is to a particular article from the same domain, the evidence is labelled as strong. If the reference is to a piece of legislation which contains articles from the same domain as well as other domains, the evidence is labelled as weaker.

- the *Riferimenti* (References) shows references from the legislation under consideration listed on the Normattiva website. The references on the Normattiva website were created manually. They can help the knowledge engineer find explicit references that were missed by the automatic reference detection tool because the textual pattern differs from the norm.
- the *Parole chiave* (Keywords) shows a list of all articles containing each term in the ontology found in the legislation, and hovering on an item in the list brings a text box containing the text of the relevant article. For the lawyer who may seek clarification on meaning and usage of terminology, a list providing all contexts in which the terms are used within the legislation under consideration can be most useful. For the knowledge engineer in charge of terminology management, the list can be useful for finding any new definitions or usage that needs to be recorded in the ontology. The list of terms can also indicate which domains are relevant for the legislation in question.
- the *Leggi simili* (Similar laws) provides a sorted list of the most similar legislation in the database. This can be most useful for finding legislation implicitly modified by later legislation. A list of similar laws is also useful for the lawyer to obtain an overview of the context of the legislation.

3.2 Classifying Laws into Topic-Based Categories

In cases where legislation cover a number of norms for various domains, it is useful for the lawyer to be able to view only articles relevant to the domain in which (s)he is interested. This requires that every relevant article and paragraph in every relevant piece of legislation is tagged as belonging to a particular domain. After new legislation is downloaded, and the Eunomos system has generated an alert message informing knowledge engineers, each article and paragraph will need to be classified by a knowledge engineer. The list of similar legislation can give a good indication of the domains that are relevant for the new legislation. The list of key terms can also be useful for identifying relevant domains. The knowledge engineer then proceeds to look at relevant domains one by one, and select the articles that are relevant to that domain. Within the Eunomos ontology, each term in the ontology is classified as belonging to particular domains. As (s)he looks at the article text, terms that belong to the domain in which the text is viewed are highlighted in yellow. Outgoing references to classified articles in other legislation that belong to the domain can also be a very good clue. Usually, if an unclassified norm contains a reference to another norm in previous legislation which has already been classified, the new norm belongs to the same class. The Eunomos system therefore ensures that when the user is looking at the legislation from the perspective of a particular domain, references to classified articles belonging to that domain are highlighted in yellow.

3.3 Explicit References

Lawyers need to know which piece of legislation references, modifies or overrides existing legislation. The Eunomos software contains a tool that automatically

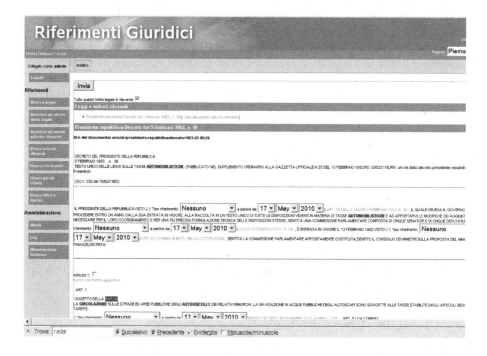

Fig. 3. Annotating legislation

finds references to articles in other legislation and creates inline hyperlinks within the legislation text. A knowledge engineer is required to look at each explicit reference and denote its type: whether it is merely a simple reference or in fact modifies or overrides existing legislation. For explicit references that were missed by the automatic tool because the expression was irregular, the Eunomos system enables the knowledge engineer to make links manually.

3.4   Implicit References

Where legislation fails to mention which existing legislation it modifies or overrides, a knowledge engineer will need to find the connections and make a record of modified legislation. The list of similar legislation can help knowledge engineers find legislation that may be implicitly modified or overridden. Eunomos has an interface to make comments about legislation and all its paragraphs and articles. This feature is especially useful for annotating elements that have been implicitly modified or overridden by other legislation.

3.5   Ontology

Behind every piece of legislation there is hidden information. Legal text is not written in natural language: each term, each concept, has a strict and defined meaning that can be quite different from everyday interpretations. Sometimes these terms are defined in the same piece of legislation. Sometimes their meaning can be found in other legislation or even in judicial or scholarly interpretations, because legal interpretations often gain acceptance with professionals before influencing subsequent definitions in legislation. Polysemy is a significant problem in legal terminology, because we have the added complexity that legal terms have significantly different meanings across jurisdictions, within contexts and over time. How can legal departments make explicit and manage this hidden information? The Eunomos package incorporates Legal Taxonomy Syllabus a specialist multilevel and multilegal ontology created by the Università di Torino for terminology management of European Directives and their national implementations. From the Eunomos interface, new terms and interpretations can be added to the ontology directly from the text of the law. Figure 4 illustrates the workflow of terminology management and the interrelationship between the ontology and legislation document management.

In the Legal Taxonomy Syllabus project, to properly manage terminological and conceptual misalignment, a distinction was made between *legal terms* amd *legal concepts*. The basic idea in the system is that the conceptual backbone consists in a taxonomy of unique concepts (ontology) to which any number of terms can refer to express their meaning. Eunomos contains specific interfaces for managing and viewing terms and concepts. The *Crea concetto* (Create a concept) page enables a knowledge engineer to create a new concept and add metadata, language, jurisdiction, an associated term that expresses the concept, date, description, notes and references to legislation defining the concept. The *Cerca termine* (Search terms) page is a table of terms in the ontology in alphabetical order. The table shows concepts associated with each term (with hyperlinks to the concept page), the language of the term, jurisdiction, and a textual description. Clicking on linked concepts (which are identified by their ID number), brings a page showing the concept as a hypernym and linking to hyponym concepts. Figure 5 below shows a concept tree for vehicles with the hyponyms being trolley-buses, motorcycles etc.

4 Technical Implementation

The methodology we use is to take inspiration from the technologies developed in the related fields of legislative drafting for parliaments, so called legislative XML, and legal ontologies, and export them in the context of applications for lawyers and law scholars. The technology used are PHP, Javascript, Ajax, XML, Postgres SQL and C++. All the data is stored in the PostgreSQL relational database.

The database architecture is divided into two independent parts, managing the Legal Taxonomy Syllabus ontology and the legal text repository. The ontology part of the database is saved as a table that is a repository of concepts, that are

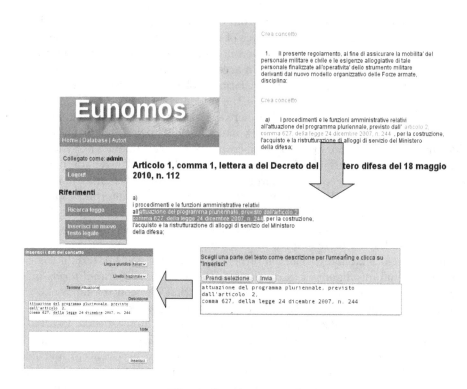

Fig. 4. Creating concepts

connected, but independent from, terms in a many-to-many relationship. The classical subject-predicate-object triple that defines the relationships between the concepts is stored in a separate table. Reconstructing transitive relations can be expensive in a relational database, so there is another cache table that stores the complete transitive closure of the ontology. The other important part of the database architecture is the repository of legal texts: each text is saved in the form of an XML document conforming to the NIR DTD and identified univocally trough their URN. Since each legislation is large, they are indexed with the PostGreSQL internal inverted index facility in order to enable fast full text searches and ranking for document similarity.

The Eunomos database of norms and legal concepts is accessible to any number of users via a web-based interface with secure login. Knowledge experts also edit the data via the web interface. The web front-end application to the system is divided into three parts:

- the pure presentation, using the Smarty[2] template engine;
- a level, implemented in a set of PHP classes, that manages the input and the output to and from the templates; and

---

[2] http://www.smarty.net

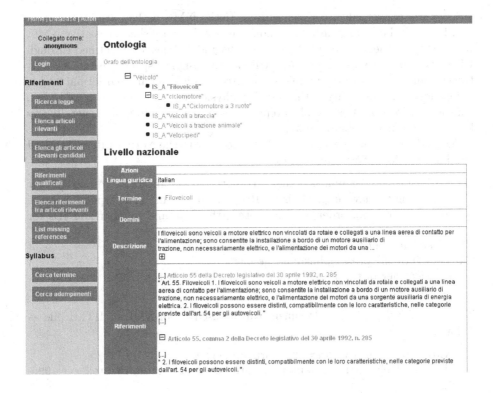

Fig. 5. Legal Taxonomy Syllabus within the Eunomos system

– the core business logic, involving another set of PHP classes that manages the input and output to the underlying database, supporting operations such as inserting a concept in the ontology or searching the legal text repository for a particular phrase in the laws of a given year.

4.1   Viewing Laws from Different Sources

Eunomos can download laws from institutional legal portals on a daily basis. Currently the software harvests the Normattiva national portal[3], the portal Arianna of Regione Piemonte[4] and a portal of regulations from the Ministry of Economy. For each legislation, Eunoms stores and time-stamps the original and most-up-to date versions. While legislative bodies may store various versions, including draft versions, of legislation using complex version control mechanism, this information is not particulary useful for our potential customers, whose primary concern is to ensure that they have up-to-date information on the law.

---

[3] http://www.normattiva.it
[4] http://arianna.consiglioregionale.piemonte.it/

Laws are converted into NormaInRete (NIR) XML if they are in pure textual format.[5] The NormeinRete standard is a well-established legislative XML. It specifies that legal documents, paragraphs and articles are identified through URNs (Uniform Resource Names). URNs are designed specifically for the Internet community to provide unique identifiers in a standardized format and is independent of availability in the network, physical location and means of access. This identifier is used as a tool to represent references - and more generally any kind of relationship - between acts.

An URN for a document constructed according to the NormaInRete standard will have the following components:

1. An ID for the original document, comprising the authority responsible for publishing the law (e.g., Ministry, Region, City, Court), the type of measure (e.g., law, decree, order, decision, etc.), the date and number and IDs for any annexes.
2. A version identifier, including the date of issue.
3. The ID of the press publishing the law.
4. An identifier of the fragment of the resource itself the URN refers to (e.g., article, paragraph, etc.).

The URN for a particular document can be used in an XML or HTML file as follows:

- XML URN: <urn valore="urn:nir:stato:legge:1996-12-31;675"/>
- HTML URN: <META name="nir.urn" content="urn:nir:stato:legge:1996-12-31;675"/>

URNs facilitate the construction of a global hypertext among the legal documents in a network environment with computer resources distributed among several publishers. It also allows the construction of knowledge bases containing the relationships between these documents. Within the Eunomos database, the URN is the ID reference number for each legislation and elements within legislation. Maintaining laws in NIR XML format makes it easier for Eunomos to extract elements such as paragraphs, articles and references so that knowledge engineers can categorise and annotate the elements, and lawyers can view relevant information.

While Eunomos uses the NormaInRete standard internally, as standards are developed for interchange between different legislative XML formats [3], it should be possible to use Eunomos in other jurisdictions. The Eunomos product can be multilingual and multilevel. This would require suitable parsers to structure laws in XML in different languages. It is already possible, however, to model EU directives and their national implementations.

4.2   Classifying Laws into Specific Domains

Classifying norms is labour-intensive. Eunomos uses natural language processing techniques to assist the user with this task. The support is based on three

---

[5] The Arianna portal already exports documents to NIR XML format.

techniques: text similarity, prevalence of domain-specific terminology, and analysis of incoming and outgoing references.

Eunomos uses a leading text similarity algorithm, the Cosine Similarity, to find the most similar pieces of legislation in the whole database. The Cosine Similarity metric uses the tf-idf measure to gauge the relative weight to be apportioned to various key words in the respective documents. The Cosine Similarity metric is particularly useful for comparing single-domain legislation. However, legislation that contain norms on different topics can introduce noise into the comparative process. We are looking to improve our similarity process by including only norms in the legislation within the same domain. This requires that each norm is classified. Our workflow ensures that norms within stored legislation will have been appropriately classified (or the classification checked) by a knowledge engineer. We have conducted preliminary experiments into automated classification of individual norms within new legislation using the Vector Support Machine algorithm, and the results are promising with an average accuracy rate of 70%.

The ontology is populated and updated manually, but once a term has been manually associated with a particular domain, the system ensures that all instances of domain-specific terms are highlighted when legislation is viewed from the perspective of the relevant domain.

The same applies to incoming and outgoing references, which are discovered automatically (see below). Where articles and paragraphs contain references to the articles and paragraphs they talk about or override, this information is used not only to link the relevant legislation via URN, but also to suggest to which category a new piece of legislation belongs. The rationale is that where paragraphs or articles contain references to classified paragraphs or articles in previous legislation, it is more than likely that the new paragraph or article belongs to the same domain. The user can check and deselect the suggested classifications.

4.3   References

Eunomos supports automated parsing of legal references, using the XML Leges Linker tool developed by The Institute of Legal Information Theory and Techniques (ITTIG). For national legislation, Eunomos also downloads a list of outgoing references from the Normattiva website. Eunomos has a specific interface for inserting explicit references missed by the parser due to irregular textual patterns whereby the knowledge engineer highlights the reference text and inserts the relevant URN. The type of reference (simple, modifying or overriding) and domain of the referenced element are inserted manually, as described in the User Workflow section above.

4.4   Ontology

The main assumptions of the Legal Taxonomy Syllabus ontology on which Eunomos is built come from studies in comparative law [8] and ontologies engineering [6]. Making a clear distinction among terms and their interlingual acceptions (or *axes*) is a standard way to properly manage large multilingual lexical databases [9,7].

5 Related Work

Our solution has some similarities with Bianchi et al. [1] in that it is designed to help users view laws and classify terms. But the scope of the Eunomos project is wider, designed as it was to address real problems in accessing and managing information by lawyers. While Bianchi et al. [1] takes XML files as input, Eunomos can download text-based laws made available in official portals and convert them into XML, where XML files are not available. Eunomos has a number of useful features for viewing and updating information, and an automatic alert messaging system on legislative updates. The downside is that Eunomos requires considerable maintenance work, as web spiders need to keep up to date with any modifications made to online legal portals, and expert users are required to verify classification and find implicit references. The use of ontology in the two systems are also quite different. Bianchi et al. [1] use the Semantic Turkey [5] ontology, where definitions can be taken from any source and arranged in any order. The Eunomos product is more careful, taking into account the strict demand for accuracy and transparency from the financial and legal sectors, encouraging the expert user to create links to definitions in legislation, judgement and official journals, and to track the evolution of terms in a systematic manner. Both Eunomos and Bianchi et al make use of statistical and reference data to help users find related norms though Bianchi combines these elements by factoring incoming and outgoing references into its statistical model. Eunomos's text similarity tool is on a legislation level, but Bianchi et al. [1]'s text similarity tool works at a paragraph and article level.

de Maat et al. [4]'s research on classification of legal sentences is also relevant, since both systems use machine learning and rule-based techniques. de Maat et al. [4] uses rules to find standardised patterns suggestive of a particular class, while Eunomos uses rules to find standardised patterns for references to classified norms in previous legislation, which provides a clue as to the classification of new norms. On the machine learning side, de Maat et al. [4] uses Support Vector Machines for text classification, while we use Cosine text similarity to find the most similar pieces of legislation, which if already annotated, suggest relevant domains as well as norms that may be overridden implicitly.

Related work on the classification of references include Mazzei et al. (2009)'s research on analysing modificatory provisions. Currently Eunomos can find most explicit references but an expert user needs to specify whether the reference is a simple reference or it modifies or overrides other legislation. Mazzei et al. (2009) have developed an automatic reference classification approach which pairs deep syntactic parsing with rule-based shallow semantic analysis and fine-grained taxonomy of modificatory provisions.

6 Conclusions

There is good research within legal informatics, knowledge management, natural language processing and artificial intelligence which can help make the legal process more effective and efficient. In this paper we illustrate the Eunomos software, a legal document and terminology management system to help lawyers manage complex information, which incorporates state-of-the art research from the above disciplines. The software has been developed with clearly-defined aims and objectives to support the work of law firms, law scholars, and in-house legal offices in financial institutions and public sector organisations. Eunomos offers users an integrated environment that makes laws easier to navigate, annotate and understand, using automatically generated hyperlinks to referenced legislation, an extensible and updatable ontology which provides current and previous definitions for norms and concepts within any specific context, and an alert system that highlights existing legislation affected by new legislation. By connecting ontologies and legislation with an XML database framework, Eunomos provides a powerful knowledge base for keeping up to date with legal changes.

The support mechanism for classification and eventual extensions of the system have been discussed in [2].

Eunomos is being developed as a commercial software part of a wider suite distributed by Nomotika s.r.l., a spinoff of Università di Torino. Eunomos has a clear business model: a combined software and services package that effectively means that legislation monitoring is outsourced. The roles, permissions and technologies have been carefully selected to address real business needs. The software and related services will be provided by experts with sound technological and business expertise.

References

1. Bianchi, M., Draoli, M., Gambosi, G., Pazienza, M.T., Scarpato, N., Stellato, A.: ICT tools for the discovery of semantic relations in legal documents. In: Proceedings of the 2nd International Conference on ICT Solutions for Justice (ICT4Justice) (2009)
2. Boella, G., Humphreys, L., Martin, M., Rossi, P., van der Torre, L.: Eunomos, a legal document management system based on legislative xml and ontologies. In: Legal Applications of Human Language Technology (AHLTL) at ICAIL 2011 (2011)
3. Boer, A., Winkels, R.: What's in an interchange standard for legislative XML? I. Quaderni 18, 32–41 (2005)
4. de Maat, E., Krabben, K., Winkels, R.: Machine learning versus knowledge based classification of legal texts. In: Proceeding of The Twenty-Third Annual Conference on Legal Knowledge and Information Systems, JURIX 2010: The Twenty-Third Annual Conference, pp. 87–96. IOS Press, Amsterdam (2010)
5. Griesi, D., Pazienza, M.T., Stellato, A.: *Semantic Turkey*: A Semantic Bookmarking Tool (System Description). In: Franconi, E., Kifer, M., May, W. (eds.) ESWC 2007. LNCS, vol. 4519, pp. 779–788. Springer, Heidelberg (2007)

6. Klein, M.: Combining and relating ontologies: an analysis of problems and solutions. In: Workshop on Ontologies and Information Sharing at IJCAI 2001 (2001)
7. Lyding, V., Chiocchetti, E., Sérasset, G., Brunet-Manquat, F.: The LexALP information system: Term bank and corpus for multilingual legal terminology consolidated. In: Proc. of the Wokshop on Multilingual Language Resources and Interoperability at ACL 2006, pp. 25–31 (2006)
8. Rossi, P., Vogel, C.: Terms and concepts; towards a syllabus for european private law. European Review of Private Law (ERPL) 12(2), 293–300 (2004)
9. Sérasset, G.: Interlingual lexical organization for multilingual lexical databases in NADIA. In: Proceedings of the 15th Conference on Computational Linguistics (COLING), pp. 278–282 (1994)

Axioms on a Semantic Model for Legislation for Accessing and Reasoning over Normative Provisions

Enrico Francesconi

ITTIG-CNR, Florence, Italy
francesconi@ittig.cnr.it
http://www.ittig.cnr.it

Abstract. In this paper an approach to the Semantic Web in the legal domain is presented: it is obtained by modelling legislative document semantic profiles using a model of normative provisions. An implementation of this approach through RDF/OWL, describing provisions and their relations, is proposed. In particular, a pattern able to implement Hohfeldian legal fundamental relations between provisions using OWL-DL expressivity is described within a case study involving duty and right provisions. An example of advanced access and reasoning over provisions using the proposed approach is shown.

Keywords: Legislative Document Semantics, Provision Model, Domain Knowledge, Hohfeldian Reasoning, OWL-DL, SPARQL.

1 Introduction

The availability of advanced information retrieval and reasoning services is particularly desirable in the legal domain, for the complex nature of the legal document workflow, as well as for the peculiarities of legal users' information needs. In this paper an approach to the Semantic Web, which is able to provide effective retrieval and reasoning services for legislation is proposed. The approach aims firstly to identify (Section 2) and describe (Sections 3, 4, 5) legislative documents semantic profiles by a model of normative provisions. Moreover, based on such modelling, an implementation of the Hohfeldian fundamental relations [7,8] between provisions is proposed (Section 5). In particular, in Section 6 an example of how this approach can support Hohfeldian inferences for improving provisions accessibility is presented with respect to a European directive case-study. Finally in Section 7 some conclusions are discussed.

2 Semantic Profiles of Legislative Documents

According to [4] the entire body of laws and regulations may be seen as a set of *provisions*, carried by speech acts [11], namely sentences endowed with meaning [10]. In this perspective a legislative text can be viewed according to two different *profiles*:

M. Palmirani et al. (Eds.): AICOL Workshops 2011, LNAI 7639, pp. 147–161, 2012.
© Springer-Verlag Berlin Heidelberg 2012

- a structural or *formal profile*, representing the traditional legislator habit of organizing legislative texts in chapters, articles, paragraphs, etc.;
- a *semantic profile*, representing a specific organization of legislative text substantial meaning; a possible description of it can be given in terms of normative provisions [4].

Following this perspective, fragments of a legislative text are, at the same time, paragraphs and provisions, according to whether they are seen from a formal or semantic view-point.

In a provision-centric view, it is possible to identify three sub-profiles according to which the semantics of a legislative text can be perceived:

- a *functional profile* representing the organization of the legislative texts structure in terms of *provision types*, namely a sequence of provisions (as *Term definition, Procedure, Duty, Right, Permission*, as well as more technical ones as *Insertion, Abrogation, Substitution*, etc.) and related *attributes*[1] (for example the *Bearer* of a *Right*), reflecting the lawmaker directions;
- a *thematic profile* representing the relations between the aspects of the reality described in legislative texts, subject to the regulative activity of the legislator; such aspects of the reality can be formally expressed as values of the *provision attributes* (for example 'Consumer' as the *Bearer* of a specific *Right*);
- a *logic profile* representing the relations between provisions (types and attributes), able to describe not only the explicit normative positions contained in a legislative text, but also to get implicit ones to emerge.

Provision types and attributes can be considered as a sort of metadata model able to analytically describe fragments of legislative texts, hence the name of *Provision Model* [4]. For example, the following fragment (article 5, paragraph 1) of the European Directive 2002/65/EC, concerning the distance marketing of consumer financial services:

The supplier shall communicate to the consumer all the contractual terms and conditions and the information referred to in Article 3(1) and Article 4 on paper or on another durable medium available and accessible to the consumer in good time before the consumer is bound by any distance contract or offer.

besides being considered part of the formal profile (a *paragraph*), can also be viewed as a component of the semantic profile of a legislative text (a *provision*) and qualified as a *Duty*, whose attributes are:

hasBearer: "Supplier"
hasObject: "Contractual terms and conditions ..."
hasAction: "Communication"
hasCounterpart: "Consumer"

Possible relations between provisions can also be identified, like the Hohfeldian fundamental relations regarding the following pairs of provisions: *Right/Duty, Liberty/No-right, Power/Liability, Immunity/ Disability*, as well as relations between provision attributes within the same Hohfeldian framework, as for example

---

[1] Also called *arguments* in [4].

the relation between the *Duty* of a subject (duty *Bearer*) towards a *Counterpart*, which can be viewed as an implicit *Right* of the duty *Counterpart* towards the duty *Bearer*.

A provision-oriented description of such profiles allows advanced access services over provisions. A typical example can be a service able to implement the previously mentioned Hohfeldian reasoning by accessing the rights of a subject, either explicitly expressed or inferred. The following sections present a possible RDF/OWL implementation of the Provision Model, how it can be used to describe legislative texts semantic profiles, as well as a Hohfeldian reasoning scheme on such a model using an OWL-DL reasoner and SPARQL.

3   The Functional Profile

The functional profile of legislative texts can be described in terms of *provision types* and *attributes* [4]. In the Provision Model provision types are organized in two main families: *Rules* (introducing entities or expressing deontic concepts) and *Rules on Rules* (different kinds of amendments). *Rules* are provisions which aim at regulating the reality considered by the including act. Adopting a typical law theory distinction, well expressed by Rawls, they consist in:

- *Constitutive rules*: they introduce or assign a juridical profiles to entities of a regulated reality;
- *Regulative rules*: they discipline actions or the substantial and procedural defaults (remedies).

On the other hand, *Rules on Rules* can be distinguished in:

- *Content amendments*: they modify literally the content of a norm, or their meaning without literal changes;
- *Temporal amendments*: they modify the times of a norm (come-into-force and efficacy time);
- *Extension amendments*: they extend or reduce the cases on which the norm operates.

A taxonomy of provisions can be represented using RDF/ OWL standards. A graphical representation of taxonomy top classes is shown in Fig. 1 ("**prv**" represents the namespace of the Provision Model).

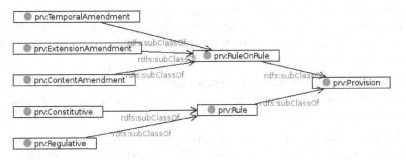

Fig. 1. Provision Model top classes

Provisions top classes are further on specialised: for example *Regulative* provisions can be distinguished into *Rules On Actions* (Right, Duty, Prohibition and Permission) and *Remedies* (Violation, Redress)[2] (Fig. 2). A complete view of provision taxonomy can be found in [4].

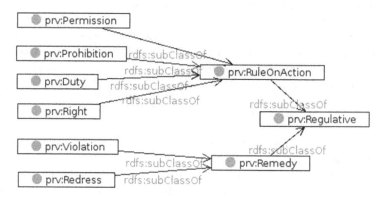

Fig. 2. Taxonomy of Regulative Provisions

Each provision type has specific *attributes* describing roles of entities (for example hasBearer and hasCounterpart are attributes of Duty and Right). For a complete view of provision attributes, the reader can refer to [4].

As discussed in [3], the Provision Model has been conceived as a metadata model able to describe legislative textual fragments, representing the domain of the discourse pertaining to normative provisions. In fact, in [3] three interpretation levels of a legislative text are distinguished:

1. Linguistic interpretation (*Provisions*): it simply requires the knowledge of the text language and consists in highlighting the semantics of legislative texts through their linguistic interpretation;
2. Dogmatic interpretation (*Rules*): it is a more systematic level of interpretation of the textual content conveyed by a provision. It requires the availability of a model of rules, and consists in highlighting and systematizing the semantics of legislative texts by a linguistic and legal-dogmatic interpretation;
3. Legal interpretation (*Norms*): it is the level of interpretation of the legal practitioners, aiming to identify the norms to be applied to specific cases. Such interpretation is either linguistic or dogmatic, and it requires also an extra-textual knowledge about the legal order, which the right sense of a norm can be derived from.

In this view the Provision Model highlights a linguistic interpretation of legislative texts. Therefore the Provision Model is not a theory of normative concepts, concerned with a formal analysis of the meaning of textual fragments (ex: what does

---

[2] Hereinafter, in the text of the paper, provision types as OWL classes (starting with capital letters) and provision attributes as OWL properties (starting with lowercase letters) are written in serif font. The prv: namespace is omitted for simplicity.

it mean to have a right?) [4] and their relations, but it is a metadata model (*provisions*), organized in terms of *rules* ("Rules" or "Rules on Rules" (amendments)).

4   The Thematic Profile

The thematic profile of a legislative text is the collection of the relationships between the aspects of the reality, described in such text, which are subject to the regulative activity of the legislator. In the Provision Model they are represented by the values of *provision attributes*; they can be expressed by lexical units, or by concepts derived from thesauri/ontologies, able to provide additional information on the entities of the regulated domain [2] [6].

An example of an ontology dealing with a domain regulated by national and EU legislations, as the consumer protection one, has been developed within the DALOS project[3] [1]. It has been implemented as an extension of the Core Legal Ontology (CLO)[4] developed on top of DOLCE foundational ontology and on the "Descriptions and Situations" (DnS) ontology [9] within the DOLCE+ library[5].

In this knowledge architecture the role of a core legal ontology is to bridge the gap between domain-specific concepts and the abstract categories of formal upper level or foundational ontologies such as, in this case, DOLCE, providing concepts belonging to a general theory of law (e.g. LegalRole, LegalSituation, etc.)

Domain-specific concepts are classified according to more general notions, imported from CLO, as LegalRole and LegalSituation. An example of concepts described in the consumer EU law (as CommercialTransaction, Consumer, Supplier, etc.) and their specific roles ([9]) is given in Fig. 3.

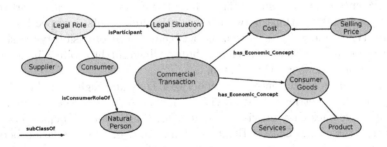

Fig. 3. Excerpt of the DALOS domain ontology

DALOS aims to describe the entities, and related relations, of the reality which the consumer law deals with. Therefore DALOS concepts can be used to describe the thematic aspects of a legislative text dealing with the consumer protection domain. The conjunction of the Provision Model and a domain ontology like DALOS is able to highlight both the functional and thematic profiles of a legislative text. An example of such use is reported in Section 6.1.

---

[3] www.dalosproject.eu

[4] www.loa-cnr.it/ontologies/CLO/CoreLegal.owl

[5] DOLCE+ library, http://dolce.semanticweb.org

5   The Logic Profile

The logic profile of a legislative text is represented by the set of relations involving provisions types and attributes. In this study Hohfeldian relations between provisions have been added as axioms to the RDF/OWL implementation of the Provision Model. Let's consider hereinafter the Hohfeldian relation between *Duty* and *Right* as an example to show our approach. In terms of Provision Model, a duty of A towards B can be expressed as follows:

Duty(hasBearer='A', hasCounterpart='B')

corresponding to

Right(hasBearer='B', hasCounterpart='A')

and vice-versa.

For example article 5 paragraph 1 of the European Directive 2002/65/EC reported in Section 2, can be considered a provision of type *Duty* involving 'Supplier' and 'Consumer', so that:

Duty(hasBearer='Supplier', hasCounterpart='Consumer')

corresponding to

Right(hasBearer='Consumer', hasCounterpart='Supplier').

These Hohfeldian relations underline an equivalence between *Duty* and *Right*, as long as the values of the duty *Bearer* and *Counterpart* are swapped, assuming symmetric roles in the *Right* provision, therefore involving equivalence relations between provision types and attributes.

However, describing these relations in the Provision Model by establishing the equivalence relations Duty ≡ Right [4] and hasBearer ≡ hasCounterpart would imply equivalence relations between *any* duties and rights, irrespective to the attribute types and values, as well as between all the provision types sharing equivalence relations between such attributes, which might produce inconsistent results in a provisions retrieval system.

In particular an equivalence relation between Duty and Right would imply that a query aiming to retrieve Right provisions having Right(hasBearer = 'Supplier'), would also give back Duty provisions having Duty(hasBearer = 'Supplier') because they satisfy the axiom Duty ≡ Right. Similarly an equivalence relation between hasBearer ≡ hasCounterpart would imply that the previously mentioned query would retrieve back Right provisions having Right(hasCounterpart = 'Supplier'), since they satisfy the axiom hasBearer ≡ hasCounterpart.

5.1   Extension to the Provision Model

To avoid these problems, while relying on Description Logic expressivity as implemented in OWL-DL, an extension of the Provision Model is proposed.

Firstly an extension which specifies provision attributes according to the related provision types can be implemented. Therefore *hasBearer* and *hasCounterpart* relations are distinguished in terms of hasDutyBearer and hasDutyCounterpart as properties of Duty, and hasRightBearer and hasRightCounterpart as properties of Right. The RDF/OWL syntax of the previous relations, limited to Duty (the same holds for Right), results as follows:

```
<owl:ObjectProperty rdf:about="prv:hasDutyBearer">
    <rdfs:domain rdf:resource="prv:Duty"/>
    <rdfs:range rdf:resource="owl:Class"/>
</owl:ObjectProperty>

<owl:ObjectProperty rdf:about="prv:hasDutyCounterpart">
    <rdfs:domain rdf:resource="prv:Duty"/>
    <rdfs:range rdf:resource="owl:Class"/>
</owl:ObjectProperty>
```

The specified Duty and Right attributes are shown in Fig. 4.

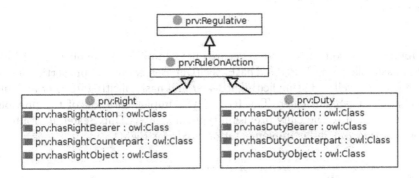

Fig. 4. Attributes of Duty and Right provisions

A model extension can also be provided by observing that a *Right*, in Hohfeldian correspondence with a *Duty*, is actually not explicitly expressed in the text, but represents an implicit provision, basically a different view of the *Duty* itself, where the values of the related bearer and counterpart attributes are swapped. Therefore the Provision Model can be extended in terms of Duty and Right implicit and explicit disjoint subclasses, whose RDF/OWL implementation is here below reported:

```
<owl:Class rdf:about="prv:ExplicitDuty">
    <rdfs:subClassOf rdf:resource="prv:Duty"/>
</owl:Class>

<owl:Class rdf:about="prv:ImplicitDuty">
    <owl:disjointWith rdf:resource="prv:ExplicitDuty"/>
    <rdfs:subClassOf rdf:resource="prv:Duty"/>
</owl:Class>

<owl:Class rdf:about="prv:ExplicitRight">
    <rdfs:subClassOf rdf:resource="prv:Right"/>
</owl:Class>

<owl:Class rdf:about="prv:ImplicitRight">
    <owl:disjointWith rdf:resource="prv:ExplicitRight"/>
    <rdfs:subClassOf rdf:resource="prv:Right"/>
</owl:Class>
```

Moreover in order to express that explicit and implicit disjoint subclasses represent a complete covering of the related superclass (ex: ExplicitRight and ImplicitRight disjoint subclasses represent a complete covering of the Right superclass), the following further axioms can be introduced:

```
<owl:Class rdf:about="prv:Right">
    <owl:unionOf rdf:parseType="Collection">
      <owl:Class rdf:about="prv:ImplicitRight"/>
      <owl:Class rdf:about="prv:ExplicitRight"/>
    </owl:unionOf>
</owl:Class>

<owl:Class rdf:about="prv:Duty">
    <owl:unionOf rdf:parseType="Collection">
      <owl:Class rdf:about="prv:ImplicitDuty"/>
      <owl:Class rdf:about="prv:ExplicitDuty"/>
    </owl:unionOf>
</owl:Class>
```

Attributes can also be specified as regards both implicit and explicit provisions, so that hasImplicitDutyBearer and hasExplicitDutyBearer are sub-properties of hasDutyBearer, as well as hasImplicitRightBearer and hasExplicitRightBearer are sub-properties of hasRightBearer. The RDF/OWL implementation of the previous relations, limited to the Duty provision type and the hasDutyBearer attribute, is below reported (a similar implementation holds for Right).

```
<rdf:ObjectProperty rdf:about="prv:hasImplicitDutyBearer">
    <rdfs:subPropertyOf>
      <rdf:ObjectProperty rdf:about="prv:hasDutyBearer"/>
    </rdfs:subPropertyOf>
    <rdfs:domain rdf:resource="prv:ImplicitDuty"/>
    <rdfs:range rdf:resource="owl:Class"/>
</rdf:ObjectProperty>

<rdf:ObjectProperty rdf:about="prv:hasExplicitDutyBearer">
    <rdfs:subPropertyOf>
      <rdf:ObjectProperty rdf:about="prv:hasDutyBearer"/>
    </rdfs:subPropertyOf>
    <rdfs:domain rdf:resource="prv:ExplicitDuty"/>
    <rdfs:range rdf:resource="owl:Class"/>
 </rdf:ObjectProperty>
```

For each attribute (property) both *domain* and *range* are specified: *domain* specifies the type of individuals a provision attribute applies to (e.g. the individuals of the class ExplicitDuty for a provision attribute hasExplicitDutyBearer); *range* specifies the type of values of this provision attribute. Since the Provision Model is a metadata model for legislative texts, and since legislative texts can deal with any aspects of the reality, the values of a provision attributes, highlighting such thematic aspects, may belong to any class of objects. Therefore a property *range* related to a provision attribute is an individual of the generic class owl:Class.

Note that only explicit provision classes (and consequently explicit properties) will be used to mark-up textual provisions, as they are the only provisions actually (explicitly) expressed in legislative texts, while implicit provision classes act as a sort of "abstract" classes, which will be used for reasoning.

5.2   Hohfeldian Relations in the Provision Model

To represent the Hohfeldian fundamental relations between *Duty* and *Right*, firstly an equivalence relation between explicit and implicit aspects of them is established, in particular ImplicitRight ≡ ExplicitDuty and ImplicitDuty ≡ ExplicitRight. Using OWL-DL the previous relations result:

```
<owl:Class rdf:about="prv:ImplicitRight">
  <owl:equivalentClass>
    <owl:Class rdf:about="prv:ExplicitDuty"/>
  </owl:equivalentClass>
</owl:Class>

<owl:Class rdf:about="prv:ImplicitDuty">
  <owl:equivalentClass>
    <owl:Class rdf:about="prv:ExplicitRight"/>
  </owl:equivalentClass>
</owl:Class>
```

In Fig. 5 the established sub-class (Section 5.1) and equivalence relations (Section 5.2) between Duty and Right in their explicit and implicit views are summed up.

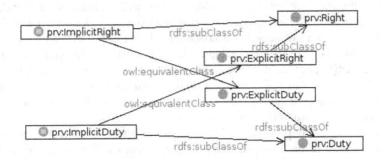

Fig. 5. Sub-class and asserted equivalence relations between Duty and Right provisions

Moreover, equivalence relations between implicit/explicit Duty and Right attributes can be established (hasImplicitRightCounterpart ≡ hasExplicitDutyBearer, hasImplicitRightBearer ≡ hasExplicitDutyCounterpart, hasImplicitDutyCounterpart ≡ hasExplicitRightBearer, hasImplicitDutyBearer ≡ hasExplicitRightCounterpart) and represented in OWL-DL as object properties as follows:

```
<rdf:ObjectProperty rdf:about="prv:hasImplicitRightCounterpart">
  <owl:equivalentProperty rdf:resource="prv:hasExplicitDutyBearer"/>
</rdf:ObjectProperty>
<rdf:ObjectProperty rdf:about="prv:hasImplicitRightBearer">
  <owl:equivalentProperty rdf:resource="prv:hasExplicitDutyCounterpart"/>
</rdf:ObjectProperty>

<rdf:ObjectProperty rdf:about="prv:hasImplicitDutyCounterpart">
  <owl:equivalentProperty rdf:resource="prv:hasExplicitRightBearer"/>
</rdf:ObjectProperty>
<rdf:ObjectProperty rdf:about="prv:hasImplicitDutyBearer">
  <owl:equivalentProperty rdf:resource="prv:hasExplicitRightCounterpart"/>
</rdf:ObjectProperty>
```

In Fig. 6 the asserted sub-property (Section 5.1) and equivalence relations (Section 5.2) between hasDutyBearer and hasRightCounterpart in their explicit and implicit views are summed up. The reader can imagine a symmetric view for the relations between a right bearer and a duty counterpart in their explicit and implicit views.

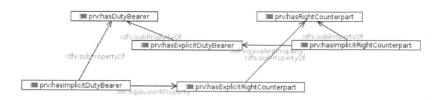

Fig. 6. Asserted sub-property and equivalence relations between hasDutyBearer and hasRightCounterpart in their explicit and implicit views

Note that the proposed patterns do not interfere with the equivalence relation between Right and Duty which still holds. In fact an individual of ExplicitDuty is also an individual of Duty, given the axiom rdfs:subClassOf(ExplicitDuty, Duty). Moreover the axiom owl:equivalentClass(ImplicitRight, ExplicitDuty) tells us that such individual is also an ImplicitRight, which is also a Right, given the axiom rdfs:subClassOf(ImplicitRight, Right). Since this is done symmetrically for explicit and implicit duties and rights, we can deduce that Right is equivalent to Duty, given that the union of the disjoint explicit and implicit subclasses covers completely the related superclass (see Section 5.1).

Therefore reasoning properties of the model are preserved, but its expressivity is improved, able to provide enhanced retrieval services. The proposed patterns in fact aim to introduce:

1. Properties equivalence, allowing direct swapping on attributes contents for addressing provision relations, without the need of using conditional statements (ex: if (hasDutyCounterpart == 'Consumer'))
2. Abstract classes (namely classes not used for mark-up, in our case "implicit" classes) so to provide different views (implicit and explicit views) on the same provision instance, as well as retrieval services able to access implicit provisions only (ex: retrieving provision instances where ImplicitRightBearer == 'Consumer');

Moreover, the proposed pattern is able to avoid inconsistent deductions which might derive from equivalence relations applied to classes and properties, as discussed in Section 5, which would bring unpleasant effects (as for example that bearers and counterparts freely mix for the same provision). In fact equivalence axioms applied on provision classes and related properties, specified according to implicit/explicit views, produce inferential deductions which keep semantic consistency. For example, given the following description of an explicit right:

a) *ExplicitRight(hasExplicitRightBearer = 'Consumer')*

given that:

ExplicitRight ≡ ImplicitDuty

hasExplicitRightBearer ≡ hasImplicitDutyCounterpart

the following consistent deductions describing the same provision instance can be obtained:

b) *ExplicitRight(hasImplicitDutyCounterpart = 'Consumer')*

c) *ImplicitDuty(hasExplicitRightBearer = 'Consumer')*

d) *ImplicitDuty(hasImplicitDutyCounterpart = 'Consumer')*

which are consistent with the semantics of the provisions: for example the "Consumer" which according to (a) is the Bearer of an ExplicitRight, is also to be consider according to (b) the Counterpart of the related ImplicitDuty. Similar considerations hold for the other deductions (c) (d).

It is worth to stress that the introduced axioms are not dealing with relations between different provisions expressed in a legislative text (which should be better described in terms of existential restrictions, as for example: 'for every explicit duty there is an implicit right where bearer and counterpart are swapped'), but they are dealing with different views (explicit and implicit views) of the *same* provision instance. In this perspective all the deductions derived from the established equivalence relations between classes, as well as the deductions derived from mixing provision qualified properties, are valid, as previously shown.

6   Hohfeldian Inference Case-Study

In this section an example of how this approach can be used for a provision retrieval system able to support Hohfeldian reasoning is shown.

6.1   Semantic Annotation

Let's first consider an excerpt of Directive 2002/65/EC, properly annotated using a CEN-Metalex [5] compliant mark-up syntax (here below).

```
<article id="art5">
 <paragraph id="art5-par1">
  1. The supplier shall communicate to the consumer all the contractual
  terms and conditions and the information referred to in Article 3(1) and
  Article 4 [...]
 </paragraph>
 <paragraph id="art5-par2">
  2. The supplier shall fulfil his obligation under paragraph 1 immediately
  after the conclusion of the contract, if the contract has been concluded at
  the consumer's request using a means of distance communication which
  does not enable providing the contractual terms [...]
 </paragraph>
 <paragraph id="art5-par3">
  3. At any time during the contractual relationship the consumer is entitled,
  at his request, to receive the contractual terms and conditions on paper.[...]
 </paragraph>
</article>
```

```
<article id="art6">
 <paragraph id="art6-par1">
   1. The Member States shall ensure that the consumer shall have a period of 14
   calendar days to withdraw from the contract without penalty and without giving
   any reason [...]
 </paragraph>
 [...]
</article>
```

According to the Provision Model and the DALOS ontology, the semantics of such document fragments, identified by document URI and specific IDs, can be summed up as in Tab. 1 (this semantic description is limited to the elements useful to demonstrate the approach, therefore conditions, actions and other involved attributes are not described).

Table 1. Semantics of Directive 2002/65/EC excerpt

Partition ID	Provison Type	Provision Attributes
art5-par1	ExplicitDuty	hasExplicitDutyBearer='Supplier' hasExplicitDutyCounterpart='Consumer'
art5-par2	Procedure	hasProcedureAddressee='Supplier' hasProcedureCounterpart='Consumer'
art5-par3	ExplicitRight	hasExplicitRightBearer='Consumer' hasExplicitRightCounterpart='Supplier'
art6-par1	ExplicitDuty	hasExplicitDutyBearer='Member States' hasExplicitDutyCounterpart='Consumer'

Having defined the following namespaces

```
xmlns:prv="http://www.ittig.cnr.it/ProvisionModel/1.0#"
xmlns:cl="http://www.ittig.cnr.it/ontologies/consumer-law/1.0#"
```

for Provision Model and DALOS consumer law domain ontology, respectively, an RDF/OWL semantic annotation of such fragments, can be the following:

```
<rdf:Description rdf:about="[URI]#art5-par1">
 <rdf:type rdf:resource="prv:ExplicitDuty"/>
 <prv:hasExplicitDutyBearer rdf:resource="cl:Supplier"/>
 <prv:hasExplicitDutyCounterpart rdf:resource="cl:Consumer"/>
</rdf:Description>

<rdf:Description rdf:about="[URI]#art5-par2">
 <rdf:type rdf:resource="prv:Procedure"/>
 <prv:hasProcedureAddressee rdf:resource="cl:Supplier"/>
 <prv:hasProcedureCounterpart rdf:resource="cl:Consumer"/>
</rdf:Description>

<rdf:Description rdf:about="[URI]#art5-par3">
 <rdf:type rdf:resource="prv:ExplicitRight"/>
 <prv:hasExplicitRightBearer rdf:resource="cl:Consumer"/>
 <prv:hasExplicitRightCounterpart rdf:resource="cl:Supplier"/>
</rdf:Description>

<rdf:Description rdf:about="[URI]#art6-par1">
 <rdf:type rdf:resource="prv:ExplicitDuty"/>
 <prv:hasExplicitDutyBearer rdf:resource="cl:MemberStates"/>
 <prv:hasExplicitDutyCounterpart rdf:resource="cl:Consumer"/>
</rdf:Description>
```

6.2   The Inferred Model

Having described the Provision Model by using OWL-DL, a provision management system can be given inference facilities through an OWL-DL reasoner able to derive a corresponding inferred model. In this case-study the Pellet[6] Java based OWL-DL reasoner has been used. The result is a Provision Model where inferences are calculated from the associated axioms.

6.3   Querying the System

An RDF triple store of provisions can be queried to retrieve specific types of provisions, involving specific entities, using SPARQL. A SPARQL query able to retrieve the rights of cl:Consumer is shown in Fig. 7, where ?par is the variable which will contain the identifier of the retrieved provision instances (usually paragraphs of legislative documents).

```
PREFIX rdf: <http://www.w3.org/1999/02/22-rdf-syntax-ns#>
PREFIX prv: <http://www.ittig.cnr.it/ProvisionModel/1.0#>
PREFIX cl: <http://www.ittig.cnr.it/ontologies/consumer-law/1.0#>
SELECT  ?par
WHERE { ?par prv:hasRightBearer cl:Consumer }
```

Fig. 7. A SPARQL query using the Provision Model and ontology concepts

Let's assume to query the Directive excerpt in Section 6.1.

In case the *non-inferred model* is queried, no provisions are retrieved since only ExplicitRight and related explicit *attributes* are used for provision annotation. To obtain the rights explicitly expressed, the query has to be specified asking for provisions whose hasExplicitRightBearer value is cl:Consumer. In this case, paragraph with id="art5-par3" is correctly retrieved.

In case the *inferred model* is queried, all the inferred provisions are retrieved, either annotated as ExplicitRight of cl:Consumer or implicitly deduced by provision relations. Since Hohfeldian relations have been implemented in the Provision Model, the result will be a Hohfeldian reasoning over provisions. By exploiting the established rdfs:subClass and owl:equivalentClass relations between provisions type and attributes, the system will act as virtually expanding the query in Fig. 7, obtaining the results as shown in Tab. 2.

Moreover, the distinction between implicit/explicit provisions and attributes allows us to select, for example, among all the *Rights* of a *Bearer*, only those which are not explicitly expressed in the text. The corresponding query will be:

```
SELECT ?par WHERE { ?par prv:hasImplicitRightBearer cl:Consumer }
```

which will provide the ExplicitDuty where hasExplicitDutyCounterpart is cl:Consumer (being hasImplicitRightBearer ≡ hasExplicitDutyCounterpart); in the

---

[6] http://clarkparsia.com/

Table 2. Virtual expansion of the query in Fig. 7, by provision axioms

Virtual query expansion	Result
?par prv:hasExplicitRightBearer cl:Consumer	art5-par3
?par prv:hasImplicitRightBearer	art5-par1
[≡ *prv:hasExplicitDutyCounterpart*]	art6-par1
cl:Consumer	

case-study of Section 6.1 the following paragraphs are retrieved: id="art5-par1",
id="art6-par1".

7   Conclusions

An approach to the Semantic Web for the legal domain through the Provision
Model and domain ontologies have been proposed. In particular Hohfeldian fundamental relations between provisions have been described through axioms on
an OWL-DL implementation of the model. To obtain this, the Provision Model
has been extended to represent provisions types and attributes, either implicitly
or explicitly expressed. An example regarding a European directive, showing how
this approach can support Hohfeldian inferences, has also been described. The
ability to provide Hohfeldian inferences, keeping the complexity of the approach
within an OWL-DL expressivity, therefore a computational tractability, represents a benefit of this approach. The possibility to represent other legal relations
using this approach is expected to be investigated in a future work.

Acknowledgement. The author would like to thank the anonymous reviewers
for their valuable comments which contributed to clarify several issues of this
paper.

References

1. Agnoloni, T., Bacci, L., Francesconi, E., Peters, W., Montemagni, S., Venturi, G.:
   A two-level knowledge approach to support multilingual legislative drafting. In:
   Breuker, J., Casanovas, P., Klein, M., Francesconi, E. (eds.) Law, Ontologies and
   the Semantic Web. Frontiers in Artificial Intelligence and Applications, vol. 188,
   pp. 177–198. IOS Press (2009)
2. Antoniou, G., Billington, D., Governatori, G., Maher, M.: On the modeling and
   analysis of regulations. In: Proceedings of the Australian Conference Information
   Systems, pp. 20–29 (1999)
3. Biagioli, C.: Modelli Funzionali delle Leggi. Verso testi legislativi autoesplicativi.
   Legal Information and Communications Technologies Series, vol. 6. European Press
   Academic Publishing (2009)
4. Biagioli, C., Grossi, D.: Formal aspects of legislative meta-drafting. In: Proceedings
   of the Jurix Conference: Legal Knowledge and Information Systems, pp. 192–201
   (2008)

5. Boer, A., Hoekstra, R., de Maat, E., Hupkes, E., Vitali, F., Palmirani, M.: Cen workshop agreement open xml interchange format for legal and legislative resources. Technical report, CEN/ISSS Workshop Metalex (2009)
6. Hoekstra, R., Breuker, J., Bello, M., Boer, A.: Lkif core: Principled ontology development for the legal domain. In: Breuker, J., Casanovas, P., Klein, M., Francesconi, E. (eds.) Legal Ontologies and the Semantic Web. IOS Press (2009)
7. Hohfeld, W.N.: Some fundamental legal conceptions as applied in judicial reasoning. I. Yale Law Journal 23, 16–59 (1913)
8. Hohfeld, W.N.: Some fundamental legal conceptions as applied in judicial reasoning. II. Yale Law Journal 26, 710–770 (1917)
9. Masolo, C., Vieu, L., Bottazzi, E., Catenacci, C., Ferrario, R., Gangemi, A., Guarino, N.: Social roles and their descriptions. In: Welty, C. (ed.) Proceedings of the Nineth International Conference on the Principles of Knowledge Representation and Reasoning, Whistler (2004)
10. Raz, J.: The Concept of a Legal System. Oxford University Press (1980)
11. Searle, J.: Speech Acts: An Essay in the Philosophy of Language. Cambridge University Press (1969)

An Open Access Policy for Legal Informatics Dissemination and Sharing

Enrico Francesconi and Ginevra Peruginelli

ITTIG-CNR, Florence, Italy
{francesconi,peruginelli}@ittig.cnr.it
http://www.ittig.cnr.it

Abstract. Scholarly communication is facing great changes due to the revolution of digital technology and the raising of new economic models for academic publishing. In the legal domain, in particular, these changes affect a scenario dominated by a rigid and centralized control of information by a few large commercial publishers. This paper analyses these changes, proposing a road map for providing a digital publishing service for legal information materials based on Open Access policies and technological implementations.

Keywords: Legal scholarly communication, Open access.

1 Introduction

It is undeniable that the Internet and the World Wide Web have brought about major changes in every sector of our lives. In legal scholarly communication these changes are very important both economically and socially, involving the whole process of it. The revolution of digital technology radically changes the economics (the system of incentives) and the law (the principles and rules) governing the production and dissemination of scientific knowledge. As a result of a number of economic and institutional factors, the traditional model of the publishing approach is broken (especially the publication of periodical articles). Due to the exponential increase of journals price, the business model is based on the idea that readers, in particular researchers of academic institutions, must pay the price of publication [1]. This leads to limitations in accessing scholarly information, as library budgets are struggling to hold the weight of more and more expensive subscriptions. Moreover, the relationship between price system and public funding of research implies a paradox. Public institutions cannot pay the cost of research multiple times: they pay researchers for their work, they acquire books, journals and databases, while subsidizing researchers in paying for accessing texts and databases[1].

In this scenario, new digital technologies play a decisive role. On one hand the digitization of scholarly information strengthens the rigid and centralized

---

[1] For a further discussion about this scenario see the paper presented by Pompeu Casanovas and Sílvia Gabarró, *New Ways of Publishing and Intellectual Property on the Web: Dialogue and Relational Law*, in AICOL 2011.

M. Palmirani et al. (Eds.): AICOL Workshops 2011, LNAI 7639, pp. 162–170, 2012.

control of information by a few large commercial publishers. On the other hand digital technologies allow publishing online articles free of charge and without any technological barriers, what precisely is the core of the Open Access (OA) philosophy. This paper tries to investigate on such an environment where Open Access is considered a spin-off of the digitization process, aiming at restoring important priorities in the area of knowledge production and dissemination [2]. In particular, it focuses on the opportunity to build an online open digital archive in the field of legal informatics in the new digital era, taking into account that the knowledge dissemination process must be subordinated to scholars and scientists requirements, and not the reverse. There are many examples highlighting national and international policies which are supporting free access to publicly funded research results[2]. Based on the relevance of these policies, the challenge of the paper is to advance on a dynamic, interactive and interdisciplinary approach to the study of the phenomenon of information control over the results of publicly funded scholarly research.

2 From OA Digital Repository to OA Journal: An Hybrid Form

The increasing concentration of the market of legal databases has led to escalating prices for legal information [3]. The contractual and market power of databases holders is strengthened by copyright laws. However, the OA movement is quickly growing in legal scholarship which is becoming more and more interdisciplinary and globalized [4,5]. Hence, the target audience interested in legal publications is very large and heterogeneous. Furthermore, in the emerging OA model on legal scholarship the major functions of publications (selecting the best works, making them accessible, publicizing and archiving works) are based on traditional participants (commercial publishers, university press, etc.) and new intermediaries (legal scholarship repositories like Social Science Research Networks, Legal Scholarship Network and Berkeley Electronic Press Legal Repository, Wikipedia, Internet search engines like Google Books and Google Scholar)

---

[2] Some of them are the following: the establishment in 2004 of the OECD *Declaration on access to research data from public funding* on behalf of OECD Committee for Scientific and Technological Policy at Ministerial Level as an incentive to develop international and national policies for free access to public funded research. The National Institutes of Health (NIH), the largest funder of medical research in the world, since 2005 asks every scientist who receives an NIH research grant, and who publishes the results in a peer reviewed journal, to deposit a digital copy of the article in PubMed Central (PMC), the online digital library maintained by NIH. PMC will then provide free online access to its copy some time after the article is published in a journal, the length of the delay to be determined by the author. The Study on the Economic and Technical Evolution of the Scientific Publication Markets of Europe published by the European Commission in 2006 has made a number of recommendations to improve the visibility and usefulness of European research outputs.

[6]. Production costs are – as in the past – born by the authors and their institutions like universities and law faculties. Dissemination costs, lower than in the past, are shared among the authors, their institutions and the traditional and new intermediaries. Open access publication policies are substantially based on two main channels and various supporting tools.

The first channel, called the "green road" to OA, relates to digital open archives, which are institutional (pertaining to academic and research organisations) or discipline-based. Institutional archives are established by universities to allow their members to self-archive their research results and products with the aim of preserve and certify them. It is a fact that nowadays almost all universities and research centres of developed countries have institutional archives [7] and there is also a number of non academic projects, like Open AIRE (Open Access Infrastructure for Research in Europe)[3], which play an important role in promoting this approach. The institutional archives containing the greatest number of documents are definitely the repositories of theses and PhD dissertations created and made available on line by universities. Theme-based archives are repositories dedicated to specific disciplines, whose aim is to provide quick communication of scientific results to scholars dealing with similar subjects. Among the most popular theme-based archives the following are worth mentioning: ArXiv (physics and informatics)[4], the first open archive created by Paul Ginsparg in 1991, containing, as of February 2011, about 580.000 open access articles; RePEC – Research Papers in Economics[5], made up of 700.000 open access articles; E-LIS[6] including library and information science material, with more than 10.000 articles. A common feature of these archives is the inclusion of texts not subject to peer review. Therefore, these repositories include pre-print and post-print documents as well as various types of scientific and educational material.

The second road, called "gold road", includes open access reviews with selected articles freely available worldwide. DOAJ–Directory of Open Access Journals[7], a portal of the Lund University, indexes current open access reviews (around 7.000 journals, 3.000 journals searchable at article level more than 62.500 articles). This approach implies a scientific committee and a peer review process.

---

[3] OpenAIRE (http://www.openaire.eu/) EU researchers, businesses and citizens can have free and open access to EU-funded research papers thanks to OpenAIRE (Open Access Infrastructure for Research in Europe), which the European Commission launched at the University of Ghent in Belgium. OpenAIRE provides a network of open repositories providing free online access to knowledge produced by scientists receiving grants from the Seventh Framework programme (FP7) and European Research Council (ERC), especially in the fields of health, energy, environment, parts of Information & Communication Technology and research infrastructures, social sciences, humanities and science in society.

[4] http://arxiv.org
[5] http://repec.org
[6] http://eprints.rclis.org
[7] http://www.doaj.org

Gold and Green roads are not alternative, but complementary: nevertheless they are interoperable by using the metadata interchange protocol OAI-PMH[8], in a way that metadata be searchable on the network. In this context legal informatics can be a pioneer field for increasing access to open legal resources. To this regard, the idea is to create an open access repository based on open access philosophy in the field of legal informatics, focusing on the following topics:

- Free access to law
- Legal information systems
- Digital law libraries
- Legal information storage and retrieval systems
- Legal literature publishing
- Legal informatics blogs
- Right to access to legal information.

The main goal is to create a repository that can become a sole point of access and dissemination of contributions on this subject, as well as an instrument for their long-term preservation. The establishment of a disciplinary repository is also in line with the legal informatics community's awareness of its responsibility to act as a research network to create, circulate and preserve scientific research outputs: in such a way new models of scientific communication serve as alternatives to the traditional paths pursued mainly by the activity of leading international publishers.

The proposed repository also intends to provide the vast community interested in legal information (legal professionals, publishers, scholars, information scientists, librarians and documentalists) with an auto-archiving tool, and support knowledge exchange in this subject area. The idea is to create a hybrid form of legal information sharing environment. As a first stage, an OA repository is developed where no peer review is performed, but simply makes an initial validation of contributions on conformity to the subject. The repository intends to collect not only new resources like technical reports, pre-prints, newly created material, but also documents already submitted for peer-review to reviews editorial committees or presented in conferences and seminars. The material is identified as peer-reviewed or not. Authors may archive their preprints without anyone else's permission. The model adopted is ID/OA (Immediate-Deposit/Optional-Access[9]), which envisages immediate archiving of publications and options to access them, decided on a case-by-case basis according to the publishers policies and the contracts signed by the authors. Closed access to the unabridged text of the document is thus allowable, although open access is preferable: immediately, if permitted by the publisher, or delayed if there are restrictions. In any case,

---

[8] The Open Archives Initiative Protocol for Metadata Harvesting (OAI-PMH) is a low-barrier mechanism for repository interoperability. *Data Providers* are repositories that expose structured metadata via OAI-PMH. *Service Providers* then make OAI-PMH service requests to harvest that metadata. OAI-PMH is a set of six verbs or services that are invoked within HTTP.

[9] http://openaccess.eprints.org/index.php?/archives/71-guid.html

bibliographic metadata will be immediately made accessible and users will be able to request the authors text.

The second stage concerns a selection of a number of contributions made by the editorial committee which will implement the OA journal. The plan is to publish one volume in two six-monthly issues, in a peer-reviewed electronic open access format, with the aim of enhancing international research on free access to law. This secures open access to contributions (authors retain copyright), peer-review by international experts and wide dissemination of published material both at national and international level, using specific appropriate IT tools. The selection of submitted articles is the task of reviewers of the editorial committee, based on their experience and competence. A double blind peer review method should be used. The review process aims at offering a competent opinion to authors on their paper, also providing suggestions, if needed, on how they can improve their works. The peer-review process can be managed with the OJS platform described in paragraph 3. Every proposal submitted for publication is read at least by two editors, for an initial review. If the paper meets editorial policies, it is sent to four reviewers for evaluation.

2.1   Organisation Policy

The proposed project is based on auto-archiving as a purely voluntary activity. This implies that multi-author works require the authorisation to deposit the contribution by all interested authors. Issues relating to authors moral rights are raised when an author's work is deposited by third parties without permission. No problems are found with public domain works or public source documentation.

All work remains the property of the author. Unless noted otherwise, authors retain copyright and other proprietary rights. Submitting authors will be responsible for ensuring the documents they archive do not have any restrictions on their electronic distribution. Authors hold the copyright for the pre-refereed pre-prints, so they can be self-archived without anyone else's permission. For the refereed post-prints, authors can try to modify the copyright transfer agreement to allow self-archiving, or, failing that, can append or link a corrigendum file to the already self-archived pre-prints. The OA Journal will be published under a Creative Commons Attribution License 3.0[10]. With the license CC-BY, authors retain the copyright, allowing anyone to download, reuse, re-print, modify, distribute and/or copy their contributions. The work must be properly attributed to its author. It is not necessary to ask further permission both to author or Journal Board. Furthermore, the journal utilizes the LOCKSS system[11] (Lots of Copies Keep Stuff Safe) to create a distributed archiving system among participating libraries. This approach ensures that these libraries create permanent archives of the journal for purposes of preservation and restoration. In such a context open archives can create a new culture dismantling old models. Authors will have to

---

[10] http://creativecommons.org/licenses/by/3.0
[11] http://lockss.stanford.edu/lockss/Home

learn how to negotiate license agreements of their rights with publishers so that
they:

- preserve their right to deposit or auto-archive;
- keep reproduction rights, (i.e. in case of educational material);
- keep translation rights into other languages;
- keep digital copyright, as in the case of transferring a work into an e-book
  in a University Press environment.

Of course all this is not an easy task. Many publishers will have difficulties to
let authors enjoy their rights; nevertheless today a number of publishers agree
on autoarchiving, realizing they cannot deny the deposit in a pre-print archive,
as documents are deposited before the rights are licensed.

Today a considerable number of Open Access support services are available.
A very useful tool is the SHERPA/RoMEO portal[12], created to facilitate researchers in managing their copyright rights in a compatible way with widespread
circulation of their products. SHERPA/RoMEO lists copyright policies of major
publishers (mainly anglophone), identified by four colours (green, blue, yellow,
white), showing whether and which version of a text is possible to auto-archive.
The above mentioned strategies can be harmonised with traditional publishing
methodologies without shaking it.

As shown in the SHERPA/RoMEO site, auto-archiving methods are compatible with policies of most publishers of major reviews. However, challenges of
OA imply far-reaching changes and adopting open access policies can lead to a
major revolution in the scientific publication environment, which is fundamental for science and society as a whole. Through their publications, scholars and
institutions acquire posts and funding, publications also serve as intermediaries
between the academia and society as a whole: citizens, students, companies,
politicians. In conclusion:

a) Newly prepared works, presentations, conference papers, articles to be submitted to reviews, etc. are deposited prior to their subsequent submission;
   later the updated version is added if the publisher agrees. Special care is to
   be put on contracts if needed, otherwise no problems arise.
b) Concerning technical papers, reports in general, contributions which are not
   submitted to reviews, i.e. grey literature which is an extremely interesting material, the open repository is just the right solution where this material can be
   placed. Careful consideration is to be given to confidential contributions, technical documentation and material of specific working groups which must be
   kept confidential within institutions and cannot be available to public access.
c) As regards works of the past, it is important to consider the contract which has
   been subscribed at the time of their production, and the authorization that
   has to be requested to publishers. All over the world even most experienced
   publishers are now granting authorization to digital reproduction for public
   access before the 3-5 years "wall" (a limit in recent years) after the publication.

---

[12] http://www.sherpa.ac.uk/romeo

More specific details about copyright issues and licenses are discussed in the paper presented by Pompeu Casanovas and Sílvia Gabarró, *New Ways of Publishing and Intellectual Property on the Web: Dialogue and Relational Law.*

3   Tool to Manage the Open Repository and the Journal

An essential pre-condition for launching a repository and a journal according to an open access distribution policy is to rely on a software platform capable to manage both back-end and front-end functionalities. Firstly such platform should manage a proper authorization scheme for the following roles:

- Author submits a contribution in the domain of interest
- Editor: manages the repository and the editorial process of the journal assignes submissions to Section Editors who organize reviews and editing, as well as issues tocs and scheduling;
- Section Editor: manages submission review and possibly Submission Editing for the assigned submissions within a specific sub-domain;
- Reviewer: is the expert invited to review contributions according to his/her specific expertise
- Repository and Journal Manager: is the technical administrator of the repository and the journal
- Copyeditor: improves grammar and clarity to the contributions, asks questions to the authors about such aspects, guarantees bibliographic references and textual styles compliance with respect to the journal styles
- Layout Editor: transforms copyedited submissions into the proper format for electronic publishing
- Proofreader: reads copyedited submissions for typographic and formatting errors.

A possible interaction schema between such roles is sketched in Fig. 1 where the phases associated to the Open Access Green and Gold Roads are coloured in light and dark gray respectively.

In this schema the Open Access Green Road can be identified in Author's self-archiving and Editor's topic selection functionalities, aiming to build up a repository of selected materials in a specific domain of interest (legal informatics for the case-study): at this stage no peer review or editorial process is foreseen. On the other hand the Open Access Gold Road involves all the roles previously mentioned, as well as their specific functionalities, as shown in Fig. 1. As discussed in the previous paragraph, OJS[13] Web platform developed within the Public Knowledge Project (PKP) is able to cope with these requirements, offering an effective and complete solution for publishing and disseminating scholarly research in an open access modality, promoting a sustainable model for academic publishing. Readiness of publication and extensive distribution are the main characteristics achieved by implementing an Open Access publication

---

[13] http://pkp.sfu.ca/?q=ojs

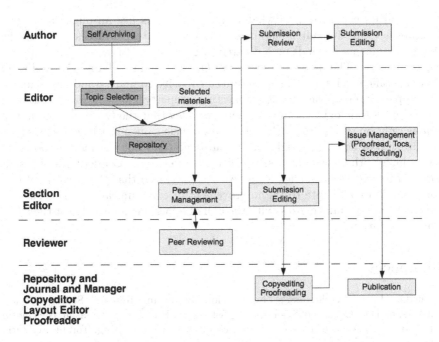

Fig. 1. Publishing workflow for an open access journal

policy, nevertheless this is not enough for an effective dissemination and evaluation of the scientific research impact. Bibliographic references have to be provided
to institutional metadata indexing services (for example Science Citation Index,
Social Science Research Network) as well as made available for bibliometric analysis able to measure and evaluate scholarly research factors. In order to cope
with these aspects, OJS implements an OAI-PMH 2.0 protocol, representing an
OAI data provider. This allows bibliographic metadata to be harvested by institutional indexers and properly disseminated for bibliometric statistics. Finally
OJS is compliant with the LOCKSS system for digital preservation, ensuring
a secure archive for the journal. It enables libraries to preserve web-published
materials by polling registered journals Web sites for new published contents.

4   Conclusion

Both in business and in science the issue of control of information arises, also
known as "access control information". However, there is a gap between the control of information that in business relies on intellectual property and what, in
the scope of the scientific community, is based on informal norms [8]. Forms of
reconciliation between informal norms of science, licensing agreements and technological standards are the key elements to create a new economic model in which
the new intermediaries of information will not repeat the threat of strict control
and centralized access to information. The stratified and competitive nature of

science and its evolution into what increasingly looks like an oligarchic power structure is quite difficult to unhinge: it is clear that open access to scientific information deeply depends on a different organization of power in an academic environment [2]. The definitive success of OA (in legal scholarship as well as in other scientific disciplines) depends on the capacity to understand and manage the complex intersection among intellectual property law, contracts, norms of science and technological standards. One may say that the OA is the revenge of authors right on the "publisher's right". Nevertheless one must be conscious that – in the digital age – copyright law is only one (and not the most important) among many other instruments, which may govern the production and distribution of information. In such a way we are convinced that the creation of this repository can contribute to the improvement of communication infrastructure in the field of legal informatics, with the aim of relocating knowledge at the heart of our civilization.

References

1. Guedon, J.C.: In Oldenburg's Long Shadow: Librarians, Research Scientists, Publishers, and the Control of Scientific Publishing. In: Proceedings Creating the Digital Future: Association of Research Libraries 138th Annual Meeting, Toronto, Ontario (2001)
2. Guedon, J.C.: Repositioning the meaning of knowledge through Open Access: implications for libraries. In: 56° Congresso AIB, Florence, November 3-5 (2010)
3. Arewa, O.: Open Access in a Closed Universe: Lexis, Westlaw, Law Schools, and the Legal Information Market. Case Legal Studies Research Paper No. 03 (2006), http://ssrn.com/abstract=888321
4. Carroll, M.W.: The Movement for Open Access Law. Law/Public Policy Research Paper No. 11 (2006), http://ssrn.com/abstract=918298
5. Hunter, D.: Open Access to Infinite Content (Or 'In Praise of Law Reviews'). Lewis & Clark Law Review 10(4) (2006)
6. Solum, L.B.: Download It While Its Hot: Open Access and Legal Scholarship. Illinois Public Law Research Paper No. 03 (2007), http://ssrn.com/abstract=957237
7. van Westrienen, G., Lynch, C.A.: Academic Institutional Repositories: Deployment Status in 13 Nations as of Mid 2005. D-Lib Magazine 11(9) (2005), http://www.dlib.org/dlib/september05/westrienen/09westrienen.html; For an updated list of OA repositories in Europe and North America see http://www.opendoar.org/countrylist.php?Continent=Europe, http://www.opendoar.org/countrylist.php?cContinent=North%20America
8. Caso, R.: Open Access to Legal Scholarship and Copyright Rules: A Law and Technology Perspective. In: Peruginelli, G., Ragona, M. (eds.) Proceedings of Law via the Internet: Free Access, Quality of Information, Effectiveness of Rights, pp. 97–110. European Press Academic Publishing, Florence (2009)
9. Guéedon, J.C.: Open Access: Contro gli oligopoli nel sapere (Internet). ETS, Pisa (2010)

Advancing an Open Access Publication Model for Legal Information Institutes

Pompeu Casanovas[1] and Enric Plaza[2]

[1] UAB Institute of Law and Technology, Law Faculty,
Universitat Autònoma de Barcelona, 08193 Bellaterra, Catalonia, Spain
pompeu.casanovas@uab.cat
[2] IIIA, Artificial Intelligence Research Institute,
CSIC, Spanish Council for Scientific Research, Campus UAB, 08193 Bellaterra,
Catalonia, Spain
enric@iiia.csic.es

Abstract. In this paper we propose an Open Access model for Legal Information Institutes (LIIs) publications in three steps: Accredited Public Archival (APA), Comment-Open Publication (COP) and peer reviewed Publication (PRP). This raises some ethical and legal issues on privacy and intellectual property which cannot be ignored. We would like to foster dialogue and discussion as the unique means to create an interactive framework among research communities, LIIs and users.

Keywords: Free Access to Law, Legal Information Institutes, Relational Law, Open Access Publishing, Privacy, Intellectual Property, Crowdsourcing.

1   Introduction

This is an updated version of the work presented at the Law via the Internet Conferences (Durban 2009, Hong Kong 2011) on the possibility of publishing a LII journal to gain visibility, transparency and interconnectivity on the web[1]. In the last few years, researchers have shown a growing interest in developing new ways of publishing and sharing their scientific works. Actually, this is a field of research in itself, going along with the developments of grid and Semantic Web technologies.

Legal Information Institutes are offering information and legal services in a free access format since 1992. They all signed the Declaration on Free Access to Law[2], and are organized around the World Legal Independent Institutes (WLII)[3]. There are several articles on the origins and short history of

---

[1] Artificial Intelligence Approaches to the Complexity of Legal Systems, AICOL III, XXV IVR-World Congress of Philosophy of Law and Social Philosophy, Goethe Universität, Frankfurt am Main, 19th August, 2011.

[2] http://www.worldlii.org/worldlii/declaration

[3] http://www.worldlii.org

M. Palmirani et al. (Eds.): AICOL Workshops 2011, LNAI 7639, pp. 171–188, 2012.
© Springer-Verlag Berlin Heidelberg 2012

the movement, which has experienced a significant grow in the last ten years [6, 7, 8, 10, 20, 39, 42]. One of its main goals is equality and social justice[4].

We will comment on two previous existing models for updating publishing: (i) the hybrid model of Open Access Repository/Journal, with a workflow interaction schema proposed by Francesconi, and Peruginelli [18] specifically for the legal field (2001); (ii) the Open Access Repository Schema developed for multiple applications in a more abstract level by the European Project LiquidPublications (2009-2011) [3][5].

We will end up with a simplified model for the LII platforms. The paper is divided into four different sections: (i) Relational law and new ways to think on intellectual property (ii) Open Access Publications and the LICT-Repository; (iii) Privacy, ethical values and free access to legal information; (iv) Steps and functions of the new process of digital publication.

2   Relational Law and New Paths to Think on Privacy and Intellectual Property

We will focus first on the culture developed through the Internet, which is changing the perception and the shape of the law. Law is now a more horizontal structure based on dialogue —with the added value of rapidity, flexibility and the immediate reaction towards particular problems— than a sole structure of rules or norms. This is what we refer to as "Relational Law" [11, 12]. This set of legal forms is not opposed to national law or to jurisprudence, but it is *superimposed* to them. In other words, dialogue is not another option but the most natural way to communicate on the Internet. As Brian Solis would put it: "People aren't lured into relationships simply because you cast the bait to reel them into a conversation. [...] Relationships are measured in the value, action, and sentiment that others take away from each conversation. Talking "at" or responding without merit, intelligence, or quality grossly underestimates the people you're hoping to befriend and influence"[6].

There are other forms to deal with relational forms of law. The underlying model of this kind of regulation, based on the developments of the Semantic Web, reuse of knowledge, and *crowdsourcing* [28] has been recently called "Metropolis" by Kazman and Chen [31]. Crowdsourcing companies [7] and humanitarian platforms (such as Ushahidi) are among the most interesting developments of governance and democracy on the web[8]. As regards the so-called "intellectual

---

[4] "The open access to law approach was developed with minimal resources; it was as a result of collaboration that legal information institutes became well established. These values and these achievements now exist for those who now want to ensure more equality and more justice in any country." [39]

[5] http://www.iiia.csic.es/en/project/liquidpub

[6] http://www.briansolis.com/2009/03/conversation-prism-v20

[7] http://compassioninpolitics.wordpress.com/2011/03/12/best-examples-of-crowdsourcing-companies/

[8] http://www.ushahidi.com

property", we think that we are moving away the discussion from the legal arena of the 20th century. This legal arena was in turn based in the rule of law of the 19th century. Even if we accept the term, we have to analyze thoroughly the concept in a non-normative way.

While Lawrence Lessig was defending the non-extension of copyright to 95 years before the Supreme Court of the United States, Dan Hunter —an expert lawyer on AI & Law— warned about the improper extension of the concept "property" concerning the intellectual products of the net [30]. He called it the paradox of the *Anticommons*, this is to say that an excessive protection of the contents may cause a weak development of the net. Protection becomes an obstacle, since the net is the result of gathering both telecommunications and information technologies, and it is not a material object with volume and consistence (such as the land). The problem has not been solved with the Web 2.0; it has become worse. Facebook's position is well known as well as the reactions it has provoke[9]. As one of the participants at the recent Hyperpublic discussion states: "Never forget... when a social media site is free (Facebook), you're not the customer, you're the product"[10]. However, there are other Web 2.0 options to share work and settings, such as Zoho, Thinkature, CiteYouLike, Scribd, IntenseDebate, DataVerse project and Swivel, among many others [3, 4, 2].

We have to take into account some well-founded criticisms against the performance of lawyers, judges and legislators coming not from Richard Stallman[11],

---

[9] In this case, Facebook policies on privacy affected the intellectual property rights of users as well. See e.g. http://www.siliconvalleyiplicensinglaw.com/blog/facebook-licensing-controversy-prompts-public-to-take-closer-look-at-social-networking-site-terms-and-conditions/ The controversial Provision (2009) was: "You are solely responsible for the User Content that you Post on or through the Facebook Service. You hereby grant Facebook an irrevocable, perpetual, non-exclusive, transferable, fully paid, worldwide license (with the right to sublicense) to (a) use, copy, publish, stream, store, retain, publicly perform or display, transmit, scan, reformat, modify, edit, frame, translate, excerpt, adapt, create derivative works and distribute (through multiple tiers), any User Content you (i) Post on or in connection with the Facebook Service or the promotion thereof subject only to your privacy settings or (ii) enable a user to Post, including by offering a Share Link on your website and (b) to use your name, likeness and image for any purpose, including commercial or advertising, each of (a) and (b) on or in connection with the Facebook Service or the promotion thereof. You represent and warrant that you have all rights and permissions to grant the foregoing licenses". See the actual policy at http://www.facebook.com/terms.php

[10] Ian Jacobs, http://twitter.com/#!/search?q=%23hyperpublic (June 10th 2011); see on the symposium http://www.hyperpublic.org/about/; see on the concept of hyperpublic ("capture local experience, organize it, and display it") http://techcrunch.com/2011/02/01/hyperpublic

[11] Stallman proposed a classification into three kinds of works according to their purpose: (i) functional works (e.g. manuals and programs should rid of copyright as open source); (ii) works that express personal position (verbatim right should apply and these works should not be modified without the author's consent); (iii) aesthetic works (the modification affects the author but may have new aesthetic uses) [3].

but from researchers of The Cooperative Association for Internet Data Analysis (CAIDA), at the University of California's San Diego Supercomputer Center[12].

They produce measurements for performance evaluation, data-sharing, broadband performance, IPv6 deployment and network analysis [15, 14]. However, they cannot monitor and map properly the evolution of the Internet due to the amount of legal obstacles that they are facing to carry out their work [13]. They point at the effects that law based on power may have to regulate the Internet and at the performance of lawyers and jurists. Tim Berners-Lee and James Hendler (W3C) have admitted also that they cannot measure the level of semantic indexing in the net [26, 11]. The present situation not only limits the access to content and knowledge, but it also limits the production of scientific knowledge about the Internet.

Therefore, it is not surprising that the scientific community has reacted according to its needs, making the most of the communication possibilities offered by the net. New perspectives on privacy, rather than on intellectual property, and ethics, rather than on legal provisions, are at stake in the scientific resilience to let the things going in the way they are going now. This constitutes the kernel of the Kenneally and Claffy's Privacy sensitive sharing framework for creative and scientific works: "We anticipate circumstances to reveal that rather than data-sharing being a risk, *not* sharing data is a liability" [32]. This is the same spirit that inspired some recent AAAI Spring Symposiums [21].

As regards publishing and scientific content-sharing, a European project of the 7FP (2009-2011) recently came to an end on this topic: *LiquidPub*, liquid publications. Its project's goals were the new ways of scientific communication that increase day by day and coexist with the peer review and the publication in journals. Wikis, blogs and virtual communities provide a discussion forum that allows progress in a particular field thanks to the contribution of specialists. There are "liquid" scientific journals (having their own problems: copy, delete or remove, and share)[13]; "liquid" journals (with copyleft), and "liquid" conferences (with irrevocable license to distribute the content). A model of "Liquid conference" is shown in Fig. 1, which understands them as virtual 'meetings' in an online environment. In this model, articles take the place of presentations and discourse follows in the form of (usually moderated) comments from system users. This enables many of the key features of 'real' conferences (detailed presentation of ideas, focused discussion and exchange) while avoiding the costs and constraints associated with bringing many people together in the same place at the same time [36].

---

[12] http://www.caida.org/home
[13] "Liquid Journals (LJs) are essentially a scientific social bookmarking service—but with a focus on making selections to share, annotate and present rather than to keep a bibliography."

In this model, Scientific Knowledge Objects (SKOs) are modeled in a single graph to monitor and control the scientific added value at each stage[14]. However, editing and managing are also crucial tasks: "The internal quality of a liquid journal depends on how good the editors are in selecting content" [4]. What the system envisages, then, is preserving and enhancing better scientific results, fostering exchange and cooperation. *Attention*, not printing, is considered the new scarce resource for scientific dissemination [2].

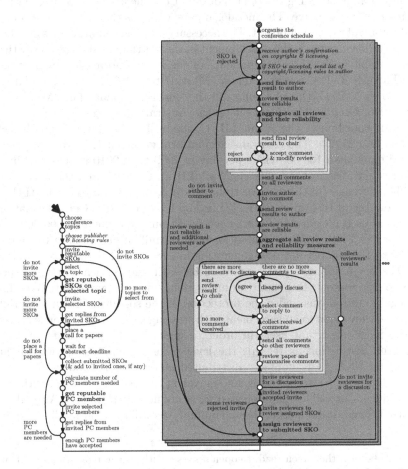

Fig. 1. "Liquid Conference" management process model. Source: [5, 36] (used with permission).

---

[14] "We see scientific contributions as structured, evolving, and multifaceted objects. Specifically, we see scientific content as something that we want to search within and help assess and disseminate by spatially representing it as scientific resources organized as a set of nodes in a graph that authors, editors, or even readers can connect or annotate. The reason for connections, and hence for modeling resources as a graph, is to capture several kinds of dependencies or relationships among them (or between resources and people or other entities)." [2]

Does this mean that editorial production does not have value anymore? Does this mean that the old concept of work's added value has to be put aside? Does this mean that any content may be used freely? We do not think so. One of the main promoters of *LiquidPub* was Springer Verlag. Publishers are very interested in these new trends. In fact, they started to allow scientific pre-prints quite a while ago[15]. There are different kinds of soft licenses in use (GFDL, CC-BY-SA) but the context has become more complex as well as the behavior of the actors. The difficulty lies in understanding the rights not only from the law perspective, but from the *metalegal* perspective. The metalegal perspective is the distance that entails the definition of a different object. The necessary dialogue among actors lies there, since the following step is the development of the net thanks to its own growth.

An example is the 15-year experience of the Legal Information Institutes. In 1992, Thomas R. Bruce and Peter W. Martin started Cornell's legal platform [6, 7, 8, 10]; in 1993, Daniel Poulin did the same in Canada (LexUM) [36, 37]; and in 1995, Graham Greenleaf set up the Australasian Legal Institute (which at present coordinates the Asian LII, CommonLII, CommonLII and Lawcite projects) [20, 25, 23]. The Australasian platform gathered 1,155 databases that received more than 100,000 visits per day in 2009[16]. In 2010 it increased in the number of databases searchable via WorldLII up to 1.205, and there were over 25 million accesses (to the non-Australian databases maintained by AustLII and located on AsianLII, CommonLII, NZLII, and LII of India)[17]. Cornell's platform is the most used platform in USA. It received from 1,500,000 to 2,000,000 visits per month in 2009, and was visited for over 14 million people in 2010, ranging from 60 to 100.000 per day and generating in excess of 71 million page views[18].

At the beginning, the platforms were based in ideals such as the universal access to free content. But countries under the Common Law soon they realized that legal material was under the protection of the so-called Crown Copyright. Therefore, they came to an agreement with state agencies, politic representatives and the most important users in order to guarantee free access to materials making no distinction as regards the level of use. The strategy was slightly different for USA, that adopted a distributed non-centralized model [6, 7].

In other words: (i) free access is not equivalent to free content; and (ii) the content producers as well as the interested users are now financing the Australasian platform because it did not get the one million Australian dollar funding required

---

[15] "Preprints form the 'green road' to open access—authors can make the text of their articles publicly available while assigning commercial rights and/or copyright to journal publishers. arXiv, the largest such preprint server, offers users a choice of licenses when submitting articles. (i) Default option: a non-exclusive and irrevocable license for arXiv to distribute the article, (ii) Compatible with most journal copyright transfer agreements, (iii) Creative Commons (CC) Attribution license. (iv) Typical open access license (PLoS, BioMed Central. Ect.), (v) CC Attribution-Noncommercial-Sharealike license., (vi) A common restrictive open access license; (vii) Public domain." [3].

[16] Communication at LII Conference held in Durban (26th 27th November 2009)

[17] Communication at LII Conference held in Hong Kong (8th -10th June 2011)

[18] Communication at LII Conference held in Hong Kong (8th-10th June 2011

to the Government in 2007. So a hybrid, collective, non-publicity but common interest based business model was thought to keep on offering the service [19]. *AustLII Foundation Limited*, a charitable company owned jointly by AustLII founding universities (UTS and UNSW) receives contributions from over 300 organizations, and keeps applying (and being successful) to competitive funding grants. Again, this business model is slightly different for the LII at Cornell, which has started accepting advertising and has formed a series of strategic business alliances with like-mined partners in the private sector[19]. LexUM has going even further. It was established in mid-nineties as a laboratory in legal informatics of the Faculty of Law (University od Montreal). In 2010, LexUM spun off and became a private company, comprising a team of 40 computer engineers, legal documentation specialists, and lawyers.[20]

As highlighted in Florence (2008)[21], Durban (2009)[22] and Hong Kong (2011)[23] Conferences, nobody questions at present the need to collaborate with companies, the need to combine business models with principles and ideals and, above all, the need of dialogue among all interested actors to go on progressing[24]. This is an example of what we have called relational law.

We think CC can move towards this same direction and, in fact, they are doing so. On 25th January 2010, the *Manifesto* of Communia in favor of the public domain seemed to open the debate [35]:

"The public domain, as we understand it, is the wealth of information that is free from the barriers to access or reuse usually associated with copyright protection, either because it is free from any copyright protection or because the right holders have decided to remove these barriers. It is the basis of our self-understanding as expressed by our shared knowledge and culture. It is the raw material from which new knowledge is derived and new cultural works are created. The Public Domain acts as a protective mechanism that ensures that this raw material is available at its cost of reproduction - close to zero - and that all members of society can build upon it. Having a healthy and thriving Public Domain is essential to the social and economic well being of our societies".

We are placed between two positions: (i) considering the public domain as the general rule and copyright as the exception; (ii) considering copyright as the general rule and the public domain as the exception. Nevertheless, we think that there is a wide space for the dialogue between these two poles.

---

[19] Communication at LII Conference held in Hong Kong (8th-10th June 2011)

[20] Communication held in Hong Kong (8th-10th June 2011)

[21] http://www.ittig.cnr.it/LawViaTheInternet/

[22] http://www.saflii.org/content/
10th-law-internet-conference-icc-durban-26-27-november-2009

[23] http://www.hklii.hk/conference/

[24] "Open access and commercial publishing can coexist. [...] Our own view is that there is room for both and that in fact both are needed. At least in Canada, commercial entities are doing a superb job publishing law" [39]; "The only realistic option for AustLII is what we could call a 'multi-contributor' model, but is really a mix of different business models. Part of its model will continue to be based on competitive grant funding [...]." [23]

As we may see with the example of the development of the Free Access to Law Movement, one of the main problems is not the launching but the *maintenance and progress* of free access to a large number of databases on daily bases. This means developing a highly specialized and time-consuming professional work. The alternative is an entirely state-base publishing legal documents, as it is the case in Civil Law countries such France, Spain or Germany, in which the state (as a legal subject, by means of the principle of the legal personality of the administration) is the source, the owner, and the publisher of most collections of legal documents. But even in this case, the public domain is not coincident with the state or administrative sphere only. It must be reconstructed through the social behavior of all the stakeholders —state agents (such as magistrates, judges and prosecutors), professionals (such as lawyers), lay people (citizens), and private companies (among them, publishers which add abstracts and indexing on the rough documentary data sold or legally delivered to them by the state). The public space is built within the interface of the state, civil society and the market.

The present paper deals with this intermediate position, focusing on Francesconi and Peruginelli's proposal of a "hybrid" publication/repository form for the platforms of the Free Access to Law Movement [18]. However, we will propose shifting from the specifically legal intellectual property domain to a more flexible structure provided by ethics and a wider conception of privacy.

3   Open Access Publications and the LICT-Repository

According to Peter Suber 4[5], "Open-access (OA) literature is digital, online, free of charge, and free of most copyright and licensing restrictions". Since the Budapest Open Access Initiative (2002), distinguishing between *self-archiving* (tools and assistance to deposit their journal articles in open electronic archives) and *open-access journals* (new generation of journals committed to open access)[25], a great deal of work has been done towards this direction[26]. Self-archiving is also known as the "green road" and Open Access Journals (OAJ) as the "gold road" to open access. It seems to us that definitions provided so far emphasize for the authors the idea of gaining control over the integrity of their works, at the same time that they make them available to a wide community of potential readers [30]. A *hybrid* way of publishing would combine these two possibilities, and

---

[25] http://www.soros.org/openaccess/read.shtm

[26] The so-called BBB common definition of "Open Access" (Budapest, 2002; Bethesda, 2003; Berlin, 2004): "By "open access" to this literature, we mean its free availability on the public internet, permitting any users to read, download, copy, distribute, print, search, or link to the full texts of these articles, crawl them for indexing, pass them as data to software, or use them for any other lawful purpose, without financial, legal, or technical barriers other than those inseparable from gaining access to the internet itself. The only constraint on reproduction and distribution, and the only role for copyright in this domain, should be to give authors *control* over the integrity of their work and the right to be properly acknowledged and cited." http://www.earlham.edu/~peters/fos/newsletter/09-02-04.htm#progress

it opens up different options offered by the main scientific publishers as well[27]. The idea advanced by Enrico Francesconi and Ginevra Peruginelli [18] follows this mixed, flexible way of conceiving and managing intellectual productions:

"The idea is to create a hybrid form of legal information sharing environment. As a first stage an OA repository is developed with the name of Legal Information and Communication Technologies Repository (LICT-Repository). It does not perform peer review, but simply makes an initial validation on conformity to the subject. The repository intends to collect not only new resources like technical reports, pre-prints, newly created material, but also documents already submitted for peer review to other editorial committees of reviews or presented in conferences and seminars. All material is identified as peer reviewed or not. Authors may archive their preprints without anyone else's permission. The model adopted is ID/OA (Immediate-Deposit / Optional-Access)[28] which envisages immediate archiving of publications and options to access them, decided on a case-by-case basis according to the publishers' policies and the contracts signed by the authors. "Closed" access to the unabridged text of the document is thus allowable, although open access is preferable: immediately if permitted by the publisher or delayed if there are restrictions. In any case, bibliographical metadata will be accessible immediately and the user will be able to request the text from the author". The Francesconi-Peruginelli workflow is shown in Fig. 2.

4 Privacy, Ethical Values and Free Access to Legal Information

We think that we can follow this original idea, but shifting from the intellectual property framework to a wider conception, more suitable for the LIIs purposes. This means changing lens: instead of viewing the publication process from the poles (the binary relationship between authors and publishers or LIIs), we might approach the same relationships stemming from the link among all the implied agents, LIIs, publishers, companies, institutions... and users (professional or lay people). In other words, to change the property perspective, where individual authors' rights and interests are the focus of the discourse, in benefit of collective trust and shared common values. Launching an OA journal means creating some kind of ties first, and securing interoperability and a more fluid and permanent communication: (i) between research communities and Legal Information Institutes, (ii) between users and LIIs, (iii) and between LIIs themselves.

Creating a community of related scientific researchers, LIIs, and users is not at all an easy task, but this social network might be at the same time a condition and a result of the Web 2.0 and Web 3.0 applications to the legal field, which is evolving along the increasing functionalities of mobile technologies and services offered on the web (legal services and, more recently, semantic services). A new balance between the increasing risks and information asymmetries of the web

---

[27] See some of the main publishing policies at
http://en.wikipedia.org/wiki/Hybrid_open_access_journal
[28] http://openaccess.eprints.org/index.php?/archives/71-guid.html

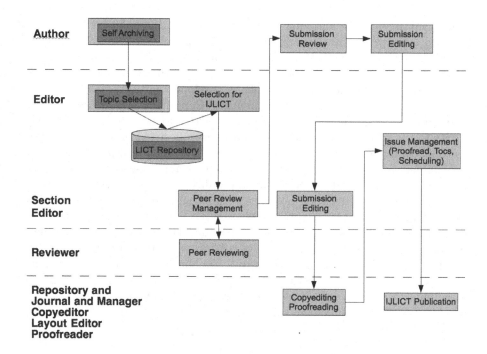

Fig. 2. Workflow interaction schema. Source: [18].

of data, and the protection offered by privacy enhancing technologies (PET) is taking place and it must be included into the policies of LIIs [16].

To do this, there is no need to break the chain value of digital intellectual property rights (in which authors, servers, publishers and sellers are equally involved as chain links). From this point of view, as shown by the Australian case in [17], protecting and gaining control over the own productions can be secured with already CC existing licenses and tools. This holds for collecting societies as well [27].

It is our contention that the legal perspective can be broadened up towards an information or computer ethics perspective [9]. The emergence of Social Network Applications and ubiquitous computing (ambience intelligence) entail other risks not covered by Privacy Enhancing Technologies (PET) [40]. Information Ethics would make sense for OA LIIs publications, because liberty and easier lay people accessibility to legal knowledge have been the main scope of the free access movement since the beginning.

Therefore, digital rights may be conceived as integrated into a broader conception of information privacy which takes into account not only the professional community needs, but the actual demands from companies and institutions within the boundaries of the market[29]. This does not mean accepting

---

[29] See the papers gathered in [21]; especially Sabah Al-Fedagli [1] on the concept of "information privacy", and his refinement of Floridi's ontological interpretation of information privacy pointing at the "Compund Personal Identifiable Information" (CPII).

these limitations, but entering into a dialogue without excluding any stakeholder. Moreover, a multicultural and pluralist approach to the different needs and cultural values of the readers seems also appropriate to make a balance between universal values and local differences and needs as well.

Global coverage and a close attention to culture and local differences is a common feature of Legal Information Institutes which have always been involved into the implementation of the Rule of Law in developing countries [18, 25, 37, 38].

This leads to a more political redefinition of "information ethics"[30], shifting from abstract entities such as SKOs, or from the ontological properties of the linked objects in the web of data, to a more user or person-centered approach[31]. *People using the web, people producing and sharing social and scientific knowledge*, and not computer objects in the web, are the center of this pluralistic approach. A minimal set of principles (or *etica minima*) affecting the collective dimension of knowledge, technical standards, and the protection of privacy should be enough [32]. And *culture* as embedded situated knowledge, provides the necessary link among creators, publishers and users which can aggregate individual decisions and values into a more collective space.

This goes into the reconstruction of a public space to transform and share individual knowledge into a collective one. This reconstruction cannot be only virtual or related to what it has been called the *infosphere*: it necessarily affects the human-computer interface that structures ethical and political actions across the global sphere. This shifting move is political in essence, deepening and strengthening the human rights dimension.

5 Steps and Functions of the New Process of Digital Publication: A Model for the LII Platforms

The process of digital publication allows separating different steps and functions that were intermingled in the old paper-based publication process. In particular,

---

[30] E.g. [34] : "Information policy is the set of strategies and actions defined at a geographical or institutional level in order to satisfy information needs expressed by people and assure development goals. With the development of information and communication technologies (ICT), new stakeholders appear, including both information producers and consumers, raising problems relative to authenticity, reliability, and evaluation of information, and also the problem of full and effective use of information technology. As information policy aims at providing access to timely information, it should attempt also to make people fluent with technology."

[31] Floridi"s perspective on Information Ethics is broadening up the field towards a substantial constructionism, *poietic* in nature. "IE suggests that there is something even more elemental than life, namely being that is, the existence and flourishing of all entities and their global environment and something more fundamental than suffering, namely *entropy*. The latter is most emphatically *not* the physicists' concept of thermodynamic entropy. Entropy here refers to any kind of *destruction* or *corruption* of informational objects (mind, not of information), that is, any form of impoverishment of *being*, including *nothingness*, to phrase it more metaphysically." [17]

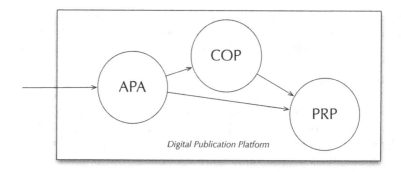

Fig. 3. Digital Publication Platform components

the low cost of "digital publication" (that is to say, making widely accessible some body of digital work) decouples the "publication" stage from the "quality certification" stages. Paper-based publication high cost made widely accessible (with a high number of physical copies) those documents that were already certified above some quality level (e.g. peer reviewing in scientific research); digital publication allows very cheap and straightforward self-publication, institution-based publication, etc., since the Internet makes any such publication widely available. Self-publication, however, lacks some institutional and social properties that are desirable. We propose now a three-component model of digital publication, namely:

APA - Accredited Public Archival
COP - Comment-Open Publication
PRP - Peer Reviewed Publication

The valid transition of a publication among these components is from the outside to 1, from 1 to 2, from 1 to 3, and from 2 to 3 (See Fig. 3). We will now describe each component in turn.

5.1   Accredited Public Archival (APA)

This component has similar functions to the way Archive.org is used by researchers in the domains of physics and mathematics. The APA component is a platform that allows the storage, indexing, retrieval, access and copy of documents. Authors of a document submit the contents of a document, and APA accredits the claim of authorship by the submitters with regard to the content of the document, and certifies the date and time stamp of the submission.

This process allows the authors to claim accreditation of authorship of the contributions at a particular date, and making it public helps their reputation and increases the transparency on research work. The authors receive a public identifier and certification, as well as the services of retrievability and accessibility provided by the platform.

The submitted document is an *archival publication*. It is a "publication" in the sense that is it is public and publicly accessible, and is "archival" in the sense that the platform commits the permanent or long-term preservation of the submitted documents. Notice, however, that this changes the common meaning of *archival publication* in scientific publication.

Submitting a research paper to a peer reviewed journal requires the authors to state that it has not yet been published in an "archival publication" it is admissible to have been published in a non-archival publication (i.e. one that does not insure permanent or long-term preservation, like workshop proceedings, or one-shot publications). The difference from our proposal is that APA is not peer reviewed, so the requirement for submission to peer reviewed publication should more exactly phrased as "a document whose content of that has not been subject *peer reviewed archival publication*". In this way, documents in APA do not fall under this requirement.

This component does not allow any kind of third-party or social commentary, or any other action, upon the APA documents; these documents are simply accessible, and their authorship claim accredited. No claim or counter claim on innovation, plagiarism, etc., takes place at this component.

5.2   Comment-Open Publication (COP)

This component receives only documents previously certified by APA. When the authors submit their publication to COP they open this publication to third-party and social commentary using the functionality provided by the platform for this purpose.

As an example, the COP platform may include functionalities like these:

– Comments of the documents by certifiably identified persons
– Comments provided anonymously
– Reputation-based mechanisms like "I like" or "I do not like"
– Endorsing mechanisms, like promoting the reading of the paper for a particular purpose, problem, or community, or recommending some particular person to read the document
– Permanent citation mechanisms, using the APA-identifier of the document to cite it, while contributing to its reputation by accumulating these citations in the platform or via citation interchange with other similar platforms.

Moreover, digital publication allows early feedback, so COP supports activities that help evolving documents. Three main evolving mechanisms are:

Refinement: which allows the authors to write a new version of the paper (but under the same title and identifier) based on the community's feedback; sometimes this may be referred to as versioning.
Superseding: the authors deprecate the document and submit to APA a new document (with a new identifier, and typically a new title) that is considered as a new take on the same issues, sufficiently different from the previous one.

Merging: authors of two (or more) documents decide to proceed, by creating a
new joint document based on their individual previous work (which becomes
superseded by the new document).

The COP papers are not peer reviewed, are open to comments and public
scrutiny, while not formally claiming a scientific contribution, and they are
deemed worthy of public debate. The reasons can be variegated: they can be
presented as food for thought, as new ideas that need elaboration, or as challenges to commonly held ideas or mores.

For instance, a paper discussing how to evaluate computer science papers
is typically not accepted as a regular paper in a computer science journal or
conference, since it is not about computer science; but it is a challenge to current
ideas or mores of the computer science community. Another example would be
State of the Art papers, which typically depend on the field, but should keep
evolving as the field evolves.

5.3   Peer Reviewed Publication

The third component is the equivalent to the usual peer reviewed archival publication, where quality is certified by a formal process. Only papers coming from
APA or COP can enter this PRP component. PRP platform can encompass one
or several "virtual journals", with each journal having a specific Editor and a
Board. PRP document identifier would be a pair (i, j), i.e. a composition of the
APA identifier i and the journal identifier j; in this way the publication aspect
(the making public stage) is decoupled from the community-certified quality (the
journal "inclusion" rather than "publication" of the paper). Once the paper has
been included in a journal it is considered a peer reviewed archival publication,
and submission to other journals it is allowed.

Finally, the PRP component does not commit to any particular process of
quality certification, although in scientific journals peer reviewing (in its different
formats) is a de facto standard.

6   Conclusions

Some years ago, Tom Bruce [6, 7] put the problem of publishing around these
alternative possibilities:

- Public-sector versus private sector
- Publication by government [or by the state] versus publication by others, be
  they private- or public-sector actors.
- Publication by centralized publisher versus self-publication by creators, regardless whether the centralized publisher is public, academic, or private.

By those years, a careful examination of the legal situation and the existing
technology leaded Graham Greenleaf to the conclusion that "privacy could now

be unduly prejudiced in favor of property" [22, 24][32]. In 2005, Dan Hunter made public the problems he had had with the California Law Review policies in storing his previous preprints in the Social Science Research Network (SSRN): "many of the top law reviews are acting as stalking horses for the commercial interests of legal database providers" [29].

CC licenses have contributed to clear up the way on how open access initiatives can protect the rights of the authors, the value added by servers and information providers, and the public access to scientific knowledge. Still, even through CC licensing, as stated recently by Lawrence Lessig [33], several problems remain to sharing and especially remixing creative works. To implement the values of justice and efficiency while keeping them under realistic terms, the implementation of open communities would be needed as well in the scientific field. We believe that it may exist a complementary attitude to the "publish or perish" research policy, which may lead to a reflexive closure of the scientific domains themselves and a misunderstanding of the effects and consequences of their implementation in social environments. Cross-fertilization means not only transparence and measurement of a fair evaluation process, but an external dialogue with specialists of other disciplines and with other stakeholders that may have some legitimate interests in the results of the research (including citizens that may discuss its applicability and side effects).

In recent times, private publishers have been forced to launch aggressive programs of Open Access Publications to survive in the highly competitive scientific market. However, authors, and therefore the private or (more often) public funds that sustain their research, are charged with the dissemination costs. They do the work and pay for it at the same time, gaining visibility or appearance on the web in exchange. In this way, authors do not participate in the market: they are the market. The hard currency is copyright: they are allowed to keep it as a negotiable value. But this is not what it was intended with Open Access programs: the so-called *hybrid* OA Journals cannot be considered free and open any more [41]. It is not visibility what it is needed, but other more fundamental values related to knowledge, free access and cross-fertilization.

In this paper, we have discussed three possible models of publishing, disseminating and improving scientific knowledge to implement the CC principles for science[33]: (i) LiquidPub Journals and Conferences (ii) the LICT-Repository; (iii) the Digital Publication Platform. Our proposal implies the reconstruction of a public space with several interfaces, to foster the recollection and dynamic rewriting of scientific works, adding a crowdsourcing perspective that maintains

---

[32] "[...] technical protection of IP in cyberspace (i.e. over networks) may protect property interests in digital works' more comprehensively than has ever been possible in physical space, and destroy many public interest elements in IP law in the process." [22]. The APEC (Asia-Pacific Economic Cooperation) economies have adopted what is now called the *APEC Privacy Framework* in 2005, which apparently has not changed very much the situation [19]

[33] http://sciencecommons.org/resources/readingroom/principles-for-open-science/ A quick reminder: (i) open access to literature, (ii) access to research tools, (iii) data in the public domain, (iv) and open cyber-structure.

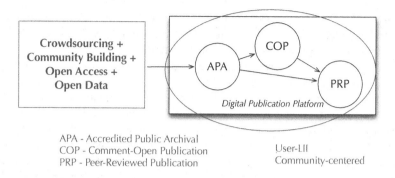

APA - Accredited Public Archival
COP - Comment-Open Publication
PRP - Peer-Reviewed Publication

User-LII
Community-centered

Fig. 4. Final resulting scheme: User-LII community-centered scheme

some balance between the private work of the authors and the collective profit that it can be produced out of it. The LiquidPub model is KSOs-centered, and therefore, editor and reputation-centered; Francesconi and Peruginelli's model is author-centered. We have tried to take the best of these two models to propose a community, user-centered model, shifting from the poles of the relation to a fast link between LIIs, users and authors, and authors themselves. The final resulting scheme is shown in Fig. 4.

Acknowledgments. Research partially funded by projects NextCBR (TIN2009-13692-C03-01) [co-funded with FEDER], ONTOMEDIA (CSO-2008-05536-SOCI; TSI-020501-2008-131) and SGR (CIRIT-2009SGR0688). The authors thank Sílvia Gabarró for her kind assistance.

References

[1] Al-Fedagli, S.: Intelligent privacy management. papers from the aaai spring symposium. In: Genesereth, M., Vogl, R., Williams, M. (eds.) Information Privacy and its Value. AAAI Press (2010)

[2] Baez, M., Birukou, A., Casati, F., Marchese, M.: Addressing information overload in the scientific community. IEEE Internet Computing 14, 31–38 (2010)

[3] Baez, M., Birukou, A., Chenu, R., Medu, M., Osman, N., Ponte, D., Sabater-Mir, J., Schneider, L., Turrini, M., Veltri, G., Wakeling, J., Xu, H.: State of the art in scientific knowledge creation, dissemination, evaluation, and maintenance (2009), http://eprints.biblio.unitn.it/archive/00001773/01/LiquidPub_D1-1_TechRep.pdf

[4] Baez, M., Casati, F., Birukou, A., Marchese, M.: Knowledge dissemination in the web era, http://eprints.biblio.unitn.it/archive/00001814/01/028.pdf

[5] Bourcier, D., Casanovas, P., Dulong de Rosnay, M., Maracke, C.: Intelligent Multimedia – Managing Creative Works in a Digital World. European Press Academic Publishing (2010)

[6] Bruce, T.R.: Tears shed over peer gynt's onion: Some thoughts on the constitution of public legal information providers. Journal of Information, Law and Technology (2) (2000)

[7] Bruce, T.: Public legal information: Focus and future. Journal of Information, Law and Technology (1) (2000)

[8] Bruce, T.: Foundings on the cathedral steps. In: Peruginelli, G., Ragona, M. (eds.) Law via the Internet. Free Access, Quality of Information, Effectiveness of Rights, pp. 411–422 (2009)

[9] Bynum, T.: Computer and information ethics. In: The Stanford Encyclopedia of Philosophy (Spring 2011 Edition). Stanford University (2011), http://plato.stanford.edu/archives/spr2011/entries/ethics-computer

[10] Carroll, M.W.: The movement for open access law - symposium. Lewis & Clark Law Review 10(4), 741–760 (2006)

[11] Casanovas, P.: The future of law: Relational law and next generation of web services. In: The Future of Law and Technology: Looking into the Future. Selected Essays, Legal Information and Communication Technologies Series, vol. 7, pp. 137–156. European Press Academic Publishing (2009)

[12] Casanovas, P., Poblet, M.: Justice via the internet: Hopes and challenges of law and the semantic web. In: Law via the Internet. Free Access, Quality of Information, Effectiveness of Rights. Series in Legal Information and Communication Technologies, vol. 5, pp. 347–359. European Press Academic Publishing (2009)

[13] Claffy, K.C.: Ten things lawyers should know about internet research (August 2008), http://www.caida.org/publications/papers/2008/lawyers_top_ten/

[14] Claffy, K.C.: The 3rd Workshop on Active Internet Measurements (AIMS-3) Report (2010), http://www.caida.org/publications/papers/2011/aims_report/

[15] Claffy, K.C.: The 2nd Workshop on Active Internet Measurements (AIMS2) Report (27010), http://ccr.sigcomm.org/drupal/files/p53-3v40n5h-polterockPS.pdf

[16] Delgado, J., Rodríguez, V. (eds.): 1st Workshop on Privacy and Protection in Web-based Social Networks. IDT Series, vol. 6. Ed. Huygens (2010)

[17] Fitzerald, A., Hooper, N., Fitzgerald, B.: The use of Creative Commons licensing to enable open access to public sector information and publicly funded research results: An overview of recent Australian developments. In: Intelligent Multimedia. Managing Creative Works in a Digital World, pp. 151–174. European Press Academic Publishing, Florence (2010)

[18] Francesconi, E., Peruginelli, G.: An open access policy for legal informatics scholarly research publishing. In: AICOL III, IVR-XXV World Conference in Philosophy of Law and Social Philosophy (2011)

[19] Genesereth, M.: APEC's privacy framework: A new low standard. annex c. to apf submission to senate legal & constitutional committee inquiry into the privacy act (1988)

[20] Greenleaf, G., Mowbray, A., King, G.: The AustLII Papers - New Directions in Law via the Internet. The Journal of Information, Law and Technology 2 (1997)

[21] Genesereth, M., Vogl, R., Williams, M. (eds.): Papers from the AAAI Spring Symposium on Intelligent Privacy Management(2010)

[22] Greenleaf, G.: Ip, phone home: The uneasy relationship between copyright and privacy, illustrated in the laws of Hong Kong and Australia. Hong Kong Law Journal 32(1), 35–81 (2002)

[23] Greenleaf, G.: AustLII's business models: Constraints and opportunities in funding free access to law. In: Law via the Internet. Free Access, Quality of Information, Effectiveness of Rights, pp. 423–437. European Press Academic Publishing (2009)

[24] Greenleaf, G.: 'Tabula Rasa': Ten reasons why australian privacy law does not exist. University of New South Wales Law Journal 24(1) (2001)

[25] Greenleaf, G., Chung, P., Mowbray, A., Chow, K.P., Pun, K.: The Hong Kong Legal Information Institute (HKLII): Its Role in Free Access to Global Law Via the Internet. Hong Kong Law Journal 32, 401 (2002)

[26] Hendler, J.A., Shadbolt, N., Hall, W., Berners-Lee, T., Weitzner, D.J.: Web science: an interdisciplinary approach to understanding the web. Commun. ACM 51(7), 60–69 (2008)

[27] Hietanen, H.: Collecting societies and creative commons licensing. In: Bourcier, D., Casanovas, P., de Rosnay, M.D., Maracke, C. (eds.) Intelligent Multimedia. Managing Creative Works in a Digital World, pp. 199–221. European Press Academic Publishing (2010)

[28] Howe, J.: The rise of crowdsourcing. Wired (2005),
http://www.wired.com/wired/archive/14.06/crowds.html

[29] Howe, J.: Walled gardens (2005), http://law.unh.edu/assets/pdf/
pierce-law-review-walled-gardens.pdf

[30] Hunter, D.: Cyberspace as place and the tragedy of the anticommons paradox. California Law Review 91, 439–520 (2003)

[31] Kazman, R., Chen, H.M.: The metropolis model a new logic for development of crowdsourced systems. Commun. ACM 52(7), 76–84 (2009)

[32] Kenneally, E., Claffy, K.: An internet data sharing framework for balancing privacy and utility. In: Engaging Data: First International Forum on the Application and Management of Personal Electronic Information. MIT (October 2009),
http://www.caida.org/publications/presentations/2009/engaging_data

[33] Lessig, L.: The architecture of access to scientific knowledge: Just how badly we have messed this up (2011), http://cdsweb.cern.ch/record/1345337

[34] Megnigbeto, E.: Information policy: Content and challenges for an effective knowledge society. The International Information & Library Review 42(3), 144–148 (2010),
http://www.sciencedirect.com/science/article/pii/S1057231710000470

[35] Network, T.C.T.: Communia: Manifesto for the public domain (2010),
http://communia-project.eu/

[36] Osman, N., Sierra, C., Sabater-Mir, J., Wakeling, J.R., Simon, J., Origgi, G., Casati, R.: Liquidpublications and its technical and legal challenges. In: Bourcier, D., Casanovas, P., de Rosnay, M.D., Maracke, C. (eds.) Intelligent Multimedia: Managing Creative Works in a Digital World, vol. 8, pp. 321–336. European Press Academic Publishing, Florence (2010)

[37] Parker, C.: Institutional repositories and the principle of open access: Changing the way we think about legal scholarship. New Mexico Law Review 37(2) (2007),
http://repository.unm.edu/handle/1928/12669

[38] Poulin, D.: Open access to law in developing countries. First Monday 9(12) (December 2004)

[39] Poulin, P.: Fifteen years of free access to law. In: Peruginelli, G., Ragona, M. (eds.) Law via the Internet. Free Access, Quality of Information, Effectiveness of Rights, pp. 15–32. European Press Academic Publishing (2009)

[40] Roig, A.: Privacy and social networks: From data protection to pervasive computing. In: Genesereth, M., Vogl, R., Williams, M. (eds.) Papers from the AAAI Spring Symposium on Intelligent Privacy Management. AAAI Press (2010),
http://www.aaai.org/ocs/index.php/SSS/SSS10/paper/view/1113

[41] Suber, P.: Nine questions for hybrid journal programs (2011),
http://www.earlham.edu/~peters/fos/newsletter/09-02-06.htm

[42] Whyte, W.F.: Advancing scientific knowledge through participatory action research. Sociological Forum 4(3), 367–385 (1989)

Combinations of Normal and Non-normal Modal Logics for Modeling Collective Trust in Normative MAS

Clara Smith[1,2], Agustín Ambrossio[1], Leandro Mendoza[2], and Antonino Rotolo[3]

[1] FACET, UCALP, Argentina
[2] Facultad de Informática, UNLP, Argentina
[3] CIRSFID, University of Bologna, Italy

Abstract. We provide technical details for combining normal and a non-normal logics for the notion of collective trust. Such combinations lead to different levels of expressiveness of the system. We give a possible structure for a combined model checker for one of the logic resulting from such combinations.

1 Motivation and Aims

Trust protection plays an important role in the law [15,21,18]. Such a protection is often and typically related to the problem of providing tools to support legally valid interactions between any kind of agents and/or to legally ground contractual transactions [15,21]. Indeed, trust protection is strongly implemented especially when agents' beliefs seem reasonable or when trustees' behaviour induces trusters' reliance. However, in multi-lateral agreement it is often the case that such reliance is mutual and this fact is relevant for trust protection. In particular, if agent x breaks group trust with regard to A, trust deception must be checked against the fact that x was supposed by the others to intend A, and x believed so.

Any computer application providing tools for detecting collective trust deception in the legal domain should require to develop

- a sound and rigorous formal analysis of the notion of collective trust,
- reasoning methods for computing when collective trust emerges and occurs in an arbitrarily large group of agents.

This general aim of this paper is to contribute to the above two research issues by significantly extending [19]'s results. In [19] Smith and Rotolo adopted [8]'s cognitive model of *individual* trust in terms of necessary mental ingredients which settle under what circumstances an agent x trusts another agent y with regard to an action or state-of-affairs, i.e. under which beliefs and goals an agent delegates a task to another agent. Using this characterization of individual trust, these authors provided a logical reconstruction of different types of *collective* trust, which for example emerge in groups with multi-lateral agreement, or which are the glue for grounding *in solidum* obligations raising from a "common front" of agents (for example, each member of the front can behave, in principle, as creditor or debtor of the whole). These collective cognitive states were characterized in [19] within a multi-modal logic based on [3]'s axiomatization for collective beliefs and intentions combined with a non-normal modal logic for the operator Does for agency. Such a combination was based on the following assumptions:

M. Palmirani et al. (Eds.): AICOL Workshops 2011, LNAI 7639, pp. 189–203, 2012.

Observation 1 (Expressiveness of the system). *A formula like* $\text{Does}_i\mathscr{A}$ *means that "agent i brings it about that $\mathscr{A}$". In this setting, the* Does *modalities are always applied to atomic propositional constants representing single behavioural actions, as in e.g.* Does_x *PayBill (which is meant to stand for "agent x pays the bill"). In the theory under study, normal operators interact with the* Does *modality in a restricted one-way manner: agents' actions always appear as innermost operators within well-formed formulas, as in e.g.* $\text{Bel}_y(\text{Does}_x$ *PayBill) (which is meant to stand for "the agent y believes that agent x pays the bill"). This means that no modality can occur in the scope of* Does.

Observation 2 (Semantics). *[19, Definition 2] proposed for the above mentioned system a semantics embedding standard multi-modal Kripke semantics for mental states into a Scott-Montague (multi-relational) semantics for* Does *[11].*

These combination and semantic embedding were assumed correct because Kripke semantics can be seen as a special case of multi-relational semantics. Although the referred concrete embedding appears to be straightforward in [19], it is worth pointing out that some basic results such as completeness and decidability are not immediately obvious and require some detailed technical machinery. This paper fills this gap by showing a simple way to prove those results, and also describes a model checking algorithm for that logic: the possibility of designing a model checker indicates that such logic can provide a feasible interpretation for norm-governed multi-agent systems and a method for computing collective trust.

The rest of the paper is organized as follows. Section 2 presents the main concepts of [19] and the proposed logical system. Section 3 reorganizes the multi-relational model in [19] as a particular combination of modal logics, which amounts to place the normal logics on top of the non-normal logics. For doing this, we first obtain two restrictions of the original logics. By exploiting results in regard to some techniques for combining logics, we prove that [19]'s system is complete and decidable. Hence, the sketch for an appropriate model checker is also outlined. Section 5 presents an independent combination of the normal and the non-normal counterparts of the base logics. This combination leads to an ontology of pairs of state-of-affairs which allows a structural basis for more expressiveness. For example, it is possible to write and test in the new ontology formulas such as $\text{Does}_i(\text{Does}_j(\text{Goal }\mathscr{A}))$. Some brief conclusions end the paper.

2   Background

There are situations where complex collective patterns are involved in social and legal interaction. Suppose that three agents x, y and z agree that some goal A should be jointly achieved. Some kind of coordination among them is of course required, but, minimally, such a multi-lateral agreement at least implies that each agent trusts that the others jointly intend to achieve A, and also believe in that. This simple agreement thus presumes a relatively elaborated collective trust background.

Collective trust and the corresponding delegation of tasks can be weak or strong [8]: weak delegation means that there are delegation situations which do not suppose any agreement, deal or promise at all, nor which yield to rights; strong delegation are the basis for promises, commitments and conventions. Since these forms of delegation support different degrees of trust intensity, different corresponding types of collective trust—joint trust, reliance, and collective trust—were introduced in [19] and can be illustrated as follows:

Example 1 (Joint trust). Suppose that agent y is at the bus stop, and there is a group G of people standing not at the bus stop but close to y, *expecting* that y will raise her hand and stop the bus.

Example 2 (Reliance). It is Mary's birthday. Her co-workers give some money to y, another co-worker who is going downtown, *relying* on y for the search and purchase of a gift. Everyone *trusts* that y will do so.

Example 3 (Collective trust). Student bands build up street-puppets filled with fireworks, which are to be burned on New Year's day. Each band builds its chosen puppet-of-the-year from scratch. The town administration institutionalized a competition and settled an award for the best figure. Bands' custom establishes that figures ought to be watched and protected day and night, this because a very common practice is to burn other bands' figures before the New Year's day by sending one band member (a saboteur). The consequence of successful sabotages is the exclusion of opponents from the competition. Assume the student band G *entrusts* its member s to burn H's puppet.

Accordingly, joint trust simply consists in the fact that all individuals in a group trust another agent for achieving their goal, reliance requires some mutual intentional coordination within the group, and collective trust assumes that the group is aware of such a coordinating effort to achieve a goal.

The following subsection outlines [19]'s logical framework, which will be our starting point for the subsequent sections.

2.1   The Logical Framework

The multi-modal language of [19] works with a finite set of agents $A = \{x, y, z, ...\}$ and a countable set of atomic propositional sentences usually denoted by $P = \{p, q, r, ...\}$. Complex expressions are formed syntactically from these in the usual inductive way using $\bot$ (*false*) and $\top$ (*true*), standard Boolean connectives, and the unary modalities we describe next.

The operator $\text{Goal}_x \, \mathscr{A}$ is used to mean that "agent x has goal $\mathscr{A}$", where $\mathscr{A}$ is a proposition. Propositions reflect particular state-of-affairs, as in [3]. $\text{Int}_x \, \mathscr{A}$ is meant to stand for "agent x has the intention to make $\mathscr{A}$ true". Intentions within the area of Cooperative Problem Solving (*CPS*) are viewed as inspiration for goal-directed activities. The doxastic (or epistemic) modality $\text{Bel}_x \, \mathscr{A}$ represents that "agent x has the belief that $\mathscr{A}$". The $\text{Does}_x \, \mathscr{A}$ operator is to be understood in the same sense given in Elgesem's account to represent successful agency, i.e. "x brings it about that $\mathscr{A}$" [5]. To simplify technicalities, the logic in [19] assumes that in expressions like $\text{Does}_x \, \mathscr{A}$ no modal operators occur in the scope of Does; therefore $\mathscr{A}$ denotes any behavioral action concerning a conduct, such as withdrawal, inform, purchase, payment, etc. We will assume the same restriction for Section 3, and we will eliminate it in Section 5 for regaining expressiveness.

As classically established [3], Goal is a K_n operator, while Int and Bel are, respectively, KD_n and $KD45_n$. The logic of Does, instead, is non-normal, it is closed under logical equivalence and amounts to the following schemata [5,11]: $\mathrm{Does}_x \, \mathscr{A} \to \mathscr{A}$, $(\mathrm{Does}_x \, \mathscr{A} \wedge \mathrm{Does}_x \, \mathscr{B}) \to \mathrm{Does}_x(\mathscr{A} \wedge \mathscr{B})$, $\neg \mathrm{Does}_x \top$, and $\neg \mathrm{Does}_x \bot$.

Remark 1. The main difference between [19]'s logic (let us call it $\mathfrak{F}$) and [3]'s system is that $\mathfrak{F}$ embeds Does and introduces new (non-primitive) operators defined on the basis of the [3]'s ones. First, $\mathfrak{F}$ defines the single-agent trust operator Trust (an agent x trusts another agent y with respect to a state of affairs ϕ) as follows:

$$\mathrm{Trust}_x^y \phi \equiv \mathrm{Goal}_x \phi \wedge \mathrm{Bel}_x \mathrm{Does}_y \phi \wedge \mathrm{Int}_x(\mathrm{Does}_y \phi \wedge \neg \mathrm{Does}_x \phi) \wedge \mathrm{Goal}_x \mathrm{Int}_y \phi \wedge \mathrm{Bel}_x \mathrm{Int}_y \phi \quad (1)$$

If G is a group of agents, the other derived operators of [19] are introduced to capture joint trust, reliance, and collective trust, respectively (see Examples 1, 2, and 3):

$$\mathrm{JTrust}_y^G A \equiv \left(\bigwedge_{i \in G} \mathrm{Trust}_y^i A \right) \quad (2)$$

$$\mathrm{Rel}_y^G A \equiv \mathrm{JTrust}_y^G A \wedge \mathrm{MInt}_G(\mathrm{JTrust}_y^G A) \quad (3)$$

$$\mathrm{CTrust}_s^G A \equiv \mathrm{Rel}_s^G A \wedge \mathrm{CBel}_G(\mathrm{Rel}_s^G A) \quad (4)$$

where the axiomatizations for MInt (mutual intention) and CBel (common belief) are those proposed in [3]:

$$\mathrm{MInt}^G A \equiv \left(\bigwedge_{i \in G} \mathrm{Int}^i(A \wedge \mathrm{MInt}^G A) \right) \qquad \mathrm{CBel}^G A \equiv \left(\bigwedge_{i \in G} \mathrm{Bel}^i(A \wedge \mathrm{CBel}^G A) \right)$$

3   Combining the Logics by Modalization/Temporalization

In this section we show how to characterize the logic of [19] as the combination of the component logics (the logic of Does and the normal component of [19]'s system) using the so-called temporalization/modalization techniques.

Before reorganizing the logic of [19] in this way, we recall some background knowledge. As is well-known, Scott-Montague semantics is a generalization of the traditional Kripke semantics [12]. Instead of a collection of worlds connected to a given world w through a relation R, consider a set of collections of worlds connected to w. These collections are the *neighbourhoods* of w. Formally, a Scott-Montague frame is an ordered pair $\langle W, N \rangle$ where W is a set of worlds and N is a function assigning to each w in W a set of subsets of W (the neighbourhoods of w). A Scott-Montague model is a triple $\langle W, N, V \rangle$ where $\langle W, N \rangle$ is a Scott-Montague frame and V is a valuation function defined as for Kripke frames, except for $\Box \mathscr{A}$: it is true at w iff the set of elements of W where $\mathscr{A}$ is true is one of the sets in $N(w)$; i.e., iff it is a neighbourhood of w.

Let us bring in the structure discussed in [19]. It is a multi-relational frame of the form [11]:

$$\mathfrak{F} = \langle A, W, \{B_i\}_{i \in A}, \{G_i\}_{i \in A}, \{I_i\}_{i \in A}, \{D_i\}_{i \in A} \rangle$$

where:

- A is the finite set of agents;
- W is a set of situations, or points, or possible worlds;
- $\{B_i\}_{i \in A}$ is a set of accessibility relations wrt Bel, which are transitive, euclidean and serial;
- $\{G_i\}_{i \in A}$ is a set of accessibility relations wrt Goal, (standard K_n semantics);
- $\{I_i\}_{i \in A}$ is a set of accessibility relations wrt Int, which are serial; and
- $\{D_i\}_{i \in A}$ is a family of sets of accessibility relations D_i wrt Does, which are point-wise closed under intersection, reflexive and serial [11].

A model based on $\mathfrak{F}$ is in its turn of the form $\langle \mathfrak{F}, V \rangle$, where V is the corresponding valuation function ([19, Definition 2]). Notice that [11] proved that Scott-Montague and multi-relational semantics are equivalent for the propositional case, so they can be interchangeably used.

Put this way, it is easy to identify two overlapping "nets" of relations over the same set W. The first net (or multi-graph) corresponds to "wires" for normal operators, the second net corresponds to the accessibility relations for the Does$_i$ modalities [1].

Following, we can assert two facts based on Definition 4.24 and Theorem 4.22 in [2] (which respectively settle how to construct a canonical model for a normal logic, and state that a normal modal logic is strongly complete with respect to its canonical model). First, that the modal similarity type built up from the normal modalities above has a canonical model; second, that this logic is complete w.r.t. its canonical model. Let us call **N** the logic with signature (Bel, Int, Goal) above (the normal modalities); hence **N** is a normal multi-modal multi-agent logic, which is complete (this proof is available in [1], we also sketch it in the Appendix).

Taking into account Observation 1 and what was stated regarding **N**, and according to the definition of temporalization given by Finger and Gabbay [9], (see also [10]) the system in [19] can be seen as a combination of logics where the normal modal machinery is placed on top of the non-normal logic. The non-normal equipment is in its turn multi-modal, as there is one Does$_i$ modality for each agent i. Indeed, [9]'s techniques were originally designed for temporalizing logics and are a special case of the modalization ones [6], which simply use the same intuition with the aim of externally applying *any* (even non-normal) modal logic to *any* generic logic system[2]. The advantage of this approach is that the resulting logic obtained from the combination is complete and decidable if both its components are, too.

Let us develop this insight.

---

[1] This definition does not include the Obl modality for obligations. Obl was originally incorporated in $\mathfrak{F}$ for dealing with the deontic connotation of an operator of the theory [19, sec. 4]. We will omit it in what follows to keep the set of modalities manageable. We come back to Obl later with the purpose to showing further possibilities for combining logics (Section 6).

[2] It has been very recently proved that modalization/temporalization techniques used in this paper are simple instances of non-iterated asymmetric importing and fibring techniques [17,16].

Consider $\mathfrak{F}$ as a split into an outer normal multi-modal frame, and inner Scott-Montague frames.

Provided this rearrangement, the intuition behind the valuation of formulas within the system is the following. When we evaluate normal operators (e.g. we parse a formula) we navigate through the outer Kripke model. When a $Does_i$ formula at any given point w is to be evaluated, we navigate through a Scott-Montague model.

The following subsection presents the technical aspects of the temporalization/modalization.

3.1 Modalization: Syntax and Semantics

Take the logic in [19]. Call **N** the restriction of $\mathfrak{F}$ to its normal part, and call Does the restriction of $\mathfrak{F}$ to its non-normal part. We can safely assume that Does is a propositional logic [11]. According to the methodology in [10], we partition the set of formulas in Does into two subsets: Boolean formulas, BDoes, and monolithic formulas, MDoes. A formula $\mathscr{A}$ belongs to BDoes if its outermost operator is a Boolean connective (e.g. $Does_x \mathscr{A} \wedge Does_x \mathscr{B}$); otherwise it belongs to MDoes (e.g. $Does_x \mathscr{A}$). It is clear that there is no intersection among the set of modalities of **N** and Does. Call **N**(Does) the modalization of Does by means of **N**.

N(Does): Syntax Let $\mathscr{L}_{Does}$ denote the language of the logic of agency (with no normal modalities and without their syntax formation rules), and $\mathscr{L}_N$ denote the language of **N** (without the Does modality and its syntax formation rule). The language $\mathscr{L}_{N(Does)}$ of N(Does)—over the set of proposition letters P—is obtained by replacing the formation rule of sentences in $\mathscr{L}_N$ that says "every proposition letter in P is a formula" by the formation rule:

every monolithic formula in $\mathscr{L}_{Does}$ is a formula

As pointed out in [9], this replacement can be matched with a process called "fuzzling" or layering: formulas in the base system can be substituted for atoms of the top system.

To formally outline the semantics for the modalization, we need a reframing of models based on $\mathfrak{F}$ in terms of the restricted models.

A modalized model for **N**(Does) has the structure:

$$\langle A, W, \{B_i\}_{i \in A}, \{G_i\}_{i \in A}, \{I_i\}_{i \in A}, V', \{d_i\} \rangle$$

where:

- A is a finite set of agents;
- W is a set of points, or possible worlds;
- $\{B_i\}_{i \in A}$ is a set of accessibility relations wrt Bel, which are transitive, euclidean and serial;
- $\{G_i\}_{i \in A}$ is a set of accessibility relations wrt Goal;
- $\{I_i\}_{i \in A}$ is a set of accessibility relations wrt Int, which are serial;
- V' is the valuation function V restricted to the normal operators, defined as follows:

1. standard Boolean conditions;
2. $V'(w, \mathrm{Bel}_i \, \mathscr{A}) = 1$ iff $\forall v \in W$ (if wB_iv then $V'(v, \mathscr{A}) = 1$);
3. $V'(w, \mathrm{Goal}_i \, \mathscr{A}) = 1$ iff $\forall v \in W$ (if wG_iv then $V'(v, \mathscr{A}) = 1$);
4. $V'(w, \mathrm{Int}_i \, \mathscr{A}) = 1$ iff $\forall v \in W$ (if wI_iv then $V'(v, \mathscr{A}) = 1$); and

– each d_i is a total function mapping, for each world w in W, for each agent i, into a multi-relational model of the form:

$$\eta = \langle W, D_i, \mathrm{v} \rangle$$

where:
– W is the (same, original) set of worlds,
– D_i is a family of sets of accessibility relations D_i wrt agency regarding agent i, which are pointwise closed under intersection, reflexive and serial [11],
– v is V restricted to the non-normal operators. That is, the valuation function for agency that says that $\mathrm{Does}_i \, \mathscr{A}$ holds in w if and only if the set of worlds where $\mathscr{A}$ is true is one of the neighborhoods of w. Formally:

1. standard Boolean conditions;
2. $\mathrm{v}(w, \mathrm{Does}_i \, \mathscr{A}) = 1$ *iff* $\exists D_i \in D_i$ such that $\forall u(wD_iu$ iff $\mathrm{v}(u, \mathscr{A}) = 1)$.

Let us call $\mathscr{K}\mathscr{L}_{\mathrm{Does}}$ the set of models for $\mathscr{L}_{\mathrm{Does}}$, then $d_i \colon W \to \mathscr{K}\mathscr{L}_{\mathrm{Does}}$.

The above semantics instantiates the construction criteria of Definition 4.2 in [9] and their generalization in [6], and so corresponds to a case of temporalization/modalization.

N(Does): Semantics Given a model $\mathfrak{M}$, given $w \in W$, given V' valuation function in $\mathfrak{M}$, and given functions d_i, the semantics for **N**(Does) is obtained by replacing the clause for **N** that says

$$\mathfrak{M}, w \models p \; iff \; p \in V'(w), whenever \; p \in P$$

with the clause:

$$\mathfrak{M}, w \models \mathscr{A} \; iff \; d_i(w) \models \mathscr{A}, whenever \; \mathscr{A} \in \mathrm{MDoes}.$$

Note here that $\mathscr{A}$ has the form $\mathrm{Does}_i \, \mathscr{B}$, as $\mathscr{A}$ is a monolithic formula.

Once a formula has entered the "Does component" it cannot come back to the top level [10]. Accordingly, we cannot test the validity of statements such as $\mathrm{Does}_i(\mathrm{Goal}_j \, \mathscr{A})$ (which can be seen as capturing a form of persuasion: "agent i makes agent j have $\mathscr{A}$ as a goal"). We address a possible solution to this drawback in Section 5.

Notice also that we combine the logics in a rather plain way: there are no bridge axioms nor intricate interactions among modal operators. Therefore, soundness and completeness results are applicable as follows. Fix a finite number of agents to prevent possible infiniteness of the system. For the normal operators, apply the results in [1](see Appendix); for the logics of agency, apply [11]. The following theorem holds [6, Theorem 3]:

Theorem 1 (Temporalization/Modalization: Transfer of Complete Logics). *If* **N** *and Does are complete logics, so is* **N**(*Does*).

Hence, **N**(Does) is complete, too.

4   Computing Collective Trust

Any possible computation model for collective trust requires that the underlying logic is at least decidable. In this section we exploit the following result [6, Theorem 4]:

Theorem 2 (Temporalization/Modalization: Transfer of Decidable Logics). *If* **N** *and Does are complete and decidable, so is* **N**(*Does*).

Hence, we show that the logic **N**(*Does*) is also decidable by simply proving that the component logics (*Does* and **N**) are decidable. On account of this result, an algorithm for model checking is subsequently outlined.

4.1   Decidability

The logic for *Does* was proved in [11] to enjoy the finite model property and to be decidable. What about the logic **N**?

Also proving that **N** is decidable is not hard. Indeed, on account of [4], proving that **N** enjoys the finite model property trivially follows. Let us adjust the following definition introduced in [4]:

Definition 1 ([4]). *A set of formulas* Σ *closed for subformulas is closed if it satisfies the following properties:*

1. *if* $\mathrm{CBel}_G\phi \in \Sigma$ *then* $\mathrm{EBel}_G(\phi \wedge \mathrm{CBel}_G\phi) \in \Sigma$
2. *if* $\mathrm{EBel}_G\phi \in \Sigma$ *then* $\{\mathrm{Bel}_i\phi | i \in G\} \subseteq \Sigma$
3. *if* $\mathrm{MInt}_G\phi \in \Sigma$ *then* $\mathrm{EInt}_G(\phi \wedge \mathrm{MInt}_G\phi) \in \Sigma$
4. *if* $\mathrm{EInt}_G\phi \in \Sigma$ *then* $\{\mathrm{Int}_i\phi | i \in G\} \subseteq \Sigma$
5. *if* $\mathrm{JTrust}_y^G\phi \in \Sigma$ *then* $\{\mathrm{Trust}_y^i\phi | i \in G\}$
6. *if* $\mathrm{Rel}_y^G\phi \in \Sigma$ *then* $(\mathrm{JTrust}_y^G\phi \wedge \mathrm{MInt}_G(\mathrm{JTrust}_y^G\phi)) \in \Sigma$
7. *if* $\mathrm{CTrust}_s^G\phi in\Sigma$ *then* $(\mathrm{Rel}_s^G\phi \wedge \mathrm{CBel}_G(\mathrm{Rel}_s^G\phi)) \in \Sigma.$

Since we omit in **N** the operator *Does*, JTrust is defined here in terms of individual beliefs, intentions and goals. Rel is the mutual intention of JTrust, and CTrust is the common belief of Rel. On account of this simple observation, we can exactly proceed as done in [4] and establish the following result:

Lemma 1 ([4]). *Given a model*

$$\mathcal{M} = \langle W, \{B_i | i \in A\}, \{G_i | i \in A\}, \{I_i | i \in A\}, Val \rangle$$

let Σ *be a closed set of formulas and*

$$\mathcal{M}_\Sigma^f = \langle W^f, \{B_i^f | i \in A\}, \{G_i^f | i \in A\}, \{I_i^f | i \in A, \}, Val^f \rangle$$

be defined as follows:

- $W^f = W / \equiv_f^\Sigma$, $Val^f(a, [w]) = Val(a, w)$;
- $B_i^f = \{([w], [v]) | \forall \mathrm{Bel}_i\phi \in \Sigma, \mathcal{M}, w \models \mathrm{Bel}_i\phi \Rightarrow \mathcal{M}, v \models \phi, \forall X_i\phi \in \Sigma, \mathcal{M}, w \models X_i\phi \Leftrightarrow \mathcal{M}, v \models X_i\phi$ *where* $X \in \{\mathrm{Bel}, \mathrm{Goal}, \mathrm{Int}\}\}$;

- $G_i^f = \{([w],[v]) \mid \forall \text{Goal}_i \phi \in \Sigma, \mathscr{M}, w \models \text{Goal}_i \phi \Rightarrow \mathscr{M}, v \models \phi, \forall \text{Int}_i \phi \in \Sigma, \mathscr{M}, w \models$
  $\text{Int}_i \phi \Rightarrow \mathscr{M}, v \models \phi\};$
- $I_i^f = \{([w],[v]) \mid \forall \text{Int}_i \phi \in \Sigma, \mathscr{M}, w \models \text{Int}_i \phi \Rightarrow \mathscr{M}, v \models \phi\}.$

The model $\mathscr{M}_\Sigma^f$ thus defined is a filtration of $\mathscr{M}$ through Σ.

From Lemma 1, it is an almost standard result to prove that the logic has the final model property and its satisfiability problem is decidable [4]. Due to the same reasons discussed in [4], also for each satisfiable formula ϕ of the logic **N** we can build a satisfying model of at most the size $O(2^{|\phi|})$, which however indicates that the following model checking algorithm has an exponential time complexity.

4.2 Model Checking

A model checker is a program that solves the model checking problem. The global model checking problem for **N(Does)** consists in checking whether, given a formula φ, and given $\mathfrak{M}$ model for **N(Does)**, there exists a $w \in W$ such that $\mathfrak{M}, w \models \varphi$. We follow the modal model checker construction of [10]. Let φ be a formula and let $\text{MM}\mathscr{L}_{\text{Does}}(\varphi)$ be the set of *maximal monolithic subformulas of* φ belonging to $\mathscr{L}_{\text{Does}}$. Let φ' be the **N**-formula obtained by replacing every subformula $\alpha \in \text{MM}\mathscr{L}_{\text{Does}}(\varphi)$ by a new proposition letter p_α. Below are the sketches of the model-checkers needed to solve the modal checking problem for **N(Does)**[3]:

Function $MC_{\mathbf{N(Does)}}((A, W, B_i, G_i, I_i, V', \{d_i\}), \varphi)$
    input: a modalized model $\mathfrak{M}$ and a formula $\varphi \in \mathscr{L}_{\mathbf{N(Does)}}$
    compute $\text{MM}\mathscr{L}_{\text{Does}}(\varphi)$
    for every $\alpha \in \text{MM}\mathscr{L}_{\text{Does}}(\varphi)$
        $i :=$ *identify the agent involved in* α
        for every $w \in W$
            $if(MC_{\text{Does}}(d_i(w), \alpha) = true)$ *then*
                $V'(w) := V'(w) \cup \{p_\alpha\}$ /*fuzzling*/

    build up φ' /* *systematically replace variables generated above* */
    return $MC_{\mathbf{N}}((A, W, B_i, G_i, I_i, V', \{d_i\}), \varphi');$/**calls to the normal checker**/

Function $MC_{\text{Does}}(d_i(w), \alpha)$
    input: a Scott-Montague model of structure η and
    a maximal monolithic sub-formula α.
    while there are neighbourhoods unchecked in $d_i(w)$
        $n_k = set \ n_i \in d_i(w)$ /**n_k iterates on the set of neighbourhoods**/
        for every $w \in n_k$
            if $\alpha \notin v(w)$ *then return false*
    return true

---

[3] To simplify the notation and have a more compact layout, we assume to work below in MC_{Does} with equivalent Scott-Montague models for Does and not with multi-relational ones. This assumption is non-problematic, since these semantics are equivalent.

$Function MC_N((A, W, B_i, G_i, I_i, V', d_i), \varphi')$
 input: a model $\mathfrak{M} = (A, W, B_i, G_i, I_i, V', d_i)$ and a formula φ'
 for every $w \in W$
  if $check((A, w, B_i, G_i, I_i, V'), \varphi')$
    return w
 return false

$Function\ check((A, w, B_i, G_i, I_i, V'), \alpha)$
 case on the form of α
  $\alpha = p_{\alpha'}$:
    if $p_{\alpha'} \notin V'(w)$
      return false
  $\alpha = \neg\alpha'$:
    if $check((A, w, B_i, G_i, I_i, V'), \alpha')$
      return false
  $\alpha = \alpha_1 \wedge \alpha_2$:
    if not $check((A, w, B_i, G_i, I_i, V'), \alpha_1)$ *or*
    or not $check((A, w, B_i, G_i, I_i, V'), \alpha_2)$
      return false
  $\alpha = \alpha_1 \vee \alpha_2$:
    if not $check((A, w, B_i, G_i, I_i, V'), \alpha_1)$ *and*
    and not $check((A, w, B_i, G_i, I_i, V'), \alpha_2)$
      return false
  $\alpha = \mathrm{Bel}_i(\alpha')$:
    for each v *such that* wB_iv
      if not $check((A, v, B_i, G_i, I_i, V'), \alpha')$
        return false
  $\alpha = \mathrm{Goal}_i(\alpha')$:
    for each v *such that* wG_iv
      if not $check((A, v, B_i, G_i, I_i, V'), \alpha')$
        return false
  $\alpha = \mathrm{Int}_i(\alpha')$:
    for each v *such that* wI_iv
      if not $check((A, v, B_i, G_i, I_i, V'), \alpha')$
        return false
  others : *return false*
 return true

The procedures should be understood as follows. Given a modalized model and a formula φ, $MC_{N(Does)}$ first computes the set $\mathrm{MM}\mathscr{L}_{Does}(\varphi)$ of maximal monolithic subformulas of φ. For each of these, the checker identifies which agent is carrying out the action. Then, the checker establishes the worlds where that action has been carried out successfully. For doing this, the MC_{Does} checker is called with the Scott-Montague model $d_i(w)$ as parameter (recall d_i has structure η). MC_{Does} is nothing but pseudo-code for the valuation function v, it tests whether there is a neighborhood of w where α holds.

If so, the new letter p_α is added to $V'(w)$ to register such successful agency. Finally, before calling the normal model checker MC_N, the new formula φ' is built without the Does modalities; these have been replaced in the former fuzzling.

5   Independent Combination of Mental States and Actions

$Does_i(Goal_j \mathscr{A})$ is a formula in which the normal modality appears within the scope of a non-normal Does. Note that, according to Observation 1, we cannot express this formula in the original system. An independent combination between a basic temporal and a simple deontic logic for MAS has been recently depicted in [20]. That combination puts together two normal modal logics: a temporal one and a deontic one.

Our aim now is to combine the normal and the non-normal counterparts of $\mathfrak{F}$ to get a new system where we can write and test the validity of formulas with arbitrarily interleaved cognitive and agency modalities.

For doing this, let us take a look to $\mathfrak{F}$ again. Consider it once more as a split into two separate substructures: one gathering the normal logics, and another one gathering the logics of agency. Again, there are two overlapping "nets" of relations identifiable over the same set W. The former is a Kripke-style cognitive ontology where goals, beliefs, intentions are interpreted, i.e., it captures internal (mental) motivational and informational aspects of agents (also the deontic aspects of the system, but recall that we do not explicitly consider them in this paper) the latter is a Scott-Montague structure which captures the external, visible, behavioral side of agents.

Now to the combination. First, duplicate and add subscripts to the elements in W to get one set of situations W_N, and another set W_D. Now build an ontology $W_N \times W_D$ of pairs (w_N, w_D).

Combination: Syntax Let $\mathscr{L}_N$ denote the language of **N** (the base logic restricted to the normal operators), and $\mathscr{L}_{Does}$ denote the language of the logic of agency. The language $\mathscr{L}_{N \times Does}$ is obtained by taking the union of the formation rules for the combination of $\mathscr{L}_N$ and $\mathscr{L}_{Does}$. Unlike the case of $\mathscr{L}_{N(Does)}$, $Does_i(Goal_j \mathscr{A})$ and $Goal_j(Does_i \mathscr{A})$ are both formulas of $\mathscr{L}_{N \times Does}$.

Combination: Semantics Assume that we have two structures: $(A, W_N, \{B_i\}, \{G_i\}, \{I_i\}, V',)$ and $(A, W_D, \{D_i\}, v)$, where to respectively test the validity of the normal modalities and the non-normal (Does) modalities. The former is a Kripke model; the latter a Scott-Montague model. Interpret $\mathscr{L}_{N \times Does}$ formulas over a combined model

$$\mathfrak{C} = (A, W_N \times W_D, \{B_i\}_{i \in A}, \{G_i\}_{i \in A}, \{I_i\}_{i \in A}, \{D_i\}_{i \in A}, V),$$

where:

- A is the set of agents;
- $W_N \times W_D$ is a set of pairs of situations;
- $\{B_i\}_{i \in A}, \{G_i\}_{i \in A}, \{I_i\}_{i \in A}$ are the accessibility relations for the normal operators (with semantics as in Section 3);
- $\{D_i\}_{i \in A}$ are the accessibility relations for the agency operators; and
- $V : W_N \times W_D \to Pow(P)$ is a function assigning to each pair in $W_N \times W_D$ the set of proposition letters in P which are true.

The definition of a formula in $\mathscr{L}_{N \times Does}$ being satisfied in a model $\mathfrak{C}$ at state (w_N, w_D) amounts to:

$$\mathfrak{C}, (w_N, w_D) \models Bel_i \; \mathscr{A} \; \textit{iff} \; \forall v_N \in W_N (\textit{if } w_N B_i v_N \textit{ then } \mathfrak{C}, (v_N, w_D) \models \mathscr{A}).$$

$$\mathfrak{C}, (w_N, w_D) \models Goal_i \; \mathscr{A} \; \textit{iff} \; \forall v_N \in W_N (\textit{if } w_N G_i v_N \textit{ then } \mathfrak{C}, (v_N, w_D) \models \mathscr{A}).$$

$$\mathfrak{C}, (w_N, w_D) \models Int_i \; \mathscr{A} \; \textit{iff} \; \forall v_N \in W_N (\textit{if } w_N I_i v_N \textit{ then } \mathfrak{C}, (v_N, w_D) \models \mathscr{A}).$$

$$\mathfrak{C}, (w_N, w_D) \models Does_i \; \mathscr{A} \; \textit{iff there exists a neighborhood } n \textit{ of } w_D \textit{ such that}$$
$$\forall v \in n \; (\mathfrak{C}, (w_N, v) \models \mathscr{A}).$$

A scan through the combined structure is done according to which operator is being tested. Normal operators move along the first component (w_N), and non-normal operators move along the second component of the current world (w_D).

Example 4 (Persuasion). The formula $Does_i(Goal_j \; \mathscr{A})$ can be seen as a form of persuasion, meaning that agent i makes agent j have $\mathscr{A}$ as goal. How do we test the validity of such a formula in a world (w_N, w_D)? The movements along the multi-graph are determined by $\mathfrak{C}, (w_N, w_D) \models Does_i (Goal_j \; \mathscr{A})$ iff $\exists$ neighbourhood n_i of w_D such that $\forall v_k \in n_i \; (\mathfrak{C}, (w_N, v_k) \models Goal_j \; \mathscr{A})$, which amounts to test $\forall v_k \in n_i$ (iff $\forall u_N \in W_N$ (if $w_N G_j u_N$ then $\mathfrak{C}, (u_N, v_k) \models \mathscr{A}$)).

6 Summary an Future Work

In this paper we have offered technical details for combining normal and a non-normal logics for modeling the notion of collective trust and for proving the completeness and decidability for the logic resulting form such a combination. Such combinations lead to different levels of expressiveness of the system by using temporalization and modalization techniques. On account of decidability results, we gave a possible structure for a combined model checker.

Let us consider three research issues for future work.

The Obl *modality.* We dealt with some of the modalities underlying the trust theory in [19]. In that work, a deontic connotation for the concept of collective trust is developed. Lawful support to collective trust is guaranteed in the theory with the schema: $(CTrust_y^G \; \mathscr{A}) \to Obl^G(Does_y \; \mathscr{A})$, which is devised with a view to reflect the lawful force of trust, relativized to groups. The schema is to be understood as a standard of (good faith) behavior that can be identified with reference to social or group norms, to correctness, or reasonableness: if the group trusts agent y with respect to $\mathscr{A}$, agent y is obliged to carry out $\mathscr{A}$.

For capturing this deontic connotation of CTrust, we must consider deontic modalities such as Obl and Obl^G. Obl is the deontic operator for generic obligations, meaning "it is obligatory that" [18,14], and Obl^G is a relativized obligation operator which is meant to stand for "it is obligatory in the interest of G that" (see e.g. [13]). If these deontic modalities have the usual accepted KD and KD_n semantics, this extension is almost trivial: it is sufficient to add appropriate accessibility relations to the frames for **N**.

Things can get more complex if we characterize the deontic operators in weaker (non-normal) systems and apply combination techniques with more than just one non-normal modal logic [7].

Complexity. The proposed logic, though decidable, is EXPTIME complete. [4] proposed some methods for reducing this complexity, such as bounding modal depth of formulas and bounding the number of propositional atoms. It is an interesting research issue to check if these techniques can be useful also in the present framework.

Further combinations. Theoretically speaking, the very idea of reasoning about time should extend the current framework. For example, a basic temporalization amounts to place the temporal machinery on top of the modalized system, just in the same spirit we placed the normal machinery on top of the non-normal one. Consider the model $(T, <, g, t_0)$. The outer frame $(T, <)$ corresponds to the temporal evolution of the system; t_0 in T is the initial point in time. The system evolves through time in the sense that new groups and generic/individual beliefs, intentions, trust relations, obligations are settled while some others become obsolete. In its turn, g is the total function that brings in a model $\mathfrak{M}$ for each point in time.

A   Completeness Proof for N

In this appendix, a completeness proof is sketched for the restriction **N**. The method used is often applied in modal logic for proving completeness with respect to finite models; is in turn inspired by the completeness proofs of mutual intentions shown by Dunin-Keplicz and Verbrugge in [3]. In fact, we adapt that to **N**, and apply Definition 4.24 and Theorem 4.22 described in [2]; these respectively settle how to construct a canonical model for a normal logic, and state that a normal modal logic is strongly complete with respect to its canonical model.

We have to prove that, supposing that $\mathbf{N} \not\vdash \varphi$, there is a model $\mathfrak{M}_\mathbf{N}$ and a $w \in \mathfrak{M}_\mathbf{N}$ such that $\mathfrak{M}_\mathbf{N}, w \not\models \varphi$. The proof has four steps:

Step 1: Closure Construct a finite set of formulas Φ called the closure of φ. Φ contains φ and all its sub-formulas, plus certain other formulas that are needed in Step 4 below to show than an appropriate valuation falsifying φ at a certain world can be defined. The set Φ is also closed under single negations.

The closure of φ with respect to **N** is the minimal set Φ of **N**-*formulas* such that, for every agent, the following hold (see also Definition 1):

1. $\varphi \in \Phi$.
2. If $\psi \in \Phi$ and χ is a sub-formula of ψ, then $\chi \in \Phi$;
3. If $\psi \in \Phi$ and Φ itself is not a negation, then $\neg\psi \in \Psi$;
4. If $\mathrm{MInt}_G(\psi) \in \Phi$ then $\mathrm{EInt}_G(\psi \wedge \mathrm{MInt}_G(\psi)) \in \Phi$;
5. If $\mathrm{EInt}_G(\psi) \in \Phi$ then $\mathrm{Int}_i\psi \in \Phi$ for all $i \in G$;

6. $\neg\text{Int}_i \perp \in \Phi$ for all $i \leq m$;
7. If $\text{CBel}_G(\psi) \in \Phi$ then $\text{EBel}_G(\psi \wedge \text{CBel}_G(\psi)) \in \Phi$;
8. If $\text{EBel}_G(\psi) \in \Phi$ then $\text{Bel}_i, \psi \in \Phi$ for all $i \in G$;
9. $\neg\text{Bel}_i \perp \in \Phi$ for all $i \leq m$;
10. $\neg\text{Goal}_i \perp \in \Phi$ for all $i \leq m$.

It should be clean that for every formula ϕ, Φ is a *finite* set of formulas (recall that the language in [19] includes: MInt, EInt, EBel).

Step 2: Canonical model. To construct a canonical model we need to define the worlds and relations between them. Each of these worlds are *maximally* **N**-*consistent* sets. To build this sets, we apply the Lindenbaum Lemma (which is proved in Lemma 4.17 [2]) over Φ *step 1*, as follows:

Let Φ be the closure of ϕ with respect to **N**. If $\Gamma \subseteq \Phi$ is **N**-*consistent*, then there is a set $\Gamma' \supseteq \Gamma$ which is maximally **N**-*consistent* in Φ.

Step 3: Build a canonical model using Definition 4.24 [2]. . This model will turn out to contain a world where $\neg\psi$ holds. Let $\mathfrak{M}_\varphi = <S_\phi, \pi, I_1, ..., I_m, B_1, ..., B_m, G_1, ..., G_m>$ be the Kripke model defined as follows:

- As domain of states, one state s_Γ is defined for each *maximally* **N**-*consistent* $\Gamma \subseteq \Phi$. Note that, because Φ is finite, there are only finitely many states. Formally, we defined $CON_\Phi = \{\Gamma | \Gamma$ is *maximally* **N**-*consistent* in $\Phi\}$ and $S_\phi = \{s_\Gamma | \Gamma \in CON_\Phi\}$.
- To make a truth assignment π, we want to conform to the propositional atoms that are contained in the maximally consistent sets corresponding to each world. Thus we define $\pi(s_\Gamma)(p) = 1$ if and only if $p \in \Gamma$. Note that this makes all propositional atoms that do not occur in φ false in every world of the model.
- The corresponding relations are defined as follows:

$$I_i = \{(s_\Gamma, s_\Delta) | \psi \in \Delta \text{ for all } \psi \text{ such that } \text{Int}_i(\psi) \in \Gamma\}$$
$$B_i = \{(s_\Gamma, s_\Delta) | \psi \in \Delta \text{ for all } \psi \text{ such that } \text{Bel}_i(\psi) \in \Gamma\}$$
$$G_i = \{(s_\Gamma, s_\Delta) | \psi \in \Delta \text{ for all } \psi \text{ such that } \text{Goal}_i(\psi) \in \Gamma\}$$

It will turn out that with this definition we get $\mathfrak{M}_\varphi, s_\Gamma \models p$ iff $p \in \Gamma$ for propositional atoms p.

Step 4: Completeness of N. If $\mathbf{N} \not\vdash \varphi$ then there is a model $\mathfrak{M}$ and a w such that $\mathfrak{M}, w \not\models \varphi$. Proof: Suppose $\mathbf{N} \not\vdash \varphi$. Take $\mathfrak{M}_\varphi$ as in step 3. Note that there is a formula χ logically equivalent to $\neg\varphi$ that is an element of Φ; if ψ does not start with a negation, χ is the formula $\neg\varphi$ itself. Now, using the Lindenbaum Lemma, there is a maximally consistent set $\Gamma \subseteq \Phi$ such that $\chi \in \Gamma$. By the Finite Truth Lemma, if $\Gamma \in CON_\phi$ then for all $\psi \in \Phi$ it holds that $\mathfrak{M}_\varphi, s_\Gamma \models \psi$ iff $\psi \in \Gamma$. Thus, this implies that $\mathfrak{M}_\varphi, s_\Gamma \models \chi$, thus $\mathfrak{M}_\varphi, s_\Gamma \not\models \varphi$. Details of the Finite Truth Lemma proof are left to the reader (see [3] and [2], Lemma 4.21).

References

1. Ambrossio, A., Mendoza, L.: Completitud e implementación de modalidades en MAS. Thesis report, Facultad de Informatica UNLP (2011)
2. Blackburn, P., de Rijke, M., Venema, Y.: Modal Logic. Cambridge Tracts in Theoretical Computer Science, vol. 53. Cambridge University Press, Cambridge (2001)
3. Dunin-Keplicz, B., Verbrugge, R.: Collective intentions. Fundam. Inform., 271–295 (2002)
4. Dziubinski, M., Verbrugge, R., Dunin-Keplicz, B.: Complexity issues in multiagent logics. Fundam. Inform. 75(1-4), 239–262 (2007)
5. Elgesem, D.: The modal logic of agency. Nordic Journal of Philosophical Logic 2, 1–46 (1997)
6. Fajardo, R., Finger, M.: Non-normal modalisation. In: Advances in Modal Logic 2002, pp. 83–96 (2002)
7. Fajardo, R., Finger, M.: How not to combine modal logics. In: IICAI, pp. 1629–1647 (2005)
8. Falcone, R., Castelfranchi, C.: Social Trust: A Cognitive Approach. In: Trust and Deception in Virtual Societies, pp. 55–90. Kluwer Academic Publishers (2001)
9. Finger, M., Gabbay, D.: Combining temporal logic systems. Notre Dame Journal of Formal Logic 37 (1996)
10. Franceschet, M., Montanari, A., De Rijke, M.: Model checking for combined logics with an application to mobile systems. Automated Software Eng. 11, 289–321 (2004)
11. Governatori, G., Rotolo, A.: On the Axiomatization of Elgesem's Logic of Agency and Ability. Journal of Philosophical Logic 34(4), 403–431 (2005)
12. Hansson, B., Gärdenfors, P.: A guide to intensional semantics. Modality, Morality and Other Problems of Sense and Nonsense. Essays Dedicated to Sören Halldén (1973)
13. Herrestad, H., Krogh, C.: Deontic logic relativised to bearers and counterparties. In: Bing, J., Torvund, O. (eds.) Anniversary Antology in Computers and Law, TANO (1995)
14. Jones, A., Sergot, M.: A logical framework. In: Open Agent Societies: Normative Specifications in Multi-agent Systems (2007)
15. Memmo, D., Sartor, G., di Cardano, G.Q.: Trust, Reliance, Good Faith, and the Law. In: Nixon, P., Terzis, S. (eds.) iTrust 2003. LNCS, vol. 2692, pp. 150–164. Springer, Heidelberg (2003)
16. Rasga, J., Sernadas, A., Sernadas, C.: Fibring as biporting subsumes asymmetric combinations. Preprint, SQIG - IT and IST - TU Lisbon, 1049-001 Lisboa, Portugal (2010)
17. Rasga, J., Sernadas, A., Sernadas, C.: Importing logics. Studia Logica (2011)
18. Rotolo, A., Sartor, G., Smith, C.: Good faith in contract negotiation and performance. IJBPIM Journal of Business Processes and Management, Special Issue on Contract Architectures and Languages 4, 154–173 (2009)
19. Smith, C., Rotolo, A.: Collective trust and normative agents. Logic Journal of IGPL 18(1), 195–213 (2010)
20. Smith, C., Rotolo, A., Sartor, G.: Representations of time within normative MAS. Frontiers in Artificial Intelligence and Applications 223, 107–116 (2010)
21. Whittaker, S., Zimmermann, R.: Good Faith in European Contract Law. Cambridge University Press, Cambridge (2000)

Software Agents as Boundary Objects

Migle Laukyte

CIRSFID, University of Bologna, via Galliera 3, 40121 Bologna, Italy
migle.laukyte@unibo.it

Abstract. Despite the wide use of agent-based applications in different areas of human activity, there hasn't been paid much attention to understand how these applications are possible, taking into account that they are build by people coming from such conceptually distant fields of study as, for example, law, artificial intelligence, and software engineering. This paper aims to fill in this gap addressing the different approaches to software agents—understood as building blocks of agent-based applications—adopted in each of these fields of study and suggesting that the way to understand how do these fields manage to work together in building a single agent-based application resides in seeing these agents as boundary objects.

Keywords: law, artificial intelligence, software agent, multi-agent system, boundary object, software engineering.

1 Introduction

Software agents[1] have been around for a while: perhaps no mistake will be made to date their appearance among computer science's topics in the late 1970s or early 1980s. From then on, software agents have been (and still are) studied, developed and discussed by the scholars of artificial intelligence, robotics, psychology, neurosciences, software engineering, and—the last but surely not the least—law.

Obviously, the nature, interests and aims of each of these scientific fields have led each of these fields to build a specific idea of what a software agent is. I am not pretending—alas!— to contribute in any way to the descriptions of agents developed in any of the fields listed above: what I instead would to like to do is to suggest to step back and take a wider perspective through which to look at these numerous agents around. This perspective, based on the idea of boundary objects, explains how it is possible to put together so many agents and involve as much many scientific fields in building a single agent-based application, and succeed in doing so. Hence, the main thesis of this paper is that the key to understand how the existence of so many agents cannot hinder, but on the contrary, can only enhance the development of a single agent-based application, is to conceive software agents as boundary objects.

To illustrate this idea I have organized this paper as follows: in Section 2, I describe how software agents are conceived in artificial intelligence (AI), while in

---

[1] I must clarify from the outset that I use software agent as a generic term covering all agents that come under this heading, thereby designating electronic agents, intelligent agents, artificial agents, and so on.

M. Palmirani et al. (Eds.): AICOL Workshops 2011, LNAI 7639, pp. 204–216, 2012.
© Springer-Verlag Berlin Heidelberg 2012

Section 3 I focus on software agents in software engineering (SE).[2] Section 4 then discusses how legal scholars understand software agents, while Section 5 offers a general introduction to the notion of boundary object used in sociology, and explores how this notion applies to our case, that is, to software agents in AI, SE and law. Finally, I conclude by explaining the benefits of considering software agents as boundary objects in developing agent-based applications.

2   Software Agents in AI

Software agents can be described in AI as rational-intentional agents based on the BDI model.[3] As AI suggests in its very name (artificial intelligence), AI agents are intended to artificially model natural intelligence. Human-intelligent agents, however, have not yet been created: what researchers have managed to develop are rational agents whose rational computing capacity is enriched with the human attributes of belief, desire, and intention, which gives the so-called Belief-Desire-Intention (or BDI) software model developed by Michael Bratman in the late 1980s ([2]).

Let us consider each of the model's three mental states in turn: beliefs are what an agent knows about the world, and are a necessary basis on which an agent figures out how to act in its own best interest. Beliefs are necessary since we couldn't work out any plan and couldn't even form intentions if we didn't believe anything to be the case. Also, an agent's beliefs need to reflect its changing environment, because that is how the agent responsively adapts to it. And if the environment is populated with other BDI agents, belief also makes it possible to infer what their beliefs and intentions might be, and hence what their behavior will be like.

Desires and intentions are close kin, in the sense that they both describe states of affairs the agent would like to come true. Desires and intentions thus enable an agent not so much to respond to its environment (because that is what beliefs are for) as to act on and modify it. Intentions then are a subset of an agent's desires: the latter are whatever an agent would wish to see in an ideal construction of the world, if only the world would correspond to its imaginings, while the former, by contrast, are those specific desires an agent has committed to.[4] They can thus be understood as

---

[2] My assumption here is that AI and software engineering are two of many disciplines in computer science. I am aware that there is no common position in this regard (see, for example, [1]), but the question as to how these disciplines are best classified falls outside the scope of this discussion, and for the purposes of this paper AI and software engineering will be sometimes be referred to by the general term of computer science.

[3] This is not to say, however, that BDI is the only agent model available in AI.

[4] This interestingly echoes the *legal* distinction between motive an intent: "intent is a state of mind [or *mens rea*] preceding or accompanying the act. Motive is the overall goal [good or bad] that prompts a person's actions" [5]. Which is to say that when we have a motive (e.g., revenge) we have not thereby also formed the intent to commit a specific act by which to satisfy that motive (a specific way to carry out the revenge, e.g., destroying so-and-so's property)—exactly as in BDI, which can thus be said to closely model the way the *law* views our decision-making. This is an example where different communities are using different names for the same concept (here the same distinction between the states of mind that move one to action). And if we only could see that the distinction between desire and intention in BDI is operationally the same as that between motive and intent in law, then we will at least have a platform on which to work in enabling the two communities to work together on future projects.

designating currently chosen courses of action [4], and that makes them deliberative in two respects, in that (i) deliberation is required to form an intention (or pick out a desire to be achieved), and (ii) once an intention has been formed, it will constrain an entire range of subsequent decisions. Intentions can thus be understood as a deliberative tool that simplifies decision-making by constraining all further reasoning and possibilities for action.

We have just seen one of the ways in which software agents can be understood in AI, namely, as entities whose rationality is enriched with beliefs, desires, and intentions. This combination gives them some autonomy, enabling them to decide on their own and act independently (at least to some extent). Clearly enough the scholars of AI study, address and develop in agents the qualities and characteristics that describe their area of research: this is why AI focuses on agent's mental states, such as intentionality or beliefs, and presents us with an idea of agent which vaguely manages to mimic some of our abilities, but is not yet ready to compete with us.

The observation to be made at this point is that even though researchers in AI work closely with software engineers, the latter do not share with the former this approach to software agents. So let us see how software engineers understand software agents.

3   Software Agents in Software Engineering

Software agents in software engineering (SE) are abstractions. I shall now have something to say about abstraction such as it relates to SE, and I will then explain how software agents fit into this concept. Then (in Section 3.1), I will illustrate these ideas by considering the use of abstractions in a specific SE context.

First of all, SE is based on abstractions: they capture content and knowledge and act as recipes on which basis to carry out different tasks, thereby simplifying them and enabling us to solve problems without requiring us to do work that has already been done. Practically this means, as engineers argue themselves, "you have to write less code and make fewer errors, and can therefore tackle more complex problems in the same time-scale" [6].

The reason for so abstracting—by simplifying and finding common elements—is to help software engineers manage the complexity involving the design and management of software systems, especially those representing a new approach in SE, namely the ones based on software agents and coined under the name of multi-agent systems (MASs) (see [7], [8], [9], and [10]). While software agents are indeed abstractions, they come in many varieties (for an overview and classification, see [11]), and this makes it necessary to investigate such agents in context: therefore we need to look at specific uses of agents to see what roles they play as abstractions in the agent-based approach to SE, or agent-oriented SE (AOSE).

3.1   Software Agents as Abstractions in AOSE

Agent-oriented SE (AOSE) is concerned with setting out approaches and methodologies on which basis to develop agent-based systems, understood as ones "in which the key abstraction used is that of an agent" [12]. An agent in such a system can be understood as autonomous and capable of interacting with other agents to satisfy the system's design objectives: these objectives then require the agents to have additional properties, such as pro-activeness (agents can act on their own accord rather than just in response to the environment), reactivity (agents perceive their environment through sensors and can appropriately respond to changes in the environment in timely fashion), situatedness (agents inhabit an physical or virtual environment), and social ability (agents never exist in isolation but always form part of a society of agents with which they interact).

Each of these properties is itself an abstraction: for example, autonomy cannot be modeled simply by matching input to output, because in this way we would limit agent to a finite set of options, and this is certainly not what we would call autonomy (an ability to make choices on one's own). So instead we need to abstract from these items and set out a model on which basis the agent can reason and plan using the information or knowledge at its disposal.

An example illustrating the role of agents in a MAS is the SODA methodology (Societies in Open and Distributed Agent Spaces, online at http://www.apice.unibo.it/xwiki/bin/view/SODA/), which brings out two ways in which agents figure as abstractions in a MAS. First, in a MAS architecture, the entire design revolves around agents, since all of its components (its non-agent abstractions, such as actions, workspaces, operations, or resources) are meant to enable agents to work (perform actions) and to interact (form into an agent society) in a way that is functional to the system's overall purpose. Agents so considered might be called functional or architectural abstractions, or again system-component abstractions.

And, second, SODA also shows a sense in which agents are abstractions as role-playing agents: since a role is itself an abstraction, it can be structured in any number of ways, but it should not identify the individuals filling that role or the specific actions that satisfy it. And since agents as roles need not specify who is acting in their respective roles, they function as abstractions. Similarly, roles may be deemed abstract as metaphors by which to say that agents act on behalf of their human users.[5] Agents so considered might be called representative abstractions, or abstractions by proxy.

In this section we considered software engineers' approach to agents, which differs from the approach of those who work in AI: software engineers see agents not as slight reproductions of humans (or at least of some human capacities), but as abstract entities which are part of a overall interactive and abstract whole of elements, each driven by its goal and all aiming to achieve a single objective. But as much as these two approaches may differ, this difference will not be as big as the one that exists between these two approaches (AI and SE) and the legal one. Hence, the next section is dedicated to describe how law conceives software agents.

4   Software Agents in Law

As already mentioned briefly in the introductory part of this paper, software agents are studied not only in computer science, but also in social sciences: here we will be concerned with the law's approach to software agents, making the case that software agents in law can hypothetically be understood as entities whose actions may bring

---

[5] This is another instance where we can strike an analogy with the legal understanding of the concept we are dealing with, here that of agency, defined in law as "a legal relationship in which one person represents another and is authorized to act for him or her" [4].

about legally relevant consequences. For this reason it has become a matter of discussion how such agents ought to be classified and treated. In fact, legal scholars[6] seem to no longer doubt *whether* autonomous and intelligent agents will be among us, and have accordingly begun to prepare for the moment *when* this will happen. The issues to be solved are many, but they prominently include that of the rights and liabilities agents should have. But let us take the whole question from the beginning.

From the legal point of view, agents can be considered in either of two ways: (a) as goods, that is, software programs (standard approach), or (b) as legal persons, that is, as entities recognized as having a capacity to act in legally relevant ways. This means that the agents have rights and duties, such as the right to sue and be sued, the right to enter into contracts, and the right act on another's behalf, just as if they were real persons (nascent speculative approach).

On the current approach (a), the accent in the discussion on software agents is placed on the word *software* rather than on the word *agents*, meaning that software agents are treated as software products: they are accordingly protected by intellectual property law, and in particular by either copyright or patent law, depending on the characteristics of the application they are used in.[7]

But we will be concerned here with the second approach (b), on which the accent falls on the word *agent* (rather than on *software*), meaning that software agents would be treated not as products, but as creatures that can deliberate about what to do and can act on that decision. This, of course, describes a fully human agent (a self-conscious one), and it is because software agents cannot yet be described as an artificial equivalent of agency so conceived that I am calling this approach speculative.[8]

So the first question, as we enter into this speculative approach, is how do legal scholars conceive software agents and how they translate the technological complexity of agents into legal terminology? We can do this by running through a selection of definitions and then considering what they all have in common.

For [13], software agents are electronic forms of real human agents, and just like human agents in law, they do things for us. This analogy is premised on three similarities: (i) both types of agents can acquire and retain knowledge; (ii) both can perform a given task; and (iii) both can communicate. Specifically, software agents can "react autonomously to changes in their environment and solve their tasks without any intervention of the user" (ibid). On this basis these two authors analogize software agents more broadly to legal persons.

---

[6] When I refer to legal scholars, I don't have in mind only legal professionals (such as attorneys), but also people, who—having studied law—actively participate in building software-agent based applications for legal or business domains (for example, in knowledge acquisition or knowledge representation phases of the application's development), or are in any other way interested in the impact that such applications (might) have on the legal issues.

[7] Furthermore, they are also regulated by norms on consumer protection.

[8] We might also call this a what-if approach: what if software agents *were* like human agents in every respect except for the fact that they do not have a biological life? For a discussion based on the claim that such fully autonomous agents *are* already with us, see [21].

[14] argues that "something is a person [i.e., an agent] iff it has states whose interactions appropriately mimic our rational architecture."

[15] similarly presents the idea of an electronic person, a legal construct on which basis to specifically consider the personhood of software agents in law. An electronic person would be recognized as having limited liability: on the one hand, this would limit its owner's liability, and on the other hand, the contracting party would be able to check a registry of agents to see whether the agent is solvent before entering into a contract.[9]

[16] presents software agents as "digital entities capable of executing autonomously the mandates assigned to them," and he ascribes to them cognitive stances (beliefs, desires, and intentions) on which basis to qualify their actions in legal terms (in e-commerce, for example).

For [17] software agents are agents "capable of independent action rather than merely following instructions": they "exhibit high levels of mobility, intelligence, and autonomy according to which their actions are not always completely anticipated, intended, or known by their users," while [18] describe software agents as "intelligent and autonomous electronic agents."

What we can extract from these definitions is that software agents are regarded as agents owning two characteristics: (a) intelligence and (b) autonomy. And it is in particular this latter characteristic that makes the action of software agents legally relevant: autonomy is understood as an agent's capacity (i) to learn from experience; (ii) modify its own instructions; and (iii) work out new instructions to follow [19].

Intelligence and autonomy make software agents more than just tools or electronic devices: in [20] we can proceed on this basis "to treat [these] programs as legal agents of their principals, empowered by law to engage in all those transactions covered by the scope of their authority."

A third characteristic that legal scholars view as essential in making software agents worthy of consideration as legal persons is their intentionality (this is implicit, for example, in Sartor's account of software agents as endowed with the cognitive attributes of desire, belief, and intention). And still other characteristics are their belonging to someone else—and usually acting on that someone else's behalf—along with their ability to socialize. [21] describes these as "external characteristics of an independent individual," which takes us back to the idea of autonomy and intentionality.

4.1   Expanding the List of Legal Persons

The consequence we can extract from these considerations is that we will sooner or later reach the point where it is reasonable for us to add software agents to the list of entities recognized as legal persons. We have seen this kind of expansion before: in the early 19th century, for example, Chief Justice John Marshall found that this is how we are to consider a corporation, "that invisible, intangible, and artificial being,

---

[9] The idea of an agent registry is also considered in [19].

that mere legal entity, a corporation aggregate" [22]. He used the legal fiction of the corporation as a person, and no doubt the time will come when software agents must also be understood in the same way.

Of course, conditions must be ripe before anything can be viewed as a legal person: this was true of the corporation, and I am arguing that the same goes for software agents.[10] And so, if we can observe legal scholars working on how to extend legal personality to software agents, trying to work out the appropriate legal analogies to be drawn in making the extension, there must be a compelling set of reasons for so doing. One such reason, I would argue, lies in the need to protect humans who interact directly or indirectly with software agents, and it seems that the law is already equipped for the extension: [23] argue that "in principle the law can attribute conditional legal personhood to any well-defined type of entity." Hence, the right thing to do would be to make the agents to become "well-defined."

Then, too, as [20] argues, the extension importantly rests on the ground that software agents have evolved to the point where they acquire social meaning, in that we recognize them as entities which populate our social existence, our modes of social and economic interaction. It might be some time before this prediction comes true, but this will certainly be an essential criterion. And since it is unrealistic to posit a specific moment when software agents become a significant part of our social life, we can think of this as a process, whereby software agents will initially attain a kind of legal personhood and will thereafter incrementally become legal persons in one way or another. This suggests that, in establishing criteria on which basis to determine that software agents are no longer just tools but qualify as legal entities, we do not have to at once make them into full-fledged legal persons: as [24] suggests, we could work out a kind of legal personhood specifically tailored to software agents in a sense "akin to Roman law," where they would be "not legal persons in their own right, but with power to enter into binding arrangements, and receive information, on behalf of their owners, in circumstances where their owners would be bound by those arrangements or that knowledge."

I have briefly outlined the process by which the law is working out its own understanding of software agents: it proceeds by taking into account a number of criteria centered on what it is that makes something an agent, focusing on what is prominent among these criteria, namely, the capacities that make one (or something) a practical agent, one that can engage on its own in practical reasoning about what to do; we are thus looking at an autonomous intentional agent, one that can accordingly be deemed morally responsible and whose actions carry consequences for those with whom it interacts. At first sight this may seem to have little to do with the previously considered conceptions of software agents developed in AI and SE. And yet, despite these differences, those who work in these three areas—AI, SE, and law—understand that they

---

[10] At the same time, what may also be at play in this extension of personality is what has been called Topffer's Law (named for Rodolpfe Topffer, the father of the modern comic book), stating that we are inclined to attribute personality and character to any squiggle we recognize as a face: we are "meaning-making, pattern-seeking creatures [...] we read personalities into all kinds of interactive artifacts" [25], and I submit that if we are including software agents among such artifacts, this psychological law may well have something to do with it, too.

are all dealing with the same object, namely, software agent. This means that there must be a common ground, and so I will devote the rest of this paper to explore it and to argue that this ground explains why the three approaches are actually fruitfully complementing one another in building agent-based applications.

5   Software Agents and Boundary Objects

The differences between software agents as conceived in AI and SE, on the one hand, and in law, on the other, are obvious ones indeed, but the point here is not the amount of differences, but that despite these differences, AI, SE and law have found a way to collaborate with each other in different enterprises which lead to successful agent-based applications. And the question is how they managed to do that?

This is indeed a very important point in discussing how the very sophisticated and complex systems are being build: in fact, in building such systems the heterogeneity of the scientific fields involved and fluent communication among them is something that ensures a positive outcome. Still, the heterogeneity doesn't mean communication: on the contrary, different fields might not understand each other. So how is it possible in case of agent-based applications?

The answer to these questions is that this is possible if only we see software agents as boundary objects, understood as cross-border objects used in at least two different areas of activity. So let us briefly consider what a boundary object is (Section 5.1) and then see how this notion applies to software agents (Section 5.2).

5.1   What Is a Boundary Object?

A boundary object is a term coined in 1989 by Susan Leigh Star (computer scientist and sociologist) and James R. Griesemer (philosopher), and it designs any object—whether it be abstract or concrete—that different communities of practice (or CPs)[11] use in different ways while still recognizing the object as such. In this sense it is a plastic object—for its use and meaning change depending on who is using it and for what purposes—and yet it is solid enough at its core that the different communities using it will still know they are essentially dealing with the same object.

But what is CP? And how it relates to our discussion? The CPs are characterized by three basic features [26]: mutual engagement of community members, communally negotiated goals, and a shared repertoire. Mutual engagement means that members pursue the same interest, developing mutual relationships in such a way as to increase everyone's sense of belonging to that particular community. Community goals and guidelines are broad, with much latitude for the community's individual members, in that each member can frame specific goals within the broad outline, thereby

---

[11] I use the term "communities of practice," but Star and Griesemer have first applied boundary objects to social worlds, and only then expanded their applicability to communities of practice, information worlds, electronic community systems, digital libraries, etc. The difference among these fields of application is not the focus of this paper. More about boundary objects, fields of their application and other details, see [29].

contributing to the community in an individual way. A shared repertoire, finally, is everything that gets built over the course of a CP's activities: experiences, stories, constructs, categories, tools, events, images, expertise, know-how, and more.

Practically any field of human activity (professional, cultural, social, ...) is CP: for example, lawyers, physicians, or astrologers could all be seen as members of their CP, because they all pursue their community's interests—legal issues, illness, or stars' influence on our characters—and all are engaged in discussions which permit them to work out the goals for their community, at the same time creating a shared repertoire of the practices, experiences and narratives that define each CP. Hence, in this paper AI, SE and law can also be seen as CPs. Obviously, this is a very simplified definition of CP which omits many important aspects and features, but I believe that it is still clear enough to give a reader an insight of what it is all about.

Where do boundary objects fit into this? In short, they are objects that different CPs can share and can use to interact, this owing to the ability of boundary objects to maintain a constant identity even as they travel across borders. Thus, on the one hand, different communities can recognize a boundary object for what it is (without mistaking it for something else), all the while tailoring that object to their different needs. Boundary objects have been described in this sense as having a structure at once weak and strong: weak because malleable—they can be fashioned into different "shapes"—but strong once they have been so fashioned, with each CP customizing the object for its own purposes. Or, if we flip this around, a boundary object makes it possible for each CP to use the object in its own way, all the while collaborating with other CPs, or at least communicating with them, in linking up the different types of knowledge they have each developed.

Hence, for example, we can say that astrologers and physicians share the concept of prognosis, which in medicine is attributed with a different meaning and based on different parameters than in astrology, even if in both cases it deals with hypothesizing about the future. Same applies to philosophy and SE which share the concept of abstraction, to art and civil engineering which share the concept of design and drawing, to political science and computer science which share the concept of institution, and so on and so forth. This shows that boundary objects populate different areas of human activity and interest, enabling the interaction among them and leading to different applications, results, uses and practices.

Having said that, we can move on to explain the idea of software agents as boundary objects used in three CPs, that is, in AI, SE, and law.

5.2   Software Agents as Boundary Objects

A software agent becomes a boundary object the moment we consider the use it is put to in the context of CPs of AI, SE, and law. As much as there is a large body of knowledge shared among the first two CPs—that is, AI and SE—this does not mean that any member of one CP will ipso facto be able to understand, communicate, and work with someone belonging to another CP. For example, a software engineer building an AI system may have to work with an AI researcher who is specialized in the psychology and physiology of the brain, two areas a typical software engineer will know nothing about, and yet the two may have to work together if they want to develop an intelligent system.

Even less common ground, than there might be between these two CPs, there is between these CPs and legal CP. But, again, this does not preclude their working together on common problems: evidence of this lies in the fact that we have an entire research area—legal informatics, or computer science and law—which (among other issues) is dedicated to bridge the law and computer science working out the best ways for legal scholars to contribute to the development of software applications dealing with legal problems.

And so we ask, How can a software agent understood as a boundary object enable the three CPs in question to benefit from one another's work? It can because a software agent so considered becomes a malleable entity, one that all CPs can recognize as such (as a software agent), but that each can use in its own way for its own purposes. This variety of uses explains why each CP will offer a different account of what a software agent is, but what appears to be a weakness (this lack of a shared definition) turns out to be a strength, for it shows that different CPs can approach software agents from different angles, or study them in different ways, all the while recognizing that they are, after all, software agents. Thus legal scholars are interested in seeing whether software agents exhibit properties akin to those that would make one an intelligent agent that can be held morally responsible for its own actions; software engineers seek to develop software agents with social abilities, such as a goal- or task-orientedness, interactivity, proactivity, and reactivity; and researchers in AI seek to model software agents having a capacity for rational behavior, endowing them with beliefs, desires, and intentions, and a capacity to act rationally on these mental states.

Now, these are not just different approaches, but different approaches that can be made complementary, because the CPs that pursue these different interests recognize them as revolving around the same object, namely, software agents. It is on this basis that we establish the common ground necessary to achieve the needed complementarity.

Thus, for example, we can recognize that all three communities understand software agents as performing precisely the role their name suggests, the role of agents or proxies, that which consists in doing something on someone's behalf: we saw this with the agency relationship in law (where an agent acts on a principal's behalf), and the same is true in computer science, where agents are entrusted with carrying out tasks for their users.

At the same time, software agents are understood in all three communities as autonomous agents: even when they carry out instructions for someone's benefit, they act with some latitude within that framework. Of course, each CP has its own view of what such autonomy means. Thus, for legal scholars, an autonomous agent is a practical agent that can be held accountable for the consequences of its actions;[12] for

---

[12] It must be noted, however, that the law of agency does limit or qualify this liability if (*a*) the agent is acting within the "scope of employment" (under the common law doctrine of *respondeat superior*), in which case the agent and the principal are both simultaneously liable for a tort of the agent (joint and several liability), or (*b*) the agent contracts with a third party on the principal's behalf and the principal is disclosed to such third party, in which case the agent bears no personal responsibility for the contract.

researchers in AI, an autonomous agent is one that can make rational decisions on its own using the information it receives and its data-processing logic in a virtual or "logical" environment; and for software engineers, autonomy is a software agent's interactive ability to be part of a MAS, in such a way as to handle the unpredictability the system may present on account of the way the other software agents in the system might behave. Hence, what we have here is the way in which the conceptual vagueness and malleability inherent in the idea of a software agent's autonomy can be an asset rather than a liability, because while each CP has its own idea of what it means for a software agent to be autonomous, they all understand that they are essentially dealing with the same thing, and indeed that the work each CP is doing in building autonomy into a software agent, or otherwise dealing with such autonomy, is complementary to the work of the others.

The complementarity involved can be also appreciated if we consider that while computer scientists deal with the question of what problems software agents could solve in the legal domain, legal scholars deal with the question what problems software agents could cause: in combination, these approaches can make it possible for computer scientists to develop better or improve the existing agent-based systems for domains where legally relevant consequences are possible (such as e-health or e-commerce). In e-commerce, for example, legal scholars have suggested taking into account the criteria of good faith and fair dealing in contract negotiation (see [27]). And many others have drawn developers' attention to the issue of privacy. As [24] have argued, "designers of artificial agents which access users' personal information or private communications need to be mindful of possible privacy implications: the more sophisticated their systems become, the more likely it is that corporations that deploy those agents will be attributed with knowledge of their users' personal information, possibly triggering significant legal liability." There is a twofold advantage to be gained here if computer scientists can interact with legal scholars, for on the one hand consumers can benefit from agent software that takes the legal scholars' concerns into account, and the law, for its part, would stand to benefit by forging rules stemming from a deeper understanding of what it is exactly that software agents actually do and how they behave.

6   Conclusions

We have considered three different approaches to software agents: the approach in AI, which considers software agents as rational entities guided by beliefs, desires, and intentions; the approach of software engineers, who treat agents as abstractions on which to build MAS; and the legal approach, which focuses on the practical implications the use of software agents brings about, and which works out hypothesis for their future legal personhood.

Then the question was put forward: how is it possible for all the people who work in AI, SE and law to work together on a single agent-based application when they all have different ideas on what an agent is. The answer to this question is given in the second part of this paper where I argue that the multilateral collaboration, interaction and understanding among these professionals is based on the idea that software agents are boundary objects among AI, SE and law.

What does this mean practically? What advantages (if any) do we get from considering software agents as boundary objects? This multiplicity of agents around us is an advantage: this way we can open a space in which AI, SE, and law can share knowledge and work together in developing products that can benefit business and consumers at the same time: in fact, agents are seen by software engineers as offering "a uniform conceptual space where all the findings of the AI field can be easily framed and related and can eventually find mainstream acceptance" [28]. And, by the same account, these software engineers also argue that software-agent abstractions provide for "a new, powerful approach to the construction of intelligent systems": an understanding of software agents as boundary objects only supports this vision, and legal scholars cannot but contribute to improve it.

References

[1] Newell, A.: The Knowledge Level. Artificial Intelligence 18(1), 87–127 (1982)
[2] Bratman, M.E.: Intention, Plans, and Practical Reason. Harvard University Press, Cambridge (1987)
[3] Binmore, K., Castelfranchi, C., Doran, D., Wooldridge, M.: Rationality in Multi-Agent Systems. Knowledge Engineering Review Archive 13(3), 309–314 (1998)
[4] Rao, A.S., Georgeff, M.P.: BDI Agents: From Theory to Practice. In: Lesser, V., Gasser, L. (eds.) Proceedings of the First International Conference on Multi-Agent Systems, pp. 312–319. MIT Press, Cambridge (1995)
[5] Emerson, R.W., Hardwicke, J.W.: Business Law. Barron's Educational Series, Hauppauge, NY (1987)
[6] Henderson-Sellers, B., Gorton, I.: Agent-Based Software Development Methodologies. White paper for the OOPSLA 2002 Workshop on Agent-Oriented Methodologies (2002)
[7] Bresciani, P., Perini, A., Giorgini, P., Giunchiglia, F., Mylopoulos, J.: A Knowledge Level Software Engineering Methodology for Agent Oriented Programming. In: Müller, J.P. (ed.) Proceedings of the Fifth International Conference on Autonomous Agents, pp. 648–655. ACM, New York (2001)
[8] Jennings, N.R.: An Agent-Based Approach for Building Complex Software Systems. Communications of the ACM 44(4), 35–41 (2001)
[9] Zambonelli, F., Van Dyke Parunak, H.: From Design to Intention: Signs of a Revolution. In: Proceedings of AAMAS 2002, pp. 455–456. ACM, New York (2002)
[10] Weiß, G.: Agent Orientation in Software Engineering. The Knowledge Engineering Review 16(4), 349–373 (2001)
[11] Nwana, H.S.: Software Agents: An Overview. Knowledge Engineering Review 11(3), 1–40 (1996)
[12] Wooldridge, M.J., Ciancarini, P.: Agent-Oriented Software Engineering: The State of the Art. In: Ciancarini, P., Wooldridge, M.J. (eds.) AOSE 2000. LNCS, vol. 1957, pp. 1–28. Springer, Heidelberg (2001)
[13] Wettig, S., Zehendner, E.: A Legal Analysis of Human and Electronic Agents. Artificial Intelligence and Law 12, 111–135 (2004)
[14] Pollock, L.J.: How to Build a Person: A Prolegomenon. MIT Press, Cambridge (1989)

[15] Karnow, C.E.A.: Future Codes: Essays in Advanced Computer Technology and Law. Artech House, Boston (1997)

[16] Sartor, G.: Cognitive Automata and the Law: Electronic Contracting and the Intentionality of Software Agents. Artificial Intelligence and Law 17, 253–290 (2009)

[17] Dahiyat, E.A.R.: Intelligent Agents and Liability: Is It a Doctrinal Problem or Merely a Problem of Explanation? Artificial Intelligence and Law 18, 103–121 (2010)

[18] Andrade, F., Novais, P., Machado, J., Neves, J.: Contracting agents: Legal personality and representation. Artificial Intelligence and Law 15(4), 357–373 (2007)

[19] Allen, T., Widdison, R.: Can Computers Make Contracts? Harvard Journal of Law and Technology 9(1), 25–50 (1996)

[20] Chopra, S.: Rights for Autonomous Artificial Agents? Communications of the ACM 53(8), 38–40 (2010)

[21] Willmott, S.: Illegal Agents? Creating Wholly Independent Autonomous Entities in Online Worlds (2004),
http://www.lsi.upc.edu/dept/techreps/
llistat_detallat.php?id=695

[22] Bank of the United States v. Deveaux, 9 U.S. 61 (1809)

[23] Koops, B.J., Hildebrandt, M., Jaquet-Chiffelle, D.O.: Bridging the Accountability Gap: Rights for New Entities in the Information Society? Minnesota Journal of Law, Science & Technology 11(2), 497–561 (2010)

[24] Chopra, S., White, L.: Privacy and Artificial Agents, or, Is Google Reading My Email? In: Proceedings of the International Joint Conference on Artificial Intelligence. Morgan Kaufmann, San Francisco (2007)

[25] Mishra, P., Nicholson, M.D., Wojcikiewicz, S.K.: Seeing Ourselves in Computer: How We Relate to Technologies? Journal of Adolescence & Adult Literacy 44(7), 634–641 (2001)

[26] Wenger, E.: Communities of Practice: Learning, Meaning and Identity. Cambridge University Press, Cambridge (1999)

[27] Weitzenböck, E.: Good Faith and Fair Dealing in Contracts Formed and Performed by Electronic Agents. Artificial Intelligence and Law 12(1-2), 83–110 (2004)

[28] Omicini, A.: Challenges and Research Directions in Agent-Oriented Software Engineering: Autonomous Agents and Multi-Agent Systems. Special issue, Challenges for Agent-Based Computing 9(3), 253–283 (2004)

[29] Bowker, G.S., Star, S.L.: Sorting Things Out: Classification and Its Consequences. MIT Press, London (1999)

Argumentation and Intuitive Decision Making: Criminal Sentencing and Sentence Indication

Andrew Vincent

School of Management and Information Systems,
Victoria University, P.O. Box 14490, Melbourne, 8001, Australia
vinandrew@gmail.com

Abstract. This paper investigates criminal sentencing in the Australian State of Victoria in particular the intuitive nature of the decision making and the difficulties of representing intuitive knowledge. In order for decision systems to be useful for the purposes of training novice practitioners and law students in the complex area of sentencing they must be constructed with an authentic cognitive model that faithfully represents the sentencing process and also the decision-making process. In this paper a pre-cognitive model of the sentencing process is presented.

Keywords: Argumentation, Criminal Sentencing, Knowledge Representation, Meta-Principles, Plea Bargaining.

1   Introduction

The task of sentencing criminals is one of the most difficult tasks that judges must undertake. The decision is difficult as it often involves the deprivation of liberty for the criminal. Over the years many attempts have been made to make the task either more simple for judges and/or more transparent for the casual observer. With the rise of information technology innovations, not solely limited to computers, and the development of methods to represent complex knowledge attempts have been made to apply these to the legal domain. Much has been made of the attempts to rationalize, computerize and sentence from a distance [1]. Aas indicates:

> [c]ategorizing human identity into axis grids is an act of deconstruction of subjectivity. It is an act of taking individuals apart and then putting them together according to the requirements of the system. The process does not require a narrative or communication . . . [d]ue to the distance created by procedural rules the offender is precluded from participating in the process of defining his or her own identity. [[1], p. 110]

The rise in actuarial practices in the legal arena has caused a great deal of concern [2]. Increasingly conservative governments are attempting to develop criminal justice regimes that remove social information from the sentencing and to predict who might breach bail and criminals who might re-offend. [3]. Aside from the problems with actuarial practices in policy and punishment (which are not further explored herein), there is a dearth of suitable avenues for the training of both students and novice practitioners in the area of both plea negotiation and sentencing. Training of judicial officers in the finer points of sentencing and in plea bargaining (for prosecutors and

M. Palmirani et al. (Eds.): AICOL Workshops 2011, LNAI 7639, pp. 217–234, 2012.

defenders) is usually done on an ad hoc basis. There are no programs that offer law students the chance to develop skills and expertise in negotiating over guilty pleas. Students and novice practitioners need to be able to practice and develop expertise in a manner that enables them to construct their own knowledge.

The purpose of this paper is to discuss knowledge representation and sentencing within the framework of developing skills and expertise in sentencing and plea bargaining. The unique elements of the sentencing process need to be preserved and represented faithfully in the knowledge representation process. This paper will describe the nature of sentencing in the Australian state of Victoria with particular reference to pleading guilty. The method of judicial decision making will also be discussed and then an argumentation scheme will be considered in relation to the procedure and to cognitive decision-making models. Discussion of the intuitive nature of the task is undertaken and the role that argumentation takes for constructing the framework of systems that could used in supporting the training of court staff and judicial officers in sentencing and plea negotiation is explored. An important aspect with regard to sentencing, plea negotiation and knowledge representation is the application of meta-principles of sentencing.

2   Expertise Development

It is crucial that novice practitioners and law students receive opportunities to develop strategies and practice negotiation in general but more specifically in the context of criminal cases. Many law schools provide opportunities for students to practice advocacy skills in relation to courtroom presentation in the form of moot courts. They seldom provide chances for students to practice advocating for defendants in relation to plea negotiation or as prosecutors presenting information to judges. At the other end of the expertise spectrum judges appointed to the courts are usually of high standing within the legal profession and in the Supreme Court are often barristers. Even though before becoming judges, they may well have had a great deal of experience with criminal cases and sentencing it is wrong to assume that they can just slot into a criminal court and conduct a case through to final sentence. Crucial to the success of training systems is the ability to both present cases and provide feedback to students both during the negotiation process and at the conclusion [4–6]. In the process of receiving feedback it is crucial to student learning and so also to the learning of novice practitioners that learners are given an opportunity to reflect on their own decisions and their processes in light of the decision of an expert [7]. There are significant issues with providing novice practitioners access to expert opinion. Opportunities for training practitioners usually takes the form of rather ad hoc arrangements where a senior staff member holds an impromptu discussion about the facts of a case. In Victoria the Judicial College of Victoria takes on the role of training judges and magistrates and this is usually conducted by way of an initial induction program and ongoing professional development via self-direct learning techniques; in that judges and magistrates are responsible for attending workshops of relevance to their practice and interest at their own leisure.

It is important to represent accurately the knowledge and information that would be utilized by both judges in the sentencing process and novice lawyers in the plea negotiation process since poorly represented and consequently poorly presented knowledge can have detrimental effects on the ability of learners to transfer their knowledge and skills to new situations. In order to reach the level of expert it takes in the order of ten years and individuals must engage in practice [8]. The system that will result from this research will be utilized to train novice professionals and law students and provide them opportunities to practice in a manner that faithfully represents the decision making requirements and practices of the judiciary.

3   Plea Bargaining and Sentencing

In the Australian state of Victoria and in other Australian states sentencing is governed by acts of parliament. In Victoria sentencing is conducted under the *Sentencing Act 1991*. The act does not define crimes (crimes are defined in various other acts, primarily in the *Crimes Act 1958*) but rather the procedural rules and guidelines that are incumbent on judges when sentencing convicted criminals. This differs from the manner that sentencing is conducted in civil law jurisdictions.

In civil law countries, legislation is often the primary source of law and consequently courts base judgments on the requirements of both statutes and codes. These generally provide solutions that can be derived and applied in a particular case. On the basis of general rules and principles contained in the various codes Courts have to reason, usually drawing analogies from legal provisions to fill voids and to achieve rationality. In common law systems, cases are the primary source of law, while statutes are only seen as providing boundaries to common law and are thus narrowly interpreted. In the Netherlands for example, a Civil law country the penal code provides restrictive rules on aggravating circumstances and contain one general mitigatory factor [11]. That said the Dutch judiciary are given wide discretion in sentencing and while there are few statutory rules which are expressed in general terms, they do not limit the court in their choice about the type and severity of sanction [11]. It is a similar case in Germany where sentences are individually indicated for each offense in the German Penal Code (Strafgesetzbuch). The judiciary is granted significant leeway in their sentencing discretion because the statutory penalty ranges are quite broad [12].

In Australia, sentencing judges are granted broad discretionary powers in determining the severity of a sentence. This stands in contrast to other jurisdictions where judges have their discretion much more restrained, methods for this take the form of mandatory sentencing regimes and gridline sentencing which are particularly popular in various states of the United States of America and the American Federal Sentencing regime [13–16].

Sentences in Victoria are not defined in terms of minimum periods of incarceration but rather maximums.[1] Judges in Victoria are delegated the responsibility by the state for the selection of the appropriate purpose for sentencing: these are [[17], s. 5(1)]:

---

[1] Sentences are classified into nine levels of severity each relating to a *maximum* penalty. A person guilty of armed robbery (*Crimes Act 1958*, s. 75A), for example, is liable to a level two imprisonment, which corresponds to a twenty-five year maximum. There are *no* minimums specified. There is a statutory requirement for the judiciary to set non-parole periods (the period of time the offender must spend in prison before being eligible for release on parole), which should not be considered minimum sentences.

(a)   to punish the offender to an extent and in a manner which is just in all of the circumstances; or

(b)   to deter the offender or other persons from committing offences of the same or a similar character; or

(c)   to establish conditions within which it is considered by the court that the rehabilitation of the offender may be facilitated; or

(d)   to manifest the denunciation by the court of the type of conduct in which the offender engaged; or

(e)   to protect the community from the offender; or

(f)   a combination of two or more of those purposes.

Judges are also then required to take in to account the following matters [[17], s. 5(2)]:

(a)   the maximum penalty prescribed for the offence; and

(b)   current sentencing practices; and

(c)   the nature and gravity of the offence; and

(d)   the offender's culpability and degree of responsibility for the offence; and

(daa) the impact of the offence on any victim of the offence; and

(da)   the personal circumstances of any victim of the offence; and

(db)   any injury, loss or damage resulting directly from the offence; and

(e)   whether the offender pleaded guilty to the offence and, if so, the stage in the proceedings at which the offender did so or indicated an intention to do so; and

(f)   the offender's previous character; and

(g)   the presence of any aggravating or mitigating factor concerning the offender or of any other relevant circumstances.

It can be seen from the two lists why wide discretion has attracted fierce detractors, especially regarding the purposes of sentencing [18, 19]. Indeed as Ashworth has indicated there has for a long time existed a:

> kind of cafeteria system, in which judges and magistrates have been encouraged to choose a rationale from several ... with relatively little constraint in the choice. [[20], p. 331]

There can be no doubt that sentencing in Victoria favours the individualization of sentence that is tailored to the particular circumstances of the offence and offender.

4   Intuition

The High Court of Australia in 2005 determined in the landmark case *Markarian v The Queen* that the only acceptable method of judicial decision making in relation to sentencing is instinctive or intuitive synthesis [21]. In Victoria the idea of instinctive or intuitive synthesis has been the accepted methodology of judicial decision making since 1975. It was fully articulated in a Court of Criminal Appeal judgment, where it was stated [22]:

> ultimately every sentence imposed represents the sentencing judge's instinctive synthesis of all the various aspects involved in the punitive process ... it is profitless ... to attempt to allot to the various considerations their proper part in the assessment of the particular punishments presently under consideration

The sentencing method in Australia has fallen in to one of two camps. On the one hand there is the "logical", "rational" approach that has stages and levels in argument which would be made transparent with full and proper reasons [4], this is known as the two-tiered approach and was ultimately rejected as the method for arriving at criminal sentences. On the other hand is the instinctive or intuitive synthesis camp. The instinctive synthesis method involves, as indicated above, the weighing of all the circumstances of the offence and offender to arrive at an appropriate sentence. All the factors are evaluated in reaching the decision but no one is given priority or weighted more than another.

The "two-tiered" approach is where on the first tier the judge considers the objective circumstances of the offence (factors associated to the gravity of the criminal activity) in order to gauge the seriousness of the offence. On the second tier the judge considers the subjective factors which usually relate to the offender (both aggravating and mitigating) and then the sentence is decided. Judges will often suggest a tariff they regard as proportionate to the crime and then adjust the tariff by specific amounts by reference to particular factors. It is obvious that this method should encourage the judiciary to be more explicit in their reasoning by declaring the weight given to individual factors. This procedure is overtly more mathematical than the so-called "black box" approach and more likely to be able to be predicted and especially modelled for the construction of decision support systems. This though cannot be further from the truth. This approach also has its critics, whose argument usually revolves around the "mathematical" nature of the process. One of the main criticisms that seems to be overlooked by commentators is that there is still an intuitive decision to be made, this involves the selection of an appropriate starting point for the various calculations of aggravation and mitigation. It has been suggested by Traynor and Potas that:

> we see no harm in retaining a two-tiered sentencing methodology wherein the sentencing judge or magistrate determines the upper, and sometimes lower, limits of an appropriate sentence based on the notion of offence seriousness or objectivity (the outer range of proportional punishment) and then proceeds to fine-tune the sentence by reference to other considerations. [23]

Given that sentences in Australia are bounded only by maximums there is another instinctive or intuitive decision required about what the starting point should be. It is further suggested that in some cases the quantum for individual factors could be disclosed. This then leads to more intuitive decisions about when to disclose specific discounts and then also explanations as to the reasons. The guilty plea is the most common mitigating factor leading to sentence reduction that is indicated by judges in their reasons and the judicial annunciation of the specific discount in a particular case could lead to further legal wrangling about the appropriateness of that figure.

There can be little doubt that the instinctive synthesis method that judges advocate as correct most accurately reflects their decision making practices. Intuition plays a very large part in the sentencing process. Mackenzie in her important study of judges in Queensland indicates that many judges describe the process as an intuitive one [37]. This is an important finding as it adds to the list of studies that have found that judges use their intuition to assist in their decision making [58]. Brest suggests that intuition and analysis are involved in a complicated dance and it is all but the most simple decisions that are analytical in nature [59].

4.1   Intuition and Expertise

In a very useful discussion of the importance role of intuition and decision support, Sauter [60] has suggested the paucity of information required for very complicated decisions means that detailed analyses by decision makers are unfeasible. In most jurisdictions, judges are under increasing pressure with respect to their case loads and given the often parlous state of information available on the effectiveness of the various sentencing options available to them, there is a great reliability on intuition in determining the appropriate sentence.

Intuition is variously described as "a sense of feeling of pattern or relationship" [60] or intuitive responses "are reached with little apparent effort, and typically without conscious awareness" [61]. It usually involves a decision that is complicated, all the information required might not be present and the outcome is not certain. In sentencing, intuition is educated by experience and by providing information relating to the interactions of the factors that can be taken in to account judges can hone their intuition. The provision of recent and relevant information to decision makers is also of critical importance. There is a growing body of literature that suggests that the ability to make intuitive decision is one of the hallmarks of expertise [24]. Intuition involves a process that matches some environmental stimuli with deeply held (non-conscious) pattern or feature. This equates to the non-conscious mapping of stimuli on to cognitive schemata [24]. Tata likens the skill of sentencing with the development of skill in craftwork [25]. This captures the notion of expertise but unfortunately confuses the development of expertise by equating it to the development of artistic skill and not making the links with decision making and problem solving. There is no doubt that experts can make highly accurate intuitive decisions [26, 27].

4.2   Building Decision Support Systems

Building systems to support the various parties involved in the sentencing process is fraught with difficulties. Tata [25] has detailed the effort in the construction of the Scottish Sentencing Information System and discusses some of the reasons why judicial decision support systems are not well received by the judiciary they are made to support. One of the primary reasons for judicial ambivalence towards decision support is that most systems do not accurately reflect the manner in which judges reach their decision or are so over complicated that they are virtually useless. In the remainder of the paper, the pre-cognitive model is presented and the preliminary on constructing a decision aid to provide expert opinion to novice practitioners will be presented.

5   Cognitive Model of Sentencing and Argumentation

[28] has presented a generalizable model of the decision-making processes of judges determining criminal sentences that consists of, at present, a set decision trees and associated argument trees based on [29]. There is also a partial instantiation of a working system in the form of a computer program. In producing a pre-cognitive model of the sentencing task, the decision trees that have been produced are being reconstructed in accordance with the pre-cognitive model that will be presented below. [28] describes the process of knowledge elicitation to produce both decision and argument trees.

Unlike many other decision-making domains, such as some of those in accounting (notably going-concern judgments and auditing decisions), sentencing knowledge does not have to be deduced or elicited from experts via interview or survey [30][31]. In Victoria there is a requirement that judges produce written sentencing remarks. These sentencing remarks contain the reasons for the sentence passed on the criminal for a particular set of circumstances. There is no generalization required. The process used by [28] initially makes the distinction as indicated by [32] that there is a difference between academic and experiential knowledge. Academic knowledge is that which is contained in legal statues, legal texts, commentaries and precedents, while experiential knowledge is that which is developed by practice. The other kind of knowledge that has been suggested is that of structural knowledge [33]. Structural knowledge is the link between declarative knowledge and procedural knowledge. This implies that there are different sets of principles that can be deduced from written sentencing remarks. There are principles that relate to statutes and others that relate to more practical matters. The practical principles are the meta-principles of sentencing, they are principal themes that synthesize around a cluster of pragmatic considerations. The pragmatic nature of meta-sentencing principles requires the identification of principles that are more practical than theoretical; those principles that are focused on what is achievable rather than ideal. The meta-principles are those that judges use to make an assessment based on their own experience and understanding of the possible effects. This idea is partially captured by [34] who suggests that there exists a tension between the sentencing purposes (deterrence, punishment and the like) and the principles that can operate to influence future behavior of the criminal.

5.1   Reasons for Sentence

The one source of legal knowledge in sentencing is the artefact produced as the result of criminal sentencing process; the reasons for sentence. In Victoria the written decision represents the reasoning processes of the judge. The sentencing remarks can be understood as structural knowledge. In the area knowledge elicitation, written records are seldom used as there is a belief that due to the amount of time that decision-makers have to prepare their answers they are able to obscure their reasoning processes [35]. While this may be true in some cases there is evidence that suggests that the longer the length of time that is spent on these types of decisions the greater the clarity in the explanations especially in terms of reasoning and argument [24, 36]. Judges in Australia seem to have an excellent understanding of their reasoning processes and are able to articulate the manner in which they decide very clearly [37]. In the Markarian case mentioned earlier the majority judgment indicates that [21]:

> [e]xpress legislative provisions apart, neither principle, nor any of the grounds of appellate review, dictates the particular path that a sentencer, passing sentence in a case where the penalty is not fixed by statute, must follow in reasoning to the conclusion that the sentence to be imposed should be fixed as it is. The judgment is a discretionary judgment and, as the bases for appellate review reveal, what is required is that the sentencer must take into account all relevant considerations (and only relevant considerations) in forming the conclusion reached. As has now been pointed out more than once, there is no single correct sentence. And judges at first instance are to be allowed as much flexibility in sentencing as is consonant with consistency of approach and as accords with the statutory regime that applies.

Judges are required, according to the common law tradition, to produce written sentences [38]. Judges are not required to account for every single issues raised at the plea stage or reveal every step taken in arriving at the sentence. In *R v Giakis* it is stated [39]:

> A sentencing judge is not obliged to express in his reasons for sentence every single matter that he has taken into account, nor to give reasons for every single step that he takes.

A number of major public policy purposes are served by the provision of reasons. Supplying reasons enables parties to see the extent to which their submissions have been both understood and accept by the sentencing judge. Reasons increase judicial accountability, which is especially important for the public good. The provision of written reasons serves as a base for which like case can be discerned and then future decisions predicted. It is critical to the effectiveness of the appellate process that the ability of appeal judges to examine whether or not a judgment is erroneous is not hindered by lack of reasons [10].

6   Operationalizing a Cognitive Model

The Judicial College of Victoria suggests that there is a sequence of considerations that should be followed and that the order outlined below ensures that a judge will consider all relevant matters in a logical order, as opposed to a random selection of factors [40]. The ultimate synthesis is the result of balancing and weighing all relevant matters. An ordered approach does not impinge upon or restraint the exercise of judicial discretion. The sequence of events is detailed below. The sequence of considerations represents the pre-cognitive model as it represents cognitive (mental) processes (functions) involving acquisition, maintenance and usage of knowledge. A cognitive process is considered as a process of human information processing. Individuals take information and following a process that might include, deconstruction, hypothesizing, analyzing, reconstruction, reorganizing, judging and reasoning, make conclusions, plans and decisions, and take action [41]. The sequence of events is detailed below and also the individual factors that could be relevant under each of the categories.

6.1   Assessment of Offence Gravity

Judges must firstly make an assessment of the offence gravity. A judge should start with a consideration of the relevant material and an assessment of offence gravity is the foundation of the sentencing process. All other factors that affect the sentence are measured against gravity and the more serious the offence the less weight may be given to the circumstances of the offender, the personal mitigating factors. There are other important elements that combine to produce an overall impression as to the gravity of the offence [38].

Crucial to the determination of offence gravity is a consideration of the maximum penalty set for the offence. This is determined by the legislative ranking of the offence which points to the general offence gravity but the nature of the specific offence depends on the nature of revelant factors, such as method of the execution of the crime and the role of other participants, the extent of victimization. Factors that are suggested by Fox and Freiburg [38] as key to the gravity of the offence (such as the mental state of the offender) are considered, in this model, according the Judicial College of Victoria, in the sequence of considerations under the circumstances of the offender category. The *Sentencing Act* suggests as indicated earlier that the court must have regard to "nature and gravity of the offence." [17]. This legislative link means that the two categories are firmly connected together in the sentencing task and exert a large influence of the final sentence. It is important to realize that offence gravity will generally extend beyond the rudimentary facts of the case and the evaluation of the "objective circumstances" will be influenced by similar evaluations by earlier courts, either by judicial remarks or sentences imposed. This represents an oblique reference to the idea of *stare decisis*. Some previous work has examined the influence of *stare decisis* in the modeling of discretionary decision-making domains [42][43] but in criminal sentencing "the principle of *stare decisis* also appears to have little operation" [44]. The issue of *stare decisis* is part of a larger debate regarding consistency in sentencing. The most recent Australian Law Reform Commission investigation into sentence consistency for Federal offenders did not mention the term *stare decisis* but the report does stand testimony to the significant importance that government places on achieving consistency of sentencing in Australia [45].

6.2   Assessment of Circumstances of the Offender

A judge should move from the objective factors indicated above to a contemplation of the circumstances of the offender that are sometimes characterized as the subjective circumstances. When the court has come to a view regarding the offence gravity and the circumstances of the offender, conclusions regarding the moral culpability of the offender can be drawn. Moral culpability is an essential consideration and may be viewed in one sense as the subjective (to the offender) aspect of offence gravity. Many personal circumstances shed light on the offender's moral culpability, including mental capacity, antecedent history and character, and state of addiction. Beyond the assessment of moral culpability, a determination of the purposes for sentence cannot be made without regard to the offender's personal circumstances. Fox and Freiburg suggest that a purely retributive approach to sentencing would require that the seriousness of the crime should be the primary if not the only determinate of the final sentence [46].

6.3   Evaluation of Policy Considerations

Some of the issues that are relevant for the court cannot be readily classified as relating to either the offence for the offender. Elements that constitute this category include co-operation with authorities, sentence indication and the guilty plea. As mentioned earlier the guilty plea is the primary method of case finalization and the practical benefits for the justice system such as avoidance of costly trial are rewarded by way of sentence discount. It is clear that once the circumstances of the offence and the offender have been evaluated these issues should be examined.

6.4   Consideration of the Purposes of Sentencing

As indicated earlier the current state of affairs with regard to the purposes of sentencing has been described as a cafeteria-style system. This is certainly the case in Victoria, but after the considerations that have preceded this one, the court may be disposed to select one or a combination of purposes for the instant case. Current sentencing practice may also be significant for the court when determining sentencing particularly in attempting to achieve that particular purpose or purposes.

6.5   Application Principles

In Victoria important sentencing principles were indicated in *R v Storey* [47]:

> Sentencing is not a mechanical process. It requires the exercise of a discretion. There is no single 'right' answer which can be determined by the application of principle. Different minds will attribute different weight to various facts in arriving at the 'instinctive synthesis' which takes account of the various purposes for which sentences are imposed - just punishment, deterrence, rehabilitation, denunciation, protection of the community - and which pays due regard to principles of totality, parity, parsimony and the like.

The appropriate time to contemplate the application of the principles of proportionality and parsimony is following the determination of the purpose or purposes appropriate for the instant case. If the court is dealing with an offender who has committed multiple offences in the instant case then the principle of totality must also be considered.

7   Factors for the Sequence of Considerations Categories

The previous sentencing modeling established decision trees based primarily on discussion with individuals involved in the sentencing process. These individuals were not in decision-making position but were however tasked with advising the sentencing judge with information for consideration in the sentencing process. The defense offers the plea in mitigation while the prosecutor provides information regarding the seriousness of the offence and usually highlights any aggravating issues. The other sources of data were the *Sentencing Act*, academic works and appellate cases [28]. Each of the decision trees was constructed using various high level concepts or factors with a number of subordinate issues that combine to indicate a level or threshold. The decision-making process is not trivial and previous sentencing studies have identified that there are a vast number of factors. [48] identified 229 factors in a study of the plea in mitigation in Magistrates' Courts in England. A study conducted by La Trobe University Department of Legal Studies in Magistrates' Courts in Melbourne identified 292 individual factors [49]. The object of the study was to collect data in a systematic manner that might reveal a variety of aspects of the courts' operations. The studies of [48] and [49] were conducted in Magistrates' courts and the complexity of the sentencing task less onerous. It is less difficult due to the severity of the offences that are determined in those courts, currently the Victorian Magistrates' Court can only sentence an offender to two years in jail and there is no requirement for the provision of written decisions.

[25] has identified that the method of assigning numerical values to each of the factors and then attempting to divine a sentencing calculus is problematic. [50] has attempted to construct a sentencing calculus for the Victorian County Court but the results of are not very accessible. The model is unable to deal effectively with offenders who have committed multiple offenses. As a method of representing the knowledge of the sentencing process the argument trees utilized by [28] are useful but there is more to the use of the argument trees than has been delivered. Other attempts at organizing sentencing knowledge include an attempt by [51] using an XML specification to allow easy transportation between various information systems.

The information judges use to sentence criminals comes primarily from submission by prosecution and defense council, judges do not use material from trial. The defense puts forward what is known as the plea in mitigation, while the prosecution canvasses available options for the sentencing judge to consider in the specific instance. The judge delivers a sentence in the form of a narrative story that has been distilled from the stories that have been presented to the court in the stage between conviction and sentence. The construction of systems that enable novice practitioners opportunities to engage constructively with the great diversity of factors and case material that can be advanced in the pre-sentence hearing will have great benefits in increasing expertise development. Enabling students this possibility will greatly enhance their ability to make arguments and advocate for the position of the clients.

8   Argumentation

Argument schemes are "conventionalized ways of displaying a relation between that which is stated in the explicit premise and that which is stated in the standpoint" [[52] p. 19-20]. These explicit premises of an argument relate to accepted truths, which in the case of legal discourse may have been negotiated and or established by legislation. Argumentation is always a defense of a point of view [53]. This definition has been built upon by van Eemeren who indicates that argumentation is:

> a verbal, social and rational activity aimed at convincing a reasonable critic of the acceptability of a standpoint by advancing a constellation of propositions justifying or refuting the proposition expressed in the standpoint [[52] p.11].

Argumentation is used here as a method of knowledge representation. Stranieri and Zeleznikow [54] suggest that the argumentation structure suggested by Toulmin is useful as the basis for knowledge representation within artificial intelligence for the following five reasons, it:

- reflects practical reasoning,
- captures many types of inferencing,
- is linked with explanations,
- captures plausible reasoning, and
- combines to form a chain of reasoning.

Argumentation has been used by artificial intelligence researchers in two distinct ways; to structure knowledge and to model dialectical reasoning. The sentencing task is not a form of dialectical reasoning. It is not concerned with the critical examination of the truth of an opinion.

The approach to modeling the discretionary and intuitive domain of sentencing is based on the model of argument proposed by Toulmin [55]. It is a method of structuring an argument that is not mathematical in nature. The Toulmin model is jurisprudential and consequently concerned with showing that logic can be seen as a kind of philosophy of law rather than science. The procedural nature of the Toulmin argument structure is useful for structuring argument after it has been articulated and beneficial for learners as it is able to be used to justify decisions and is often used as the structure for systems that provide explanations [6]. It is able to capture arguments regardless of content. The procedural nature and simplicity of the Toulmin model means that argument chains can be constructed by linking together single argument units. The claim of one argument can be used as the data item for the next. According to Toulmin, an argument is made up of a combination of five components: a claim, some data (grounds), a warrant, some backing, and a qualifier. Claims are ideas that the arguer would like the audience to believe. The data lends support to the claim and makes it more likely that the audience will believe it. The warrant, on the other hand, is the logic of the argument: the rules of inference that lead the claimant to conclude the claim, given one ground or a set of grounds. Backings usually give reasons why the audience should believe the warrant. Modal qualifiers modify the claim by indicating a degree of reliance on, or scope of generalization of, the claim, given the grounds, warrants, and backing available. Rebuttals are the possible exceptions to the conditions under which a claim holds.

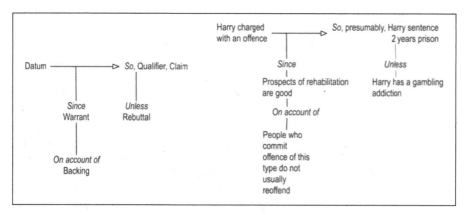

Fig. 1. Toulmin argument structure

The Toulmin argument structure offers those interested in knowledge engineering a method of structuring domain knowledge. It also enables the reasoning behind certain claims to be made explicit. There are a number of benefits that stem from utilizing Toulmin structures; decision makers often need to provide reasons especially so as to ensure transparency of processes and fulfill procedural justice principles. Yearwood and Stranieri [56] present a list of other attempts to use the Toulmin argument model as a method of structuring reasoning and modeling discourse. The approach is very beneficial in attempting to model the sentencing domain due to the ability for a resultant representation to be used in a computer-based system to provide information in a number of different rubrics. It might be possible to store information can be stored around clusters of arguments. An example of one of the argument trees from [28] is presented below.

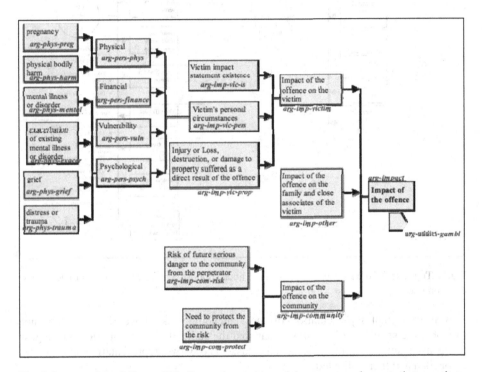

Fig. 2. Impact of the Offence. This figure shows some of the common elements that constitute the impact of the offence.

To use the argument trees from [28] it is necessary for the majority of them to be re-constructed based on the cognitive model has been presented earlier. The figure above (Fig. 2) according to the cognitive model must be incorporated into the argument tree that represents the circumstances of the offence.

9   Knowledge Use and Building Systems

The sentencing information system framework is non-hierarchical and allows judges to navigate the through the various elements as determined by their requirements [28]. As has been detailed above the Judicial College of Victoria have described an approach that indicates the order in which matters should be considered. The figure below (Fig. 3) represents the previous attempt to represent the knowledge of the sentencing task.

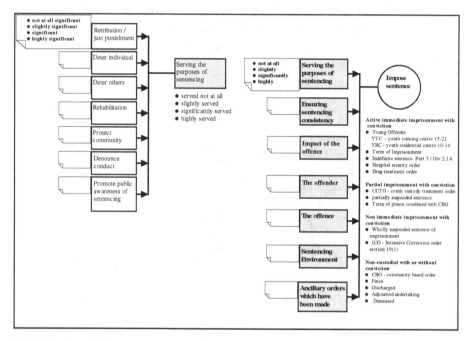

Fig. 3. Top-level of the argument tree with the node (Impose Sentence) the referent back to the decision tree

It is clear that representation of the sentencing task above (Fig. 3) does not resemble the sequence of considerations that the Judicial College of Victoria suggest to be most suitable for the sentencing task. These two representation need to be reconciled and then validated.

This approach is of critical importance to the usefulness of any information system relating to the sentencing task especially for the task of training novice practitioners and law students. It has been suggested previously that the method of presenting the information required to make an evaluation of each of the arguments that link together to make reach the main contention, that is the final sentence, is strained at best. The approach that was taken previous has resulted in a system that does not adequately reflect the cognitive model of sentencing. The system, the initial screen of which is displayed below, provides the decision maker too much freedom and needs to be made more structured. It should present the decision maker with the sequence of considerations and then facilitate the progression through the sequence.

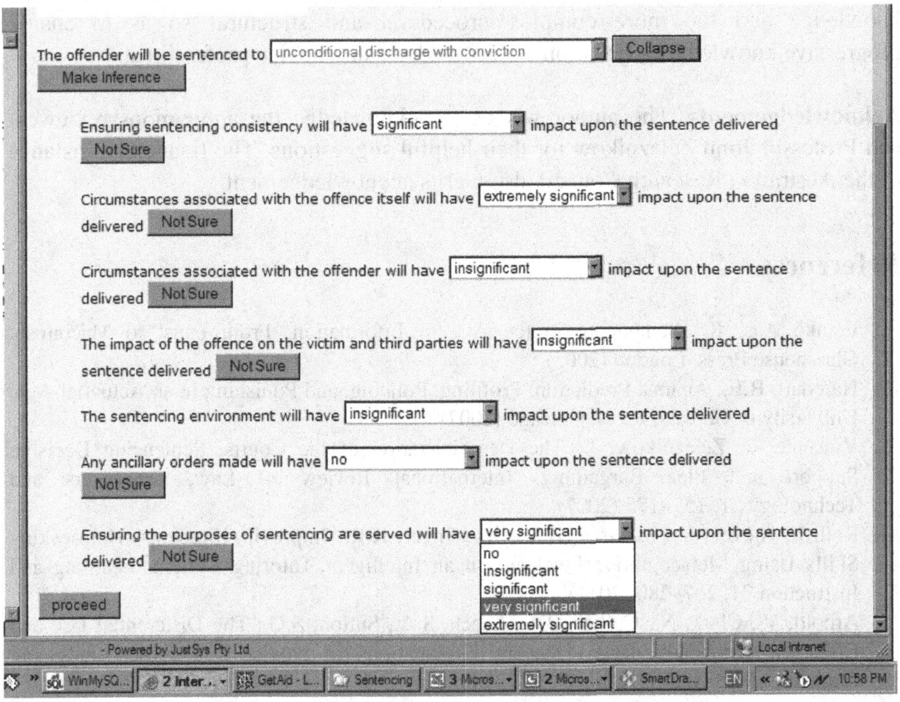

Fig. 4. Initial screen of the original proto-type sentencing system

As can be seen from the above diagram the user of system can proceed and explore the system and thus avoid the sequence of considerations that represents the cognitive model. It also suggests that the meta-sentencing principles are inconsequential. Requiring students to work through the sequence of considerations, that are fundamental to the decision making of judges, is a crucial part of the process of enabling students and novice professional to become familiar with the cognitive model and to internalize the process.

9.1   Future Work

There are several outstanding issues with respect to the sentencing model discussed in this paper. Firstly, the representation needs to validated. This task is not as simple as it might appear. The judiciary in Victoria are very reluctant to provide their expertise in these types of endeavours and rarely even engage with the senior academics. It has been suggested that this reluctance to discuss their decision-making roles and sentencing more generally is a result of usually harsh treatment they receive in the media [57]. The lack of judicial enthusiasm is not uncommon [58], especially regarding knowledge repositories that could possibly be used to expose even more starkly the lack of consistency which detractors of the current sentencing environment in Victoria often highlight. Any system that results from the cognitive model also needs to be able to provide explanations both in terms of the simple declarative knowledge and the more complex procedural and structural so as to enable progressive knowledge acquisition by law students and novice professionals.

Acknowledgements. The author wishes to acknowledge the anonymous reviewers and Professor John Zeleznikow for their helpful suggestions. The financial assistance of the Australian Research Council also merits acknowledgement.

References

1. Franko Aas, K.: Sentencing in the Age of Information: From Faust to Macintosh. Glasshouse Press, London (2005)
2. Harcourt, B.E.: Against Prediction: Profiling, Policing, and Punishing in an Actuarial Age. University of Chicago Press, Chicago (2007)
3. Vincent, A., Zeleznikow, J.: The Desocialization of the Courts, Sentencing Decision Support and Plea Bargaining. International Review of Law, Computers and Technology 21, 157–175 (2007)
4. Roll, I., Aleven, V., Mclaren, B.M., Koedinger, K.R.: Improving Students' Helpseeking Skills Using Metacognitive Feedback in an Intelligent Tutoring System. Learning and Instruction 21, 267–280 (2011)
5. Arnold, V., Clark, N., Collier, P.A., Leech, S.A., Sutton, S.G.: The Differential Use and Effect of Knowledge-Based System Explanations in Novice and Expert Judgment Decisions. MIS Quarterly 30, 79–97 (2006)
6. Arnold, V., Clark, N., Collier, P.A., Leech, S.A., Sutton, S.G.: Explanation Provision and Use in an Intelligent Decision Aid. Intelligent Systems in Accounting, Finance and Management 12, 5–27 (2004)
7. Arnold, V., Collier, P.A., Leech, S.A., Sutton, S.G., Vincent, A.: Incase: Simulating Experience to Accelerate Expertise Development by Knowledge Workers. Intelligent Systems in Accounting, Finance and Management (forthcoming)
8. Ericsson, K.A.: Deliberate Practice and the Acquisition and Maintenance of Expert Performance in Medicine and Related Domains. Academic Medicine 79, S70–S81 (2004)
9. Victorian County Court: County Court of Victoria Annual Report 2008-2009. Victorian County Court, Melbourne (2009)
10. Fox, R.: Victorian Criminal Procedure: State and Federal Law. Monash Law Book Co-operative, Clayton (2005)
11. Tak, P.J.: Sentencing and Punishment in the Netherlands. In: Tonry, M.H., Frase, R.S. (eds.) Sentencing and Sanctions in Western Countries, pp. 151–187. Oxford University Press, Oxford (2001)
12. Weigend, T.: Sentencing and Punishment in Germany. In: Tonry, M.H., Frase, R.S. (eds.) Sentencing and Sanctions in Western Countries, pp. 188–221. Oxford University Press, Oxford (2001)
13. Roberts, J.V.: Mandatory Sentences of Imprisonment in Common Law Jurisdictions: Some Representative Models. Research and Statistics Division, Department of Justice Canada (2005)
14. Hoel, A., Gelb, K.: Sentencing Matters: Mandatory Sentencing. Victorian Sentencing Advisory Council, Melbourne (2008)
15. Machin, D.: Sentencing Guidelines Around the World. The Sentencing Commission for Scotland (2005)

16. Guerra Thompson, S.: Sentencing Guidelines in the US: A Primer. Reform. 86, 45–48 (2005)
17. Sentencing Act, Victoria (1999)
18. Frankel, M.E.: Criminal Sentences: Law without Order. Hill and Wang, New York (1973)
19. Bagaric, M., Edney, R.: What's Instinct Got To Do With It? A Blueprint for a Coherent Approach to Punishing Criminals. Criminal Law Journal 27, 119–141 (2003)
20. Ashworth, A.: Sentencing and Criminal Justice. Cambridge University Press, Cambridge (2005)
21. Markarian v The Queen. High Court of Australia 25 (2005)
22. Williscroft, R.V.: Victoria Reports 292 (1975)
23. Traynor, S., Potas, I.: Sentencing Methodology: Two Tiered or Instinctive Synthesis. Sentencing Trends and Issues 25, 1–22 (2002)
24. Dane, E., Pratt, M.G.: Exploring Intuition and its Role in Managerial Decision Making. Academy of Management Review 32, 33–54 (2007)
25. Tata, C.: Sentencing as Craftwork and the Binary Epistemologies of the Discretionary Decision Process. Social and Legal Studies 16, 425–447 (2007)
26. Dreyfus, H.L., Dreyfus, S.E.: Mind Over Machine: The Power of Intuition and Expertise in the Era of the Computer. The Free Press, New York (1988)
27. Klein, G.: The Power of Intuition: How to Use your Gut Feelings to Make Better Decisions at Work. Doubleday, New York (2003)
28. Hall, M.J.J., Calabro, D., Sourdin, T., Stranieri, A., Zeleznikow, J.: Supporting Discretionary Decision-Making with Information Technology: A Case Study in the Criminal Sentencing Jurisdiction. University of Ottawa Law and Technology Journal 2, 1–36 (2005)
29. Zeleznikow, J., Hunter, D.: Building Intelligent Legal Information Systems: Representation and Reasoning in Law. Kluwer Law and Taxation Publishers, Deventer (1994)
30. Leech, S.A., Collier, P.A., Clark, N.: Modeling Expertise: A Case Study Using Multiple Corporate Recovery Experts. Advances in Accounting Information Systems 6, 85–106 (1998)
31. Leech, S.A., Collier, P.A., Clark, N.: A Generalized Model of Decision-Making Processes for Companies in Financial Distress. Accounting Forum 23, 155–174 (1999)
32. Susskind, R.E.: Expert Systems in Law. Clarendon, Oxford (1987)
33. Jonassen, D.H., Beissner, K., Yacci, M.: Sturcutral Knowledge: Techniques for Representing, Conveying, and Acquiring Structural Knowledge. Lawrence Erlbaum Associates, Hillsdale (1993)
34. Thomas, D.A.: Principles of Sentencing: The Sentencing Policy of the Court of Appeal Criminal Division. Heinemann, London (1979)
35. Ericsson, K.A., Simon, H.A.: Protocol Analysis: Verbal Reports as Data. MIT Press, Cambridge (1993)
36. Jonassen, D.H.: Learning to Solve Problems: A Handbook for Designing Problem-Solving Learning Environments. Routledge, New York (2011)
37. Mackenzie, G.: How Judges Sentence. The Federation Press, Annandale (2005)
38. Fox, R., Freiberg, A.: Sentencing: State and Federal Law in Victoria. Oxford University Press, Melbourne (1999)
39. Giakis, R.v.: Victorian Reports 973 (1988)
40. Judicial College of Victoria: Victorian Sentencing Manual,
    http://www.justice.vic.gov.au/emanuals/VSM/default.htm

41. Niu, L., Lu, J., Zhang, G.: Cognition-Driven Decision Support for Business Intelligence: Models, Techniques, Systems and Applications. Springer, Berlin (2009)
42. Stranieri, A., Zeleznikow, J.: The Role of Open Texture and Stare Decisis in Data Mining Discretion. In: Jurix 1998, pp. 101–112 (1998)
43. Stranieri, A., Yearwood, J., Meikle, T.: The Dependency of Discretion and Consistency on Knowledge Representation. International Review of Law, Computers and Technology 14, 325–340 (2000)
44. Bagaric, M.: Punishment and Sentencing: A Rational Approach. Cavendish, London (2001)
45. Australian Law Reform Commission: Same Crime, Same Time: Sentencing of Federal Offenders. Australian Law Reform Commission, Canberra (2006)
46. Fox, R., Freiberg, A.: Sentencing: State and Federal Law in Victoria. Oxford University Press, Melbourne (1999)
47. Storey, R.v.: Victorian Reports 359 (1998)
48. Shapland, J.: Between Conviction and Sentence: The Process of Mitigation. Routledge and Kegan Paul, London (1981)
49. Legal Studies Deprtment, L.T.U.: Guilty, your Worship: A study of Victoria's Magistrates' Courts. Legal Studies Department, La Trobe University, Bundoora (1980)
50. Lovegrove, A.: Judicial Decision Making, Sentencing Policy, and Numerical Guidance. Springer, New York (1989)
51. Apistola, M., Mommers, L., Lodder, A.R.: A Knowledge Management Exercise in the Domain of Sentencing: Towards an XML Specification. In: Winkels, R., van Engers, T., Bench-Capon, T. (eds.) Proceedings of Second International Workshop on Legal Ontologies, pp. 48–57. The University of Amsterdam, Amsterdam (2001)
52. van Eemeren, F.H.: The State of the Art in Argumentation Theory. In: van Eemeren, F.H. (ed.) Crucial Concepts in Argumentation Theory, pp. 11–26. University of Amsterdam Press, Amsterdam (2001)
53. van Eemeren, F.H., Kruiger, T.: Identifying Argumentation Schemes. In: van Eemeren, F.H., Grootendorst, R., Blair, J.A., Willard, C.A. (eds.) Argumentation: Perspectives and Approaches (Proceedings of the Conference on Argumentation 1986), pp. 70–81. Foris Publications, Dordrecht (1986)
54. Stranieri, A., Zeleznikow, J.: A Survey of Argumentation Structures for Intelligent Decision Support. In: Burnstein, F. (ed.) Proceedings of 1999 International Society for Decision Support Systems, pp. 1–23. Monash University, Melbourne (1999)
55. Toulmin, S.E.: The Uses of Argument: Updated Edition. Cambridge University Press, Cambridge (2003)
56. Yearwood, J., Stranieri, A.: The Generic/Actual Argument model of practical reasoning. Decision Support Systems 41, 358–379 (2006)
57. Mackenzie, G., Stobbs, N., O'Leary, J.: Principles of Sentencing. Federation Press, Annandale (2010)
58. Ashworth, A., Genders, E., Mansfield, G., Peay, J., Player, E.: Sentencing in the Crown Court: Report of an Exploratory Study. Centre for Criminological Research, Oxford (1984)
59. Brest, P.: The critique of pure reason: The role of intuition in judgement and decision making. In: International Conference on Artificial Intelligence and Law (2005) (keynote paper)
60. Sauter, V.L.: Intuitive decision-making. Communications of the ACM 42, 109–115 (1999)
61. Hogarth, R.M.: Educating Intuition. University of Chicago Press, Chicago (2001)

Application of Model-Based Diagnosis to Multi-Agent Systems Representing Public Administration

Alexander Boer and Tom van Engers

Leibniz Center for Law, University of Amsterdam, The Netherlands
{A.W.F.Boer,T.M.vanEngers}@uva.nl

Abstract. In public administration, legal knowledge in the form of *critical incidents* and, for want of a better word, *noncompliance storylines* is important for monitoring and enforcement, but has no natural place in traditional forms of legal knowledge representation such as normative rules or legal argument schemes. In this paper we present a model-based diagnosis view on the complex social systems in which large public administration organizations operate. The purpose of diagnosis as presented in this paper is to identify problematic agent role instances in a multi-agent system (MAS). We propose the model-based diagnosis problem as an explanation of the driving forces behind requests for change to the IT, business process design, and policy making departments in public administration. This makes model-based diagnosis an important potential legal knowledge acquisition frame for public administration.

1 Introduction

Although few would argue that large organizations are designed, with *agents* as the components, little has been done to address complex social systems as a model-based diagnosis problem. Localization of fault is however an important part of a sterotypical legal problem: we look for *someone* responsible for norm violation. Enforcement involves a search for a reasonable *responsibility assignment* to one or more agents, and subsequent punishment of, or remedial action by, those agents. Problems that threaten public administration also involve localization of agents responsible for performance problems, and design-based remedies, like resource reallocation, redesign of software, revision of guidelines, etc, can be characterized as *component repairs*.

In [1], we have characterized the problem of adapting business processes and decision support software in public administration, in response to changing society and government policy, as a design & diagnose problem solving cycle, operating on agent roles as components. In public administration, legal knowledge in the form of *critical incidents* and, for want of a better word, *noncompliance storylines* used for monitoring and enforcement has an important function in knowledge management. A central place in these stories is taken by stereotypical agent intentions like the intention to evade income taxes. These stereotypical agent intentions tend to resurface even if changes are made to the specific legal rules to stop abuse. This knowledge has no natural place in traditional forms of legal knowledge representation such as normative rules or legal argument schemes.

M. Palmirani et al. (Eds.): AICOL Workshops 2011, LNAI 7639, pp. 235–244, 2012.
© Springer-Verlag Berlin Heidelberg 2012

In this paper, a sequel to the more general presentation in [1], we evaluate the efficacy of Reiter's model-based diagnosis frame [2], explained in section 3, applied to the complex social systems in which large public administration organizations operate. The purpose of diagnosis as presented in this paper is to identify problematic agent role instances in a multi-agent system (MAS). To this end, we introduce in section 4 the model-based diagnosis problem of a diagnostic agent having a diagnostic model of an observable multi-agent system in its environment. We introduce the *agent role* as component of the multi-agent system, and agent role descriptions as models of normal, functional components and abnormal, faulty components. The diagnostic agent is *part* of the multi-agent system, has only limited access to the information exchanged in the multi-agent system, limited ability to control and change the multi-agent system, and actions to obtain information change the state of the system.

The model-based diagnosis problem, based on executable agent role descriptions, is a model for the acquisition of diagnostic compliance knowledge in public administration. In the paper we use the domain of real estate property transactions, from the point of view of a tax administration, as an simple example, in section 4.2.

2   Related Research

Fig. 1 presents the functional classification of generic tasks in public administration that we proposed in [1, 3]. The organization principle for this classification is functional dependency between knowledge roles, inspired by the typology of problems and views on problem solving presented two decades ago in [4]. The *suite of problem types* in [4] was based on an analysis of the problem and task decompositions found in then-state-of-the-art knowledge-based system literature [5–7]. The suite of problem types presents us with a generic problem solving cycle, and two different vocabularies for describing it, depending on the type of model of the domain that is available:

1. *Model → Design → Implement → Monitor → Diagnose*
2. *Classify case → Plan → Execute → Monitor → Assessment*

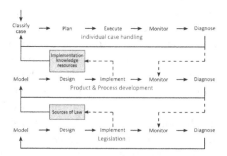

Fig. 1. The case handling, development, and legislation problem solving cycles

Design in public administration can be approached from the first perspective, suggesting a multi-agent model-based diagnosis problem type. The principles of model-based diagnosis have been explained well by [2]. In this paper we follow the problem formulation given there, because Reiter's approach to the problem can be straightforwardly mapped to more recent logics used for representation of law.

The problem of multi-agent diagnosis has been addressed in different contexts. Generally, there are several interesting ways to combine multi-agent systems with model-based diagnosis that have been explored in the literature. One approach focuses on distributed diagnosis of (generally non-agent) systems [8, 9], for instance the diagnosis of systemic failure of distributed sensors in a complex system. The systems being diagnosed are of the traditional kind, and external to the agents, but the diagnostic problem solver is a multi-agent system. This approach addresses the information and coordination problems that arise in diagnosis of complex multi-agent systems by participants of the system. The diagnostic agents only have local access to a greater system, and need to coordinate the diagnostic hypothesis generation and testing process. The centralization of diagnostic information in a single agent that discriminates between hypotheses and orders tests is theoretically the most efficient solution [8].

Another approach is to diagnose *multi-agent plans*. A diagnostic hypothesis from this point of view is an identification of failing parts of a plan [10]. Although this approach treats a multi-agent system as the subject of model-based diagnosis, the agents are not the components of the system of interest. This problem formulation has some similarity with the one addressed in this paper, but the setting to which it applies is a fully cooperative setting. It shares with [8, 9] the assumption that problems are at the root caused by different or false beliefs about the system, and not by competing interests, as in our problem setting.

Lastly, one could consider the *normative assessment* approach found in the normative multi-agent system problem formulation [11] to be a kind of diagnosis approach to mullti-agent systems, but the classical normative assessment problem formulation does not make a central issue out of localization of faults, which is an essential characteristic of model-based diagnosis. Although related in function, it is a specialized case of the simpler assessment problem formulation in the typology of problems.

3   Model-Based Diagnosis Problems

The model-based diagnosis setting can be viewed as a problem of a single diagnostic agent having a model of a system to be diagnosed as its environment. Diagnosis presumes that a system can be decomposed into small components with well-understood behaviour models. Effectiveness of model-based diagnosis approaches in the literature is largely determined by the extent to which it is possible to obtain observations on the states implied by the model of the system, and the possibility to test components independently from the rest of the system. How to do this is of course not directly obvious for non-trivial social systems.

Following [2], a description of a system can be characterized as a pair $(SD, COMP)$, where SD, the system description, is a set of first order sentences, and $COMP$ a finite set of constants identifying components. Typically, a system description describes how a system normally behaves, and it often distinguishes a description of structure from a description of function of the components. The functional model causally relates input and output terminals of components, and terminals of components are connected.

An observation of a system is a set of first order sentences OBS on the events happening at component terminals. If OBS and SD are inconsistent with the assumption that all components are normal, certain of the components behave abnormally. A diagnosis is a hypothesis that components $AB \subset COMP$ in $(SD, COMP, OBS)$ are abnormal and the rest normal. The diagnostic process usually involves making additional observations as evidence for ruling out hypotheses (measurement). Diagnostic reasoners may use the functional model for both causal and evidential reasoning.

Diagnoses can be generated on the basis of component fault modes. The diagnosis in this case conjectures alternative behaviour descriptions for the components that are behaving abnormally. The set of fault modes of a component may be complete: in this case the component must be behaving according to the health mode or one of the fault modes. Alternatively there may be unknown fault models.

Default reasoning about normality of component behaviour can be modeled with some predication of (ab)normality, here n, and a normal default theory of the form $DT = (\{ Mn(c)/n(c) \mid c \in COMP\}, SD \cup OBS)$ [2]. For this default theory, Reiter's default logic extensions are exactly those of the generic diagnosis problem for $(SD, COMP, OBS)$ directed towards a *minimal* set of abnormal components [2]. In SD a complete set of fault modes $f_1, f_2, \ldots, f_n$ can be expressed by the first order axiom $t(x) \wedge \neg n(x) \supset f_1(x) \vee f_1(x) \vee f_2(x) \vee \ldots \vee f_n(x)$, and $\neg(f_i(x) \wedge f_j(x))$ for any different $1 \leq i, j \leq n$, [2], where t is the component type to which the partition into behaviour modes applies. In analogy, health modes $n_1, n_2, \ldots, n_n$ can be distinguished. In addition, we may rule out diagnoses on grounds of impossibility, faults in components may imply faults in other components, and a probability distribution of fault modes may be known that guides selection of hypotheses.

Reiter's default logic extensions are stable extensions in terms of Dung's argumentation frameworks [12]. This makes the notion of a diagnosis concrete enough to implement it with logic programming [12]. in multi-agent programming environment Jason [13], which supports logic programming. The main problem is conceptual: to represent the multi-agent system as a set of components to be diagnosed, and messages as opportunities for observation.

4   Agent Roles as Model-Based Diagnosis Components

In [1] we argued for the use of agent role instances rather than agents as the knowledge components in a simulation of noncompliance storylines. The agent role construct is associated to *reflective function*. By attributing beliefs, desires, and intentions to others, we make the behaviour of others meaningful. Agents are able to flexibly activate, from a collection of self-other representations organized by prior experience, the one(s) best suited to the circumstances [14]. We have similar collections of representations of the self in relation to the affordances of an environment, for instance for the use of tools. Agent role knowledge distinguishes itself from other self representations by the potentiality of agent role inversion: we can imagine being in the other agent's shoes, and reflect on the beliefs, desires, and intentions we would have. For agent roles such as buyer–seller we literally have the option of experiencing the transaction from the other point of view.

In MAS literature we find various proposals for adding agent roles to MAS in order to model social systems [15]. In the conventional view, the most important operations in MAS role dynamics are understood to be are enact and deact, which means that an agent starts and finishes to occupy an agent role, and activate and deactivate, which means that an agent starts executing actions belonging to the role and suspends the execution of the actions [15].

We propose to drop the notion of a deliberative agent-level process that exercises control and enacts, deacts, activates, and deactives roles, and to make the coordination between agent roles part of agent role specifications. The agent role description itself specifies when it activates another agent role or deactivates itself. This means that the deliberative agent self of a natural person becomes an empty shell: nothing more than an identity marker to impute legal and moral responsibility to. In [1] we defended this choice in the context of simulation of legal storylines in a MAS. In the MAS as a model-based diagnosis problem setting, the possibility of localization of faults is another, pragmatic reason to keep planning ability in agent components simple.

The existence of social structures, and the conventional or healthy behaviours associated with them, must be accepted by the participants in a social structure. Complex combinations of agent roles, with social coordination mechanisms between them, may be considered stereotypes of organizations, other social arrangements, and individual persons, in the *society of mind* tradition [16]. If the agent role description is adequate, its behaviour in response to the events to which it is liable to respond is predictable. If the agent knows the agent role it is dealing with, and it has an appropriate matching self representation, it does not normally need to model the intentions and beliefs of the other agent. It has planning routines for all normal outcomes of interaction.

Competence at task performance allows for a certain degree of flexibility of behaviour within a role. Diagnosis occurs when events indicate a *problem* calling for reinterpretation of the others in the environment in terms of health and fault modes. System breakdowns result from participants working with different social environment models, accidentally or on purpose.

Diagnostic agents participate in the social system being diagnosed, and affect eachother. In real social systems, the participants do not always share the results of diagnosis. This observation is for instance relevant to business process design in public administration: if personnel on the work floor implements its own workarounds for perceived problems, this may lead to impoverished diagnostic information on higher levels, and therefore an invisible design problem may not be addressed. Moreover, agents operating in a fault mode may actively diagnose for signs of the fault mode being caught, and in response hide their tracks. A diagnostic agent draws attention to it if it has to ask for information, tests the trustworthiness of others by comparing reports from different sources, etc.

A final limitation is that relevant events are always messages between agents. A payment from a to b is for instance reinterpreted as an order from a to its bank, and an acknowledgement to b from its bank. This works adequately in bureacratic environments, although we recognize the limitations of this stance in principle.

A diagnostic agent does not have access to all messages exchanged between relevant agents. There is a distinction between topological connections to other agents that are real, in the sense that the agent really sends and receives messages, and topological connections that are hypothesized, of which the agent may receive information from other trusted agents, that may however behave in a fault mode. The only ways in which agent c can know of the messages from a to b are 1) notification by a, 2) notification by b, 3) a trusted carbon copy of a message sent from a to b. The diagnostic agent can only get somewhere if it trusts either a's or b's health status.

4.1   Social Structure and Agent Role Component Descriptions

In Reiter's terms, a description of a social environment is a pair $(ENV, ROLES)$, where ENV is a set of first order sentences, and $ROLES$ a set of agent role instances in the environment. All $a \in ROLES$, are assigned an agent role class $A(a)$ in ENV, which determines possible health modes $Ah_1, Ah_2, \ldots, Ah_n$ and fault modes $Af_1, Af_2, \ldots, Af_n$ for the agent role, as specified in section 3. The structural topology of the system is determined by messages $M(a_1, a_2, m)$, where a_1, a_2 are agent roles and m is a message, and agents have goals $G(a, g)$, where a is an agent role and g a goal. Behaviour descriptions in ENV appeal to agent role class and behaviour mode, e.g:

$$A(x) \wedge Ah_1(x) \wedge A(y) \wedge G(x, z) \subset M(x, y, z)$$

The following should be the case for such behaviour descriptions:

1. A message sent may be conditional on the agent role class and health or fault mode of the sender;
2. A message sent may be conditional on the agent role class of the receiver, but not of the health or fault mode of the receiver, since the sender does not know it;
3. A specific health mode an agent role operates in should not normally affect completion of goals, only the path towards it.

The pragmatic approach to the diagnosis problem we have chosen, depends on the possibility of translating between agent descriptions in a BDI-style agent representation language, AgentSpeak, and a representation in first order sentences. For this purpose others in the same project are working on a semi-automatic mapping from business rules specified in RIF to AgentSpeak in [17].

The acquisition of legal diagnostic knowledge depends on the conceptual model of the diagnosis problem, with its health modes and fault modes, rather than on its implementation in the form of model-based diagnosis algorithms as in [2]. Each mode represents a possible agent in a MAS simulation, while the first order sentences describing agents, allow for diagnostic reflection on the social environment by agents equipped with diagnostic ability implemented in logic programming constructs.

The example behaviour rule earlier for instance translates to: an x_{A,Ah_1} has the following AgentSpeak plan operator $+!z : a(Y) \leftarrow m(Y, z)$. This means that $m(Y, z)$ becomes a intention if an intention to z is added, and $a(Y)$ is the case. The partial translation reflects the fact that the logical rule, depending on perspective and the reasoning abilities of the agent to which it is added, may be implemented in different forms.

4.2   An Example Knowledge Acquisition Environment: Real Estate Tax Evasion

In this section we present an example domain for diagnosis that we work on, and an agent description in the form in which it can be acquired from an expert. The model-based diagnosis framework helps us to organize fraud storylines.

Within the tax administration, important diagnostic processes deal with tax evasion, and feedback to the legislator on new tax avoidance strategies that erode the tax base. The tax administration, and its customs and fiscal intelligence departments, play a supporting role in crime fighting.

A representative knowledge domain subject to monitoring by the tax administration is real estate transactions. The tax administration has no direct interest in buying or selling real estate, but real estate transactions do involve a lot of money and are a good vehicle for tax evasion, because they are typically not taxed, or taxed at a much lower rate than VAT, personal income taxes, etc. In the field of real estate we find a variety of types of crime:

Tax Evasion Proper: selling or buying below or beyond real market value to avoid income taxes or VAT on an amount due to the other party;

Bid Rigging: typical of foreclosure auctions, sellers unknowingly sell property below value to a buyer cartel, which distributes the profit among the cartel participants;

Hiding Kickbacks: use of real estate transfers below or beyond real market value to hide illegal kickbacks; and finally

Extortion and Theft: a buyer or seller deviating from real market value because of threat with some form of violence or blackmail, or because its agent received a kickback.

A shared feature of these scenario types is that they involve large deviations from apparent market value, or untypical quick depreciation or appreciation of real estate property value. It is not, however, trivial for an outsider to determine what the real market value of a property should be, and the market value may be misrepresented even to outsiders by appraisers that participate in a scam. The deviation itself is for a tax administration not a reason to act: to enforce, a plausible story is needed, involving a motive, and collusion between parties.

Knowledge acquisition is about eliciting these stories, with a special focus on intentions and plans of the participants, and the messages that must have been exchanged.

Health modes reflect proper intentions of the buyer and seller. A healthy buyer wants to obtain at least market value mv, and expects little more. A healthy seller wants to pay no more than market value, and expects little less. Both may refer to alternative bids to judge market value, or may trust an appraiser to estimate market value mv'. A typical healthy factor that will affect value negatively is for instance haste.

Fault modes for the buyer or seller are based on an intentional deviation $addv$ from market value. This is either the intention of buyer or seller alone, in collusion with false competitors, a fraudulent agent of the other party, or a fraudulent appraiser ($mv' = mv + addv$), or it is a coordinated intention between both, because the amount $addv$ is owed for some service, or is a gift, or is being extorted. When the buyer or seller is the victim, this will become apparent if the participants are interviewed about the transaction directly. Doing this does however tip off the participants in a coordinated transaction, and allows them to hide their tracks.

The following is a generic AgentSpeak plan for a fault mode seller who intends to transfer an amount $addv$ to the buyer (with predicates in natural language, and variables of central importance emphasized, to enhance readability):

- ! transfer $addv$ to recipient
- +! transfer $addv$ to recipient ← propose to sell a property worth mv for $mv - addv$.
- + Plan to sell a property worth mv for $mv - addv$ accepted ← propose a property worth mv.
- + Property is worth mv is accepted ← ! secure property for mv; ! sell property for $mv - addv$.

In this form, agent mode descriptions can be directly written down during an interview with experts, to be worked out later in code. As we consider these scenarios in the context of model-based diagnosis, the number of relevant differentiated agent roles and modes tends to grow, because we rely on interception of (coordination) messages to verify the fault mode, and to provide us with a motive for the behaviour.

For a tax administration, various sources of information (appraisers, notary lawyers, competitors, colleagues of an agent representing buyer or seller, straw men, patsies, etc) are available to test the hypothesis that it is dealing with this type of buyer. The buyer leaves a conspicuous evidence trail if he has to acquire the property and then immediately sells it for a loss. An appraisal mv' before the property sells for $mv' - addv$ is for instance conspicuous, just like the buyer immediately selling it again for mv. These transactions attract attention to the relationship between buyer and seller. One way to hide a relationship between participants, and therefore the motive of the transaction, is to make transactions part of larger packages of transactions: a for instance sells property p_1 to foreign party b for $mv - addv$, who then sells property p_2 to c for $mv - addv$, the right holder for a payment of an amount $addv$ towards a, who then cashes $addv$ by selling p_2. Much more complicated arrangements exist, and generally $addv$ can remain smaller as a percentage of the whole, and therefore be less conspicuous, by increasing the mv of the property package. But participants in a complicated scam may for instance have trouble trusting eachother, leading them to sign all deeds at the same time in the presence of a notary lawyer. This makes the notary lawyer an important monitor of transactions, but some notary lawyers often involved in setting up such complicated transactions are suspicious themselves.

In [18], a more dense version of this example, with competing healthy and faulty explanations, is worked out in more detail.

5   Conclusions

Central to our knowledge acquisition approach is a focus on social arrangements, motives, and messages that prove coordination, a perspective that is often hard to distill from case stories from experts. The agent role components with their health and fault modes help experts to explore a dense diagnostic hypothesis of fault modes, and similar health modes being mimicked, and focus attention on the problem of discriminating between suspects.

This approach is a model-driven alternative to ad hoc approaches to embedding tacit knowledge about noncompliance in the business rules and business processes of the organization. Although there are good reasons to believe that social systems are too complex for application of model-based diagnosis in Reiter's terms, a diagnosis-based methodology for compliance knowledge representation:

- distinguishes knowledge of conventional behaviour, and intended, designed behaviour, from common incidents and noncompliance masquerading as conventional behaviour in a systematic manner;
- permits the modeling of misunderstandings between network partners arising from different views of the social environment;
- encourages early simulation of abuse of new policy by stereotypical policy abusers, based on known design patterns; and
- encourages systematic collection of diagnostic knowledge for enforcement and redesign purposes.

An alternative description of our diagnostic approach, not based on Reiter's concepts in [2], is found in [18].

An aspect of our approach that needs more development is the use of generic agent coordination design patterns, to deal with the variety of approaches of coordinating the same plan among multiple agents. The approach of [19], and later work based on it, specifically attracts our attention, and combines well with our view on the use of normative rules in law. Standardization of coordination modeling is especially relevant because coordination also takes the place of a higher order deliberation process within an agent, and because normative positions play an important role in coordination in this domain.

References

1. Boer, A., van Engers, T.: An agent-based legal knowledge acquisition methodology for agile public administration. In: ICAIL 2011: Proceedings of the 13th International Conference on Artificial Intelligence and Law. ACM, New York (2011)
2. Reiter, R.: A theory of diagnosis from first principles. Artificial Intelligence 32(1), 57–95 (1987)
3. Boer, A., van Engers, T.: Knowledge Acquisition from Sources of Law in Public Administration. In: Cimiano, P., Pinto, H.S. (eds.) EKAW 2010. LNCS, vol. 6317, pp. 44–58. Springer, Heidelberg (2010)
4. Breuker, J.: Components of Problem Solving and Types of Problems. In: Steels, L., Schreiber, G., Van de Velde, W. (eds.) EKAW 1994. LNCS, vol. 867, pp. 118–136. Springer, Heidelberg (1994)
5. Chandrasekaran, B., Johnson, T.R.: Generic tasks and task structures: History, critique and new directions. In: David, J.M., Krivine, J.P., Simmons, R. (eds.) Second Generation Expert Systems. Springer (1993)
6. Clancey, W.J.: Heuristic classification. Artificial Intelligence 27, 289–350 (1985)
7. Steels, L.: Components of expertise. AI. Mag. 11(2), 30–49 (1990)
8. Kalech, M., Kaminka, G.A.: On the design of coordination diagnosis algorithms for teams of situated agents. Artificial Intelligence 171(8-9), 491–513 (2007)

9. Roos, N., ten Teije, A., Witteveen, C.: A protocol for multi-agent diagnosis with spatially distributed knowledge. In: Proceedings of the Second International Joint Conference on Autonomous Agents and Multiagent Systems, AAMAS 2003, pp. 655–661. ACM, New York (2003)

10. Witteveen, C., Roos, N., van der Krogt, R., de Weerdt, M.: Diagnosis of single and multi-agent plans. In: Proceedings of the Fourth International Joint Conference on Autonomous Agents and Multiagent Systems, pp. 805–812. ACM, New York (2005)

11. Boella, G., Pigozzi, G., van der Torre, L.: Five guidelines for normative multiagent systems. In: Proceeding of the 2009 Conference on Legal Knowledge and Information Systems, pp. 21–30. IOS Press, Amsterdam (2009)

12. Dung, P.M.: On the acceptability of arguments and its fundamental role in nonmonotonic reasoning, logic programming and n-person games. Artificial Intelligence 77(2), 321–357 (1995)

13. Bordini, R., Hübner, J., Vieira, R.: Jason and the golden fleece of agent-oriented programming. In: Weiss, G., Bordini, R., Dastani, M., Dix, J., Fallah Seghrouchni, A. (eds.) Multi-Agent Programming. Multiagent Systems, Artificial Societies, and Simulated Organizations, vol. 15, pp. 3–37. Springer US (2005), 10.1007/0-387-26350-0-1

14. Fonagy, P., Target, M.: Attachment and reflective function: Their role in self-organization. Development and Psychopathology 9(04), 679–700 (1997)

15. Tinnemeier, N., Dastani, M., Meyer, J.J.: Roles and norms for programming agent organizations. In: Proceedings of The 8th International Conference on Autonomous Agents and Multiagent Systems, AAMAS 2009, vol. 1, pp. 121–128. International Foundation for Autonomous Agents and Multiagent Systems, Richland (2009)

16. Singh, P.: Examining the society of mind. Computing and Informatics 22, 521–543 (2004)

17. Gong, Y., Overbeek, S., Janssen, M.: Integrating semantic web and software agents: Exchanging rif and bdi rules. IJSSOE 2(1), 60–76 (2011)

18. Boer, A., van Engers, T.M.: Implementing compliance controls in public administration. In: Atkinson, K. (ed.) Legal Knowledge and Information Systems - JURIX 2011: The Twenty-Fourth Annual Conference, University of Vienna, Austria. Frontiers in Artificial Intelligence and Applications, December 14-16, vol. 235, pp. 33–42. IOS Press (2011)

19. Jennings, N.R.: Commitments and conventions: the foundation of coordination in multi-agent systems. Knowledge Engineering Review 8(3), 223–250 (1993)

Semantic Annotation of Legal Texts through a FrameNet-Based Approach

Marcello Ceci[1], Leonardo Lesmo[2], Alessandro Mazzei[2],
Monica Palmirani[1], and Daniele P. Radicioni[2]

[1] University of Bologna, CIRSFID, Italy
{marcello.ceci,monica.palmirani}@unibo.it
[2] University of Turin, Computer Science Department, Italy
{lesmo,mazzei,radicion}@di.unito.it

Abstract. In this work we illustrate a novel approach for solving an information extraction problem on legal texts. It is based on Natural Language Processing techniques and on the adoption of a formalization that allows coupling domain knowledge and syntactic information. The proposed approach is applied to extend an existing system to assist human annotators in handling normative modificatory provisions –that are the changes to other normative texts–. Such laws 'versioning' problem is a hard and relevant one. We provide a linguistic and legal analysis of a particular case of modificatory provision (the *efficacy suspension*), show how such knowledge can be formalized in a linguistic resource such as FrameNet, and used by the semantic interpreter.

Keywords: Knowledge modelling, semantic interpretation, NLP.

1 Introduction

Legal systems are dynamic by nature, since they change over time. Modifications affect legal texts, their temporal properties, and even the meaning of the norms expressed in those texts. Many efforts have been invested in the last ten years towards the digitalization in the legal domain. Researches to produce updated collections of legal documents on the Web are being conducted with multiple aims, such as intelligent indexing, querying, searching, filtering, retrieval of documents or of meaningful parts, and to help managing changes in the legal content, through the so-called *consolidation process*. The digitalization process requires solving two sorts of problems: defining (XML) file formats to conveniently encode the texts, and designing systems to assist human experts in the annotation of the legal texts according to a format devised at the previous step. Much work has been done in both directions. Various initiatives have been established at the national and international levels to devise XML standards for describing legal sources and schemas to identify legal documents [8]. Also, systems have built that automatically identify and classify structural portions of legal documents and their intra- and inter-references [2,12]; the problem of semantic analysis is currently being investigated[15]. Unfortunately, due to the

M. Palmirani et al. (Eds.): AICOL Workshops 2011, LNAI 7639, pp. 245–255, 2012.
© Springer-Verlag Berlin Heidelberg 2012

'natural language barrier' (i.e., the problem of translating a sentence into some form of semantic interpretation [11]), this is still an open problem. This paper is concerned with this problem, that is the extraction of modificatory provisions and their annotation.

One main aspect makes legal texts suitable for applying information technologies commonly used to deal with hypertexts. Legal texts contain references to other legal texts or to other parts within the same document, so that a legal text can be naturally considered as a particular case of hypertext.[1] To have a sound example, let us consider the following example of *consolidation* problem: a legal document A contains a reference to document B. Say, e.g., that A contains a locution such as 'the first article of the law number 9 in the document B is suspended until January 29, 2011'. Unfortunately, a person interested in inspecting the validity of the norms in B could encounter some problems to figure out whether the norms in B are still valid, in force, etc., because B contains no reference to A (that is, it is not possible to add backward pointers to existing legal documents). Under this perspective, legal systems can be seen as tangled webs. It should be noted that at least for some normative systems –such as the Dutch and the Italian ones (see [9] and [15,3,4], respectively)–, the consolidation problem is a relevant one. In fact, the uncertainty on the effects of normative modifications would undermine the certainty of the law, making it hard to clearly understand which one of several versions of a provision counts as law. Automating the process of semantically annotating modificatory clauses and provisions would be of great help in simplifying the legal system and in consolidating texts of law,[2] because the human annotation process is expensive and error-prone. From a practical perspective, the consolidation process involves identifying the main elements of the modificatory provisions, annotating them in the legal text according to a given DTD (we adopt the NIR standard, but in principle in a different context another standard could be adopted), and generating a set of metadata that compactly describes the considered modification.

In past works we detected some regularity in the linguistic structure of modificatory provisions [14], and we showed how this regularity, coupled with a XML markup [12] can be used by automated tools to qualify a modificatory provision [13]. In particular, our approach relies on a tree-matching technique to put together deep parsing and shallow semantic interpretation [7]. In the present work we extend our approach by devising a specialised version of FrameNet [1] to cope with modificatory provisions. In particular, we model the efficacy suspension modificatory provisions. The paper is structured as follows: we first illustrate the considered problem of automatically annotating XML files with information describing the modificatory provisions. We then consider the case of the efficacy suspension, which is by far more complex on a linguistic perspective, and argue that it requires enhanced modelling efforts with respect to integration,

---

[1] According to WordNet, hypertext is a "machine-readable text that is not sequential but is organized so that related items of information are connected", http://wordnetweb.princeton.edu/perl/webwn?s=hypertext

[2] *Consolidated text* is the updated version of a normative text, embodying the changes.

substitution and repeal modifications, that have been previously considered. Finally we illustrate how the novel approach can be integrated in the implemented system by considering how a sentence containing a suspension is extracted from the XML input format, processed in the syntactic analysis phase, and then how the main elements of the suspending modification are individuated by the semantic interpreter, based on the FrameNet formalization.

2   Suspension Analysis

The Suspension is the action by which a textual provision interrupts the efficacy of a legal text (or fragment thereof) for a given period. It is important to deal with suspensive modifications because from a linguistic perspective they are more complex and rich than other temporal modifications. Furthermore, suspension is a relevant modification in that it is often used as a *legislative drafting technique* for introducing a *temporary* law. This need stems from for two main reasons: *i*) when the topic is so complex but urgent that it is necessary to have a temporary solution (e.g. Genetic Law); and *ii*) when some time is needed to fully apply the new dispositions (e.g. Euro Law in 1999). The rationale underlying the Suspension of Efficacy is that some norms so strongly affect their addressees (citizens, businesses, social actors) that some time is needed to tune them up. One ambitious and long-term goal is to track this rationale over time. Recognizing the suspension process even if it is fragmented across several intervals of efficacy will allow unveiling that each macro-suspension is driven by a normative principle.[3]

The suspension can be either *explicit* or *implicit*, depending on the language of the provision in question. And, temporally, it can be either *defined* or *undefined*. A suspension is *defined* when the period during which a norm efficacy is interrupted is explicitly stated in the text, with the suspending provision clearly indicating a beginning and an end (or an initial and a final event). By converse, a suspension is *undefined* when this time interval is not explicitly set out in any part of the suspending provision. This class of suspensions includes at least three subclasses as follows: (*i*) *sine die* (that is, with no ending date) suspension; (*ii*) suspension conditioned by an external event (e.g., "Article 5 is suspended for a six-month period starting from entry into force of the Treaty [...]"); and (*iii*) suspension intervals described with a set of other parameters such as the duration (e.g., "Article 5 is suspended for four months starting from January 31, 2011)". In these cases, extracting the correct values may be a complex task.

Suspension modificatory provisions are themselves subject to *modification*. A suspension can be reflexive, with the law introducing the suspension being the same as that affected by the suspension. However, it is rather habitual that later provisions are compiled to modify that suspension for the same reasons that led to its introduction. For example, the Decision 2000/185/EC (Article 3) said

---

[3] Such as the principle that all norms on the use of human embryonic material will remain suspended until a coordinated regulatory framework is in place. Ordinanza 30 maggio 2003 (GU n. 158 del 10/07/2003).

that the decision itself "shall apply from January 1, 2000 to December 31, 2002", thus limiting the document efficacy. Later, the Decision 2002/954/EC modified the second subparagraph of Article 3 by replacing "December 31, 2002" with "December 31, 2003". Then, finally, a third Directive again changed the term, from '2003' to '2005'. The rationale guiding this suspension remains the same, and it is important to grasp it by first detecting the arguments that characterize the suspension modification—so to identify and adjust the main suspension of efficacy—and then describing the phenomena in their atomicity. Another type of suspension provision is the *disapplication*. When a document "disapplies" another document, the latter is "frozen," its efficacy being suspended.[4]

3   Extraction of Suspension Modifications

The annotation of modificatory provisions is a three steps process. Although this process has been illustrated in previous work (full details are provided in [10]), we briefly recall them in order to make the paper more complete and readable. We then show how the FrameNet formalization is used in the semantic interpretation process, pointing out the benefits due to encoding the knowledge about modifications in declarative form.

3.1   System Architecture

In the first step we look for the possible location of a modificatory provision within the document, and we simplify the input sentences, so to prune text fragments that do not convey relevant pieces of information (*input preprocessing*). In the second step we perform the syntactic analysis (*parsing*) of the retrieved sentences; in the third step (*semantic interpretation*) we semantically annotate the retrieved provisions through a tree matching approach. We briefly recall the first two steps and then focus on the annotation phase and on the semantic interpreter design.

The input to the system is encoded in the NormeInRete (NIR) XML standard format for Italian Legal Text. The NIR format encodes the structural elements used to mark up the main partitions of legal texts, as well as its atomic parts (such as articles, paragraphs, subparagraphs, and lettered and numbered items) and any non-structured text fragment. Additionally, the NIR standard includes in its Document Type Definitions a part describing modifications, to implement this model in XML. Based on the XML structure, we retain the text excerpts contained between some meaningful tags (e.g.,⟨corpo⟩, which is the Italian word for body, where the modifications may be found). The text tagged by ⟨rif⟩ (Italian abbreviation for reference) and ⟨virgolette⟩ (Italian word for quotes) is then rewritten with the IDs of the corresponding tags. For example, given the XML encoding of a sentence such as "L'efficacia del decreto ministeriale 17 novembre

---

[4] Disapplication may be motivated by various legal phenomena, such as the aim at resolving conflicts of laws between regional and national law or between national law and European regulations.

2006 è sospesa fino alla data del 30 aprile 2007" (*The efficacy of the Ministerial Decree is suspended until the date of April 30th, 2007*), we rewrite the sentence we rewrite the sentence like "L'efficacia del RIF12 è sospesa fino alla data del 30 aprile 2007". This sentence, which is much simpler to analyze with no loss of information, is then given in input to the parser.

The TULE parser is a broad coverage rule based parser for Italian [6]: it returns a dependency tree that represents the syntactic analysis of the source Italian sentence. It relies on a morphological dictionary of Italian (about 25,000 lemmata) and on a rule-based grammar that describes dependency structures. Let us consider again the sentence: "L'efficacia del RIF12 è sospesa fino alla data del 30 aprile 2007". After two preliminary steps (the morphological analysis and part of speech tagging), necessary to recover the lemma and the part of speech (PoS) tag of the words, the words sequence goes through three phases: chunking, syntactic analysis of the coordination, and verbal subcategorization. The parser produces in output a dependency tree that makes explicit the structural syntactic relationships occurring between the words of the sentence. Each word in the sentence is associated with a node of the tree, as depicted in Figure 1.[5] The nodes are linked via labeled arcs that specify the role of the dependents with respect to their governor (the parent). In the considered example, "efficacia" (*efficacy*) is the subject of the verb "(è) sospesa" ((*is*) *suspended*), while "è" (*is*) is the auxiliar, marked with *aux*. A special node "trace" is framed by a dashed line and labeled *t*: it specifies that the deep subject of the suspension (the agent, in terms of roles) is not expressed. Finally, the temporal argument is in a dependent that is labeled as a *modifier*, tagged as *RMOD* in Figure 1.

3.2 The Interpretation Process

Modifications are represented by means of semantic frames, composed by slots [5]. Retrieving a modificatory provision amounts to choosing the frame describing that modification, and to filling its slots with the correct arguments. Alternatively, annotating a modificatory provision means that given a modification description we are able to recognize it in a sentence. The task of the semantic interpreter is twofold. First it consists in inspecting the dependents of the verb on the one hand, and in inspecting the frames and the available syntactic and semantic information on the other hand. Then the semantic interpreter is charged to find the frame that best fits to current setting. Secondly, once the appropriate frame has been individuated, the related set of rules is applied to retrieve the fillers for the frame slots. The information stored in the FrameNet formalization is thereby fundamental, since it provides a necessary interface between the syntactic and the semantic levels. Additionally, it allows formalizing syntactic and semantic knowledge about modificatory provisions in a *declarative* (as opposed to *procedural*) manner. That is, the FrameNet formalization allows illustrating the rationale underlying and governing the application of rules, since it puts

---

[5] Actually, the nodes include further data (e.g., the gender and number for nouns and adjectives and verb tenses) which do not appear in the figure for space reasons.

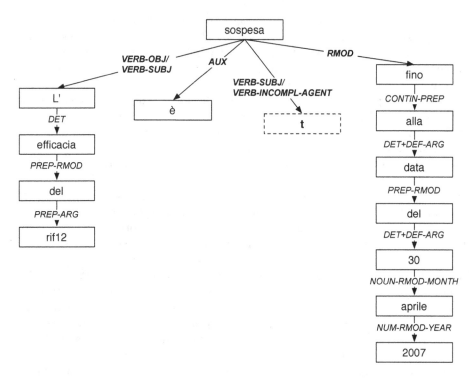

Fig. 1. The (simplified) *dependency tree* structure for the input sentence "L'efficacia del RIF12 è sospesa fino alla data del 30 aprile 2007" (*The efficacy of the rif12 is suspended until the date of April 30th, 2007*)

together both the information about the modification, and their grammatical and syntactical possible realizations.

FrameNet Encoding. FrameNet is a lexical database that represents concepts related to events, relations and states in terms of *semantic frames* [1]. Some features make FrameNet particularly well-suited to our modeling purposes. Frames can be thought of as concepts, composed by sub-elements (called *frame elements*, FEs), that act as semantic roles. Words meaning is encoded through *lexical units* (LUs) that are the FrameNet counterpart of words *senses* in a traditional dictionary.[6] Moreover, which is perhaps more relevant to our present ends, for each such lexical unit an annotation is provided, where the possible realizations of that LU are mapped onto a syntactic structure. The annotated component of FrameNet is of the highest relevance to computational approaches to linguistics (be them based on hand-crafted rules, or acquired through machine learning techniques), in that it provides fully analyzed working examples for each lexical unit. FrameNet retains also information on parts of speech (PoS) such as verbs, adjectives, nouns, etc., so that it can be exploited at these levels.

---

[6] This implies, e.g., that polysemous words are represented by different lexical units.

We are currently developing some FrameNet-like frames (ideally extending the original FrameNet) to encode the legal knowledge needed for recognizing the main features of suspensive provisions (and their variants, such as the modified suspension and the disapplication). To provide an account for suspension modifications it is possible to start by devising two frames, the *Efficacy_Inclusion* and *Efficacy_Exclusion* frames. Such frames are composed of the elements illustrated below and are part of a *Main_Suspension* frame, which can be thought of as an *abstract class* to be implemented through the *Efficacy_Inclusion/Exclusion* frames (Table 1).

Table 1. The basic frames for *Efficacy_Inclusion* and *Efficacy_Exclusion*

frame (Efficacy_Inclusion)
frame_elements (Passive_Norm, Period_Start, Period_End)
scene (*Passive_Norm* has efficacy from *Period_Start* to *Period_End*)
frame (Efficacy_Exclusion)
frame_elements (Passive_Norm, Period_Start, Period_End)
scenes (Passive_Norm is suspended from Period_Start to Period_End, *Passive_Norm* has not efficacy from *Period_Start* to *Period_End*, *Passive_Norm* has efficacy until *Period_End*)
frame (Main_Suspension)
frame_elements (Passive_Norm, Suspension_Start, Suspension_End)
scenes (*Passive_Norm* is suspended from *Suspension_Start* to *Suspension_End*, *Passive_Norm* has efficacy from *Suspension_End* to *Suspension_Start*)

Also, a *Suspension_Modification* frame can be used to describe provisions modifying a suspension previously introduced by another norm (please refer to the analysis of suspensions, Section 2). It is fairly easy to distinguish between the two kinds of provisions, since they are textual modifications lacking a term that evokes an *Efficacy* frame and contains some *Change_event_time* frame. In order to properly interpret the modification, there needs to be a comparison between the *Suspension_Modification* and the *Main_Suspension* (contained elsewhere). For this reason, the *Suspension_Modification* element is presented without any semantic specification of its content, since the exact interpretation of the provision is entrusted to the semantic interpreter.

We have collected a set of relevant terms, that evoke either the *Efficacy_Inclusion* or the *Efficacy_Exclusion* frame. The '.n', '.v', etc. notation reports about PoS information for nouns, verbs, adjectives, and so forth (Table 2). The *TemporalArguments* of the shift in efficacy is captured by the *Period_Start* and *Period_end* Frame Elements (FEs), and the target norm is marked as *Passive_Norm*. Frame Element Groups (FEGs) represent the occurrence of FEs in the examined provisions (P=*Passive_Norm*, S=*Period_Start*, E=*Period_End*). Some typical examples of annotated suspensions are provided in Table 3.

The *Main_Suspension* frame is modelled by inheriting the *Process* frame. The *Suspension* is therefore treated as a process, with a "target" represented by the *Passive_Norm* and whose state is affected by one or more events: it starts with the *Suspension_Start* event and/or ends with the *Suspension_End* event.

Table 2. Terms relevant, *evoking* the efficacy of suspension

Efficacy_Inclusion
(efficacia.*n*, efficace.*adj*, applicarsi.*v*, valido.*adj*, validità.*n*, effetto.*n*, applicazione.*n*, vigore.*n*)
Efficacy_Exclusion
(sospendere.*v*, disapplicare.*v*, cessare.*v*+efficacia.*n*, non.*adv*+Efficacy_Inclusion)

Table 3. Example of annotated sentences containing efficacy suspensions

FEG	Annotated Example
P, S, E	[L'obbligo di cui all'articolo 51, comma 1, della legge 27 dicembre 2002, n. 289]$_P$, è sospeso [dalla data di entrata in vigore del presente decreto]$_S$ [fino al 31 dicembre 2006]$_E$.
P, E	[Le disposizioni del presente provvedimento]$_P$ hanno efficacia [sino a tutto il 7 maggio 2007]$_E$.
P, S+E	[Le disposizioni della legge 29 dicembre 1988 n. 554]$_P$ si applicano [negli anni 1989 e 1990]$_{S+E}$

Moreover, the start of the process can be advanced or postponed by another norm, and the same can be done to its end. These events are represented by specific frames, subclasses of the *Suspension_Modification* frame that are presently not reported for lack of space.

The FrameNet model described above is designed to deal with legislative texts encoded in XML format, with some elements already annotated, in a supervised manner. A parser called Norma-Editor automatically detects *references*, *dates*, and allows adding metadata in legislative texts [14]. Norma-Editor is employed to convert legal texts in a XML format based on Legal XML standards (such as Akoma Ntoso and NiR, [2]). The XML file is then given in input to the TULE parser. The FrameNet modelling helps us clearly investigate and understand the possibile linguistic realizations of suspensions and how such information can be exploited by a syntactic interpreter. Efficacy-evoking terms help us formulate an hypothesis on the type of provision being examined: for example, if the evoking word occurs as the subject, then the prepositional phrase is marked as *Passive_Norm* (as in "Efficacy of law X"). Also, if the evoking word occurs as the predicate, the *Passive_Norm* element will be represented by the subject ("Law X is suspended"). Words and locutions expressing (the beginning or the end of) a time span are marked as *Period_Start* and *Period_End*.

Arguments Extraction. After describing how legal and linguistic knowledge is represented in FrameNet terms, we show how such knowledge is used by the semantic interpreter.

The semantic interpreter is charged to test whether the root node of the syntactic tree is a verb, and if it belongs to the taxonomy of the verbs relevant to modificatory provisions (see [10]). For example, given the parse tree in Figure 1, we take the verb lemma *sospendere* (*suspend*), search for it in the knowledge base, and find that it is a possible instantiation of a modification whose *legal-Category* property is *suspension*, together with the verbs *disapplicare* (*to cease to apply*), *applicarsi* (*enforce*), etc.. In this case we have a fundamental cue that the sentence being analyzed contains a modificatory provision and the semantic interpreter is triggered. We note that there is potentially a terminological collision, in that the frame allocated (and to be filled) is a data structure that can be thought of as an object, and has nothing to do with the frames of FrameNet. Once the frame is allocated, the main task of the semantic interpreter consists in filling its slots. To identify the main elements of the *efficacy inclusion* and *efficacy exclusion* modifications we have to retrieve the information needed to fill the following slots: *Passive_Norm*, [*Position*], *Start* and *End*.

Once discovered that the modification is probably a suspension, the appropriate set of rules is executed so to exploit the information grasped through the FrameNet formalization to retrieve the correct slot fillers from the parse tree. Filling a modification frame amounts to finding an appropriate mapping between tree dependents and frame slots. To carry on with the sentence under consideration, let us consider a typical realization for the *Efficacy_Exclusion* frame:

$$\text{Passive\_Norm is suspended from Start to End.} \tag{1}$$

By introducing the terms used above, we can rewrite the previous sentence as:

$$P \text{ is suspended from } S \text{ to } E. \tag{2}$$

In practical cases it may happen that either the *Start* or the *End* argument is lacking, therefore determining an *open* time span, where one of the two temporal arguments may be absent. Among many possible variants of the sentence in (3.2), a slightly different linguistic construction can be

$$\text{The efficacy of } P \text{ is suspended from } S \text{ to } E. \tag{3}$$

Once the semantic interpreter recognizes a particular surface realization, further relevant information can be made available and exploited, that is directly related to the syntactic structure:

$$[\text{The efficacy of } P]_{subj} \text{ is suspended } [\text{from } S]_{rmod} \, [\text{to } E]_{rmod}. \tag{4}$$

Like it is apparent from this simple example, the FrameNet formalization provides a compact description for (some of) the possible syntactic realizations of the modificatory provisions. That is, the locution "The efficacy of P" is expected to occur in a branch of the parse tree rooted under the main verb. Namely, the semantic interpreter inspects the branch containing the subject of the sentence, labeled *verb-subj*. The processing of such tree branch allows extracting the reference to the *passive norm*. Similarly, extracting both the *Start* and the *End* time will imply traversing the tree branches labeled *RMOD* (see Figure 1). As suggested in the description of the frame, the presence of words/locutions such as "a partire da" (*starting from*), "a decorrere da" (*starting day will be*) or "fino a", "sino a" (*until*) will provide precious cues about where to find the starting and ending time of the suspension time span.

Triggered by the recognition of the root verb, the set of rules related to each modification are executed to test the content of the verb arguments and the verb modifiers to fill the slots of current frame. The rules are charged to discover whether in the syntactic arguments like *subject*, *object* or in any modifier are present any meaningful locutions or constants, such as *RIF*. In this way we can conveniently map the syntactic pattern described in the FrameNet formalization onto the set of slots of a semantic frame.

4   Conclusions

In this paper we provided a linguistic account and syntactic analysis for a particular type of modificatory provision, that is efficacy suspension. A system aimed at automating the consolidation process is being developed, that extends an existing one in dealing with further sorts of modifications (in its first release we only accounted for integration, substitution and repeal). The system is designed to extract modificatory provisions from a large database consisting of about 29,000 normative documents. The system is grounded on a description of modifications paired with a full-fledged syntactic annotation of such modifications.

In this paper we described a methodology for approaching legal texts analysis, with special focus on temporal modifications. We showed how the adoption of the FrameNet approach allows to use a wealth of information about legal language phenomena, that span over different layers, such as the legal one, the grammatical one and the syntactic one.

In our view, the proposed approach benefits from a declarative description of modifications. Decoupling declarative knowledge from procedural components of the system is helpful in separating legal knowledge from its use, which is not only more convenient on a software engineering perspective, but is also helpful in extending the system coverage. Further, from preliminary tests, we are confident to be able to improve the system accuracy, that over simpler modifications (substitution, integration and repeal) is around 70% recall and over 80% accuracy. The results of the first experiments of the system seem to corroborate the approach undertaken; however an extensive experimentation is necessary to assess the approach.

Future works will involve investigating the related –though different– modification of exceptions in its connections to suspensions, in order to yield a broader coverage of the modifications handled and a deeper comprehension of legal and linguistic phenomena.

Acknowledgments. The authors would like to thank the ICT4Law project (http://www.ict4law.org/), funded by Piedmont Region, for partly supporting current research.

References

1. Baker, C.F., Fillmore, C.J., Lowe, J.B.: The Berkeley FrameNet Project. In: Proceedings of the 17th International Conference on Computational Linguistics, vol. 1, pp. 86–90. Association for Computational Linguistics, Montreal (1998), http://portal.acm.org/citation.cfm?id=980860

2. Biagioli, C., Francesconi, E., Spinosa, P., Taddei, M.: The NIR project: Standards and tools for legislative drafting and legal document web publication. In: Proceedings of ICAIL Workshop on e-Government:Modelling Norms and Concepts as Key Issues, pp. 69–78 (2003)

3. Brighi, R., Lesmo, L., Mazzei, A., Palmirani, M., Radicioni, D.P.: Towards Semantic Interpretation of Legal Modifications through Deep Syntactic Analysis. In: Jurix 2008: The 21st Annual Conference. Frontiers in Artificial Intelligence and Applications, vol. 189, pp. 202–206. IOS Press (2008), http://www.di.unito.it/~radicion/papers/brighi08towards.pdf

4. Cherubini, M., Giardiello, G., Marchi, S., Montemagni, S., Spinosa, P., Venturi, G.: NLP-based metadata annotation of textual amendments. In: Proc. of Workshop On Legislative Xml, JURIX 2008 (2008)

5. Fillmore, C.J.: Scenes-and-frames semantics. In: Zampolli, A. (ed.) Linguistic Structures Processing, pp. 55–79. North-Holland, Amsterdam (1977)

6. Lesmo, L.: The Rule-Based Parser of the NLP Group of the University of Torino. Intelligenza Artificiale 2(4), 46–47 (2007)

7. Lesmo, L., Mazzei, A., Radicioni, D.P.: Semantic Annotation of Legal Modificatory Provisions. In: Serra, R., Cucchiara, R. (eds.) AI*IA 2009. LNCS, vol. 5883, pp. 304–313. Springer, Heidelberg (2009), http://dblp.uni-trier.de/db/conf/aiia/aiia2009.html#LesmoMR09

8. Lupo, C., Vitali, F., Francesconi, E., Palmirani, M., Winkels, R., de Maat, E., Boer, A., Mascellani, P.: General XML format(s) for legal Sources - ESTRELLA European Project. Deliverable 3.1, Faculty of Law, University of Amsterdam, Amsterdam, The Netherlands (2007), http://www.estrellaproject.org

9. de Maat, E., Winkels, R.: Automatic Classification of Sentences in Dutch Laws. In: Legal Knowledge and Information Systems, Jurix 2008: The 21st Annual Conference. Frontiers in Artificial Intelligence and Applications, pp. 207–216. IOS Press (December 2008)

10. Mazzei, A., Radicioni, D., Brighi, R.: NLP-based Extraction of Modificatory Provisions Semantics. In: Proceedings of the International Conference on Artificial Intelligence and Law, ICAIL 2009, pp. 50–57. ACM, Barcelona (2009)

11. McCarty, L.T.: Deep semantic interpretations of legal texts. In: ICAIL 2007: Proceedings of the 11th International Conference on Artificial Intelligence and Law, pp. 217–224. ACM, New York (2007)

12. Palmirani, M., Brighi, R.: An XML Editor for Legal Information Management. In: Traunmüller, R. (ed.) EGOV 2003. LNCS, vol. 2739, pp. 421–429. Springer, Heidelberg (2003)

13. Palmirani, M., Brighi, R.: Model Regularity of Legal Language in Active Modifications. In: Casanovas, P., Pagallo, U., Sartor, G., Ajani, G. (eds.) AICOL-II/JURIX 2009. LNCS, vol. 6237, pp. 54–73. Springer, Heidelberg (2010)

14. Palmirani, M., Brighi, R., Massini, M.: Automated Extraction of Normative References in Legal Texts. In: ICAIL 2003, pp. 105–106 (2003), http://dblp.uni-trier.de/db/conf/icail/icail2003.html#PalmiraniBM03

15. Soria, C., Bartolini, R., Lenci, A., Montemagni, S., Pirrelli, V.: Automatic Extraction of Semantics in Law Documents. In: Biagioli, C., Francesconi, E., Sartor, G. (eds.) Proceedings of the V Legislative XML Workshop, pp. 253–266. European Press Academic Publishing (2007)

Developing a Frame-Based Lexicon for the Brazilian Legal Language: The Case of the `Criminal_Process` Frame

Anderson Bertoldi and Rove Luiza de Oliveira Chishman

Universidade do Vale do Rio dos Sinos,
Applied Linguistics Graduate Program, Av. Unisinos, 950,
São Leopoldo, 93.022-000, Rio Grande do Sul, Brazil
`andersonbertoldi@yahoo.com`, `rove@unisinos.br`

Abstract. The FrameNet database has been used for semantic tagging and multilingual lexicon development. This paper describes the initial steps in the development of a lexicographic project that aims to build a legal frame-based lexicon for the Brazilian legal language. First, we discuss the FrameNet lexical analysis methodology. Second, we present the `Criminal_process` frame for the Brazilian legal system and the methodology we adopted in the development of this frame. Third, we discuss how FrameNet frames and Brazilian legal frames differ and how the semantic tags developed in the scope of this project could be used for semantic tagging and semantic parsing.

Keywords: Semantic frames, legal lexicons, legal knowledge representation.

1   Introduction

Semantic Web technology for the legal domain has been an important topic in the last years. Semantic Web technologies involve both applications in corporate settings, such as knowledge management and intranet systems, and public information retrieval on internet [1]. In this context of semantic web and legal information management, semantic lexicons and legal ontologies have been developed to facilitate the access to legal information.

This paper describes the first step in the development of a lexicographic project that aims to build a legal frame-based lexicon for the Brazilian legal language[1]. This project applies Frame Semantics and the FrameNet paradigm to develop a frame-based lexical database for legal information retrieval purposes. In this first step, the expansion methodology was applied to expand the FrameNet `Criminal_process`

---

[1] This work was developed in the scope of the *Semantic Technologies and Legal Information Retrieval Systems Project*, supported by CAPES-CNJ (Brazil) under the rubric N°. 020/2010/CAPES/CNJ and coordinated by Prof. Dr. Rove Luiza de Oliveira Chishman. This work was also supported by the Brazilian agencies CAPES, CNPq and FINEP under the rubric N°. 001/2010 - MCT/CNPq/FINEP National Program of Post-Doctorate (PNPD).

M. Palmirani et al. (Eds.): AICOL Workshops 2011, LNAI 7639, pp. 256–270, 2012.
© Springer-Verlag Berlin Heidelberg 2012

frame from English (American legal system) to Brazilian Portuguese (Brazilian legal system).

FrameNet frames have been used for developing lexical databases and annotated corpora for different languages. Spanish FrameNet [2], Japanese FrameNet [3] and FrameNet Brasil [4] are just some examples of FrameNets created for languages other than English. Kicktionary [5] is an example of a specialized multilingual frame-based lexicon for the soccer language. SALSA project [6] uses the FrameNet semantic tags to manually annotate a German corpus and to automatically develop a frame-based lexicon of German. Other works have applied FrameNet frames for automatic development of lexicons using automatic transfer of corpus annotation in parallel corpora [7], [8], [9].

This paper (i) discusses the challenges of using expansion methodology in social-oriented areas, such as Law, (ii) presents the methodology used in the first step of this project, and (iii) presents the Criminal_process frame for the Brazilian legal language. Therefore, the remaining sections of this paper are organized as follows. Section 2 approaches the previous projects that inspired this work. Section 3 presents the methodology to create semantic frames in the FrameNet project. Section 4 presents the mismatches of using expansion methodology for FrameNets creation. Section 5 describes the FrameNet Criminal_process frame. Section 6 presents the Criminal_process frame for the Brazilian legal language and the methodology used to develop legal frames (the adaptation of the expansion methodology). Section 7 discusses the mismatches between FrameNet frames and Brazilian legal frames. Finally, section 8 points the conclusions reached from this work and the future directions that will be taken in the development of a frame-based legal lexicon for the Brazilian legal language.

2   Legal Lexicons

The lexicographic work we present in this paper is inspired by Jur-WordNet [10] and LOIS [11]. Jur-WordNet is a terminological lexicon of the legal language. Jur-WordNet is a semantic lexicon that follows the structure of WordNet [12], organizing the legal terms in synonym sets, called *synsets*. As synonymy is limited in specialized languages, in Jur-WordNet synsets tend to group terms used by specialists and general words used by non-specialists. An example is the specialized term "*locazione di immobile*" (*lease*) and the general word "*affito*" (*rent*) [12]. In general terms, synonymy in terminological areas are related to social-dialectal variation. For this reason, Jur-WordNet serves as an interface between the *common language* used by citizens and the specific terminology of legal standards [11].

LOIS (*Lexical Ontologies for Legal Information Sharing*) was an investigation project supported by European Commission within the e-Content program. The aim of LOIS was to build a European legal WordNet for legal information retrieval. The semantic relations connect terms in different languages. The LOIS architecture was based on another European project, the EuroWordNet [13]. In LOIS the different language databases were connected through an interlingual index.

3   FrameNet and Frame Development Methodology

FrameNet is a lexical database that describes word meaning according to the principles of Frame Semantics. In FrameNet lexical items are conceived as lexical units [14]. A lexical unit is the combination of a word form with a meaning. Every new meaning of a word represents a new lexical unit. Therefore, it is the lexical unit that evokes the frame, not the word. According to [15], the method of lexical analysis in FrameNet follows five steps:

Characterizing the Frames: A situation for which the language has provided lexical units is described, i.e., arresting a suspect.
Describing and Naming Frame Elements: After characterizing a frame, the frame elements are named, i.e., AUTHORITIES, CHARGES, OFFENCE, SUSPECT.
Selecting Lexical Units: the frame-evoking lexical units are identified, i.e., *apprehend.v, apprehension.n, arrest.n, arrest.v, book.v, bust.n, bust.v, collar.v, cop.v, nab.v, summons.v.*
Creating Annotations of Sample Sentences: Sample sentences collected from BNC are annotated, i.e., Are [you AUTHORITIES] arresting [me SUSPECT] [for the murder of Topaz Brown? OFFENCE].
Automatically Generating Lexical Entries: Annotated sample sentences are converted to lexical entries. The lexical entries contain the definition of the lexical unit, the syntactic realizations of each frame element and the valence patterns.

FrameNet makes a differentiation between 'core' frame elements and 'peripherical' frame elements. According to [15], the distinction between 'core' and 'peripherical' is not clear. In general, frame elements that are necessarily realized are core.

4   Expanding FrameNet Database to Other Languages

FrameNet for languages other than English has been created using the expansion methodology. Expansion methodology assumes that semantic frames stay the same and only the linguistic information is substituted to create new FrameNet for other languages. This is the methodology adopted by Spanish FrameNet [2] and Japanese FrameNet [3].

According to [16], expansion methodology may disregard differences in language lexicalization. In [16], four types of mismatches between frames in FrameNet construction are presented. These mismatches cause the creation of a new frame in a non-English FrameNet:

Semantic Frame: The first possibility is an inadequate description of the FrameNet frame when compared to the frame of the other language, i.e., the FrameNet Statement frame is represented in Japanese FrameNet using two different frames: Statement_verbal_act and Statement_verbal_trasfer [16]. The second is an inadequate coverage of a domain in the English FrameNet. i.e., FrameNet does not have a frame to cover the meaning of the Return frame in Spanish FrameNet [16].

Frame Elements: The necessity of new frame elements to describe the meaning of a semantic frame in a language other than English entails the creation of a new frame. The creation of new frame elements results in a more specific frame. Second [16], he more specific frame is a child of the FrameNet frame and could be related to the parent frame by inheritance relation.

Semantic Type and Frame Element Coreness: Semantic frames are not considered equivalents when core frame elements in FrameNet semantic frames are not core in a semantic frame of other language.

Frame Relations: Every time a semantic frame presents a different number of frame elements or these frame elements present a different semantic type or a different coreness, the relations among frame elements will change. Considering this observation, changes in frame elements relations in a semantic frame of other language will represent the creation of a new frame in the FrameNet of other language.

Considering that Law is a social-oriented creation, it varies from one society to another. For this reason, legal frames may change from one country to another, which means that the original Criminal_process frame may change when expanded to Portuguese. Changes may occur in different levels: (i) frame: the legal event represented by FrameNet frame is not equivalent to the Brazilian legal event; (ii) frame elements: the legal event is equivalent, but the legal agents are different in the USA and Brazil; (iii) semantic type and frame element coreness may differ in the FrameNet database and in Brazilian legal frames; and (iv) frame relations: as court procedures are not the equal in the USA and Brazil, frame-to-frame relations may differ from FrameNet and Brazilian legal frames, especially Subframe relations. The next sections will present the FrameNet Criminal_process frame, the Brazilian Criminal_process frame and the challenges to use expansion methodology in social-oriented areas, such as Law.

5   The FrameNet Criminal_Process Frame

In the FrameNet terminology, Criminal_process frame is a non-lexical frame. The function of non-lexical frames is to connect semantically related frames. Non-lexical frames do not present frame-evoking lexical units. They represent complex events divided in more specific frames. Criminal_process frame describes the different steps of a criminal process according to the American legal system.

In FrameNet, relations are established between frames, not words. Therefore, lexical relations, such as antonymy and synonymy, are not considered. Figure 1 shows Criminal_process frame and relations among frames. In case of complex frames, like Criminal_process, each sequence of events or states is described as a single frame, related to the complex frame through *Subframe* relations and to the other subframes through *Precedes* relation.

Criminal_process frame is divided in four subframes temporally succeeded: Arrest, Arraignment, Trial, and Sentencing. Arraignment frame is divided in three subframes: Notification_of_charges, Entering_a_plea, and `Bail_decision`. `Trial` frame also presents three subframes: `Court_examination`, `Jury_deliberation` and `Verdict`. `Trial` frame and `Try_defendant` frame are related through *Perspective_on* relation.

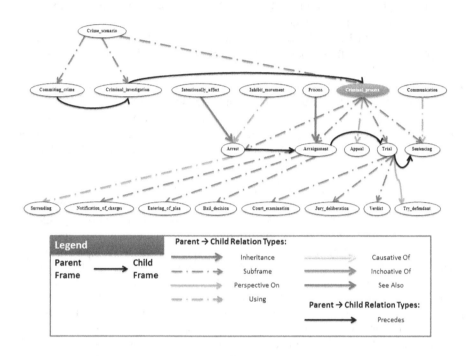

Fig. 1. FrameNet `Criminal_process` frame

According to [17], complex frames represent sequences of states and events and each sequence can be separately described as a frame. The *Subframe* relation relates the separate frames, called *subframes*, to the complex frame. In cases of complex frames, some frame elements of the complex frame may be inherited by subframes. In FrameNet, a set of frame elements is described for each frame. Frame elements are semantic roles evolved in the comprehension of the frame.

In the case of `Criminal_process` frame, each step of a criminal process is represented as a subframe in the FrameNet database: `Arrest`, `Arraignment`, `Trial` and `Sentencing`. In the FrameNet database, these subframes are related to the `Criminal_process` frame through Subframe relation. Subframes may also describe complex states and events, e.g. `Arraignment`. The `Arraignment` frame is a subframe of the `Criminal_process` frame and a complex frame by itself. The different steps of an arraignment session are represented as separate frames: `Notification_of_charges`, `Entering_a_plea`, and `Bail_decision`.

The Precedes relation connects two subframes of a complex frame. The subframes of `Criminal_process` frame are related to each other via Precedes relation. This relation specifies the sequence of steps of a complex event. In general, all the subframes of a complex frame will be connected by Precedes relation. In the figure 1, the black lateral arrows show that `Arrest`, `Arraignment`, `Trial`, and `Sentencing` frames have precedence relation.

In the FrameNet database, `Try_defendant` frame perspectivalizes the more general `Trial` frame. Second [17], the Perspective relation indicates the presence of at least two different points of view of a neutral frame. The classical example of Perspective relation is the commerce scenario example. There are at least two points of view of a commerce scenario: one of the buyer and the other of the seller. In the case of the FrameNet `Criminal_process` frame, `Trial` describes the steps of a typical criminal trial in the USA and `Try_defendant` describes the legal event of trying a defendant. Perspectivalized frames have different frame-evoking lexical units. While the lexical unit *try.v* evokes `Try_defendant` frame, `Trial` frame is evoked by *case.n* and *trial.n.*

6   The `Criminal_Process` Frame for the Brazilian Legal System

This section presents the methodology used in this work and the structure of the Brazilian `Criminal_process` frame. Starting from the FrameNet `Criminal_process` frame, the steps of a criminal process were reorganized, according to the Brazilian Code of Criminal Procedures. Some subframes of the FrameNet Criminal_process frame were maintained the same, others were adapted to the Brazilian legal system, and those subframes that described legal events that did not exist in the Brazilian criminal process were discarded.

6.1   Methodology for Frame Expansion and Frame Creation

In this first step of the project, it was used expansion methodology to create the Brazilian `Criminal_process` frame. Some `Criminal_process` subframes were expanded to Portuguese without problem. This is the case of `Arrest` and `Try_defendant` frames. In these cases, the legal event in the American legal system was equivalent to the Brazilian legal system. Therefore, only the English lexical units were substituted by Brazilian Portuguese.

When legal events are equivalents the expansion methodology works very well. Notwithstanding, legal events are not always equivalent. As legal systems are generally divergent, the challenge in using expansion methodology is placed in the cases in which legal events differ or a legal event represented in FrameNet simply does not exist in the Brazilian legal system. In these cases, all the Subframe relations must be rearranged. Therefore, complex legal events, in other words, legal events that are represented in FrameNet as complex frames and divided in many subframes, will be more difficult to expand. This is what happens to `Criminal_process` frame.

As figure 2 shows, while the American legal system has a typical procedure to try a person accused of a felony, the Brazilian legal system presents two different procedures. The special procedure (jury) is used in cases of crime against life considered intentional. The ordinary process is used to try different crimes, including crimes against life considered not intentional. The `Criminal_process` frame described in this work represents only the special process. In future steps of the frame-based legal lexicon the ordinary process will be described and contrasted with the special process.

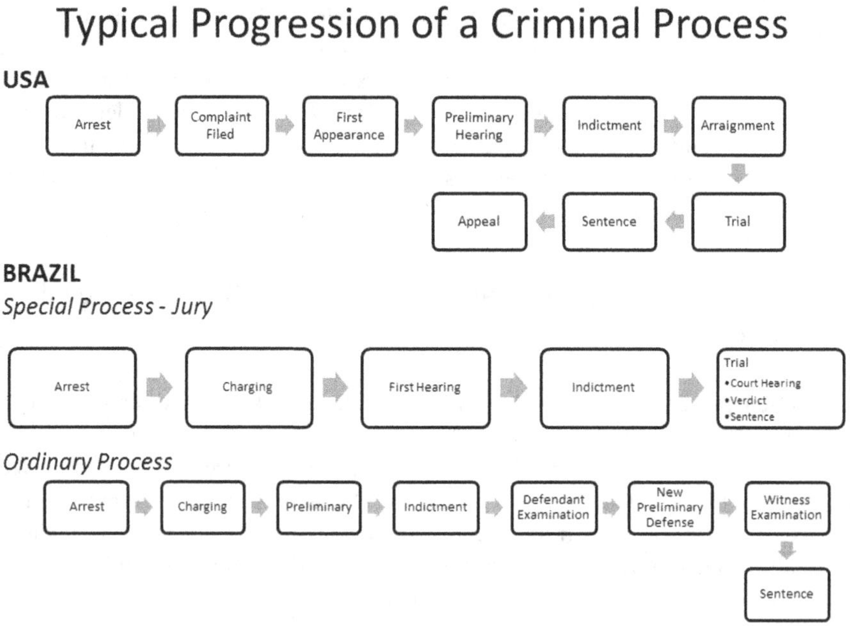

Typical Progression of a Criminal Process

Fig. 2. Typical progression of a criminal process in the USA and Brazil

Some frames did not find complete equivalence in the Brazilian legal system. In these cases, two steps were adopted. The first was to find translation equivalents in Portuguese for the English frame-evoking lexical. The second step was to find the legal frame evoked by the Portuguese lexical unit and to follow the lexicographical methodology used in FrameNet [15]. The methodology, described in section 3, is divided in five stages, but only the first four stages were adopted in this work: (i) characterizing the frame, (ii) describing and naming frame elements, (iii) selecting lexical units, and (iv) creating annotations of sample sentences. The FrameNet `Notification_of_charges` frame is a good example of a frame that does not have a complete equivalence in the Brazilian legal system. When the frame-evoking lexical units were translated to Portuguese, it was seen that those lexical units could evoke two different frames: `Charging` and `Indictment`.

6.2   Frame Structure

The Brazilian `Criminal_process` frame is divided in five subframes: `Arrest`, `Charging`, `First_hearing`, `Indictment`, and `Trial`. `Trial` frame is divided in three subframes that represent the steps in a Brazilian criminal trial:

Court_hearing, Verdict, and Sentencing. The Try_defendant frame specifies the general legal event described in the Trial frame. While the Trial frame represents the major steps in a Brazilian criminal process, the Try_defendant frame describes the event of trying a defendant.

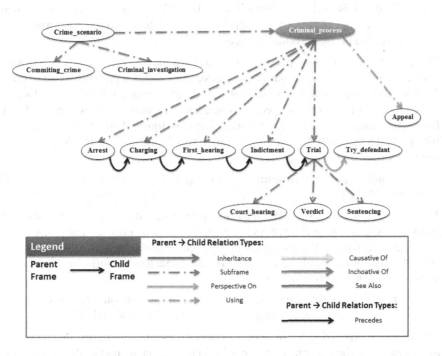

Fig. 3. Criminal_process frame for the Brazilian legal system

The Arrest frame describes a legal act in which AUTHORITIES charge a SUSPECT for a crime, the CHARGES, and take him/her into custody. The core frame elements of this frame are AUTHORITIES, SUSPECT, OFFENSE, and CHARGES. The frame-evoking lexical units are *prender* (*to arrest*), *prisão* (*arrest*), *fichar* (*to book*), *deter* (*to arrest*), and *capturar* (*to arrest*).

(1) [França AUTHORITIES] **prende** [95 suspeitos SUSPECT] [de colaboração com terror argelino. OFFENSE]

The Charging frame represents a legal event in which the prosecution, PROSSECUTION_AUTHORITY, charges the SUSPECT. The core frame elements of this frame are ACCUSED, PROSSECUTION_AUTHORITY, and CHARGES. The frame-evoking lexical units in this frame are *acusar* (*to charge*), *acusação* (*charge*), *denunciar* (*to charge*), and *denúncia* (*charge*).

(2) A partir desses documentos, [o Ministério Público PROSSECUTION_AUTHORITY] **denunciou** [os bicheiros ACCUSED] novamente e ficou comprovado que eles mantinham suas atividades mesmo de trás das grades

The FIRST_HEARING frame describes a preliminary hearing in which a JUDGE examines the SUSPECT. The core frame elements of this frame are JUDGE, ACCUSED, WITNESS, and CHARGES. The frame evoking lexical units are *interrogar* (*to examine*) and *depor* (*to testify*).

(3) [A juíza Giselle Lima e Silva Rocha, da 36ª Vara Criminal, JUDGE] **interrogou** [ontem TIME] [nove dos 14 bicheiros ACCUSED] [condenados por formação de quadrilha no ano passado. CHARGES]

The Indictment frame represents the legal event in which the JUDGE makes a first evaluation of the evidences against the suspect and writes a preliminary sentence. If the evidences against the suspect are not considered relevant, he/she will be absolved. If the evidences against the suspect are considered relevant, he/she will be declared a DEFENDANT and will face a criminal trial. The core frame elements of this frame are JUDGE and DEFENDANT. The frame evoking lexical units are *pronúncia* (*indictment*) and *pronunciar* (*to indict*).

(4) [O juiz JUDGE] deve **pronunciar** [o réu DEFENDANT] (TJSP, RCrim 71.325, RT 648 / 275).

The Trial frame describes the legal event in which the JURY decides about the guilty or not-guilty of a DEFENDANT and the JUDGE sentences the DEFENDANT. The core frame elements of this frame are JUDGE, JURY, PROSSECUTION, DEFENDANT, and DEFENSE. The frame evoking lexical units are *julgamento* (*trial*), *processo* (*suit*) e *ação penal* (*criminal proceeding*).

(5) O recurso pode provocar, em 95, um novo **julgamento** [dos acusados DEFENDANT] [pelos desembargadores do Tribunal de Justiça do Estado. JUDGE]

The Trial frame is divided in three subframes: Court_hearing, Verdict, and Sentencing. The Court_hearing frame represents the part of the trial in which the defendant is examined and the witnesses testify. The core frame elements of this frame are DEFENDANT, WITNESS, and JUDGE. The frame evoking lexical units are *testemunhar* (*to testify*), *depor* (*to testify*), and *interrogar* (*to examine*).

(6) [Principal testemunha da chacina WITNESS] **depõe** no II Tribunal do Júri reafirma denúncias e diz que Emanuel mentiu ao inocentar Côrtes.

The Verdict frame describes the findings of a trial. The core frame elements of this frame are JUDGE, FINDING, and CHARGES. The frame evoking lexical units are *decidir* (*to decide*), *considerar* (*to consider*), *absolver* (*to acquit*), *inocentar* (*to acquit*), *condenar* (*to convict*), *condenação* (*conviction*), and *veredito* (*verdict*).

(7) Quanto a essa acusação, o [júri JUDGE] **decidiu** [absolver FINDING] [o réu Alexandre Cardoso, o Topeira, DEFENDANT] e [condenar FINDING] [Sandro Baggi e André Rodrigues da Silva, o Gargamel. DEFENDANT]

The Sentencing frame represents the final step in a Brazilian criminal process, when the judge pronounces the sentence giving the punishment to the defendant. The core frame elements of this frame are DEFENDANT, COURT, OFFENSE, and PUNISHMENT. The frame evoking lexical unit is *condenar* (*to convict*).

(8) [Ubirajara DEFENDANT] foi **condenado** [a 19 anos PUNISHMENT] [para cada homicídio OFFENSE] e [a 12 anos PUNISHMENT] [pela tentativa de homicídio de Orlando OFFENSE]

The Try_defendant frame describes the trial of a DEFENDANT who is charged of a crime. A JURY is responsible for evaluating the CHARGES and deciding whether the DEFENDANT is guilty or not-guilty. The core frame elements in this frame are JUDGE, JURY, DEFENDANT, OFFENSE, and CHARGES. The frame evoking lexical unit is *julgar* (*to trial*).

(9) Para o governador, o fato de [os acusados DEFENDANT] serem **julgados** [por um júri popular JURY] é muito positivo.

Although the legal frames we have developed by now and presented in this paper have many similarities with the FrameNet Criminal_process frame, there are some conceptual differences that need to be clear. In the next subsection, we will discuss how FrameNet frames and Brazilian legal frames differ.

7   Mismatches between FrameNet and Brazilian Legal Frames

The expansion of the Criminal_process frame revealed four types of mismatches between FrameNet and Brazilian legal frames, most of them presented in section 4: (i) semantic frame, (ii) frame elements, (iii) frame relations, and (iv) lexical units. The FrameNet Try_defendant frame is equivalent to the Brazilian Try_defendant frame. They describe the same legal event: trying a defendant in a court. These frames are considered equivalents because they find correspondence of semantic frame, frame elements, frame relations, and frame-evoking English lexical units presents translations equivalents in Portuguese. Figure 4 shows the similarities between these two frames.

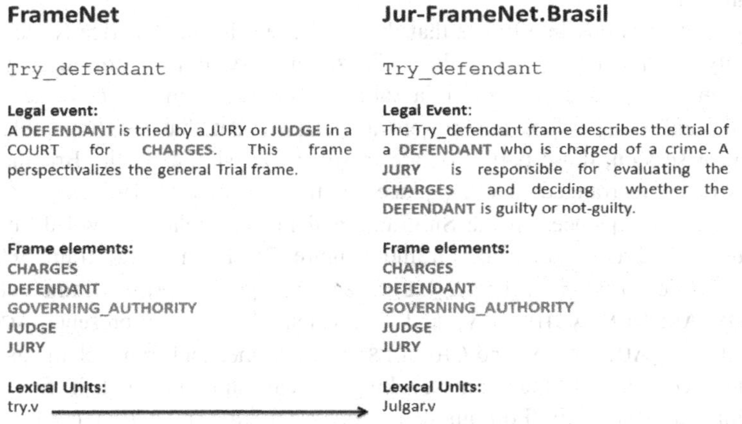

FrameNet	Jur-FrameNet.Brasil
Try_defendant	Try_defendant
Legal event: A DEFENDANT is tried by a JURY or JUDGE in a COURT for CHARGES. This frame perspectivalizes the general Trial frame.	**Legal Event:** The Try_defendant frame describes the trial of a DEFENDANT who is charged of a crime. A JURY is responsible for evaluating the CHARGES and deciding whether the DEFENDANT is guilty or not-guilty.
Frame elements: CHARGES DEFENDANT GOVERNING_AUTHORITY JUDGE JURY	**Frame elements:** CHARGES DEFENDANT GOVERNING_AUTHORITY JUDGE JURY
Lexical Units: try.v ⟶	**Lexical Units:** Julgar.v

Fig. 4. Equivalence of Try_defendant Frames in English and Portuguese

Even the frame-to-frame relations for these two frames seem to stay the same, as can be seen in the figure 5.

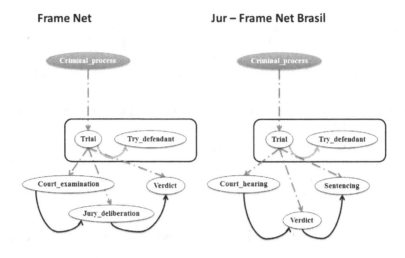

Fig. 5. Relations for `Try_defendant` frames

Some frame presented only partial equivalence. In the case of `Notification_of_charges`, it represents two different frames in Brazilian legal system: `Charging` and `Indictment`. Therefore, `Charging` frame represents only part of the `Notification_of_charges` frame, that part in which the suspect is charged by the prosecution, before being notificated of the charges against him/her. Figure 6 shows the contrast between the FrameNet and the Brazilian frames.

In figure 6, it is possible to see that the legal event in the America System is not completely equivalent to the Brazilian system. Considering that `Notification_of_charges` is a subframe of `Arraignment` frame, and the Brazilian legal system does not present any legal event that is comparable to the arraignment session, those parts that are in some way similar to the Brazilian legal system will be rearranged in other places of the Brazilian `Criminal_process` frame. As a consequence, all the Subframe and Precede relations will differ in the Brazilian `Criminal_process` frame (Figure 7). Even some frame elements change. While `Notification_of_charges` presents the frame elements ACCUSED, ARRAIGN_AUTHORITY, and CHARGES, `Charging` presents ACCUSED, PROSSECUTION_AUTHORITY, and CHARGES. If the frames and frame elements are not completely equivalents, the frame-evoking lexical units in English find perfect translation equivalents in Portuguese: *to accuse* (*acusar*), *charge* (*acusação*), *to charge* (*acusar*), *to indict* (*pronunciar*), and *indictment* (*pronúncia*).

FrameNet	**Jur-FrameNet.Brasil**
Notification_of_charges	Charging

Legal event:	**Legal Event:**
The judge or other court officer (the ARRAIGN_AUTHORITY) informs the ACCUSED of the CHARGES against him/her, i.e. the alleged actions and the relevant laws.	The Charging frame represents a legal event in which the prosecution, PROSSECUTION_AUTHORITY, charges the ACCUSED.

Frame elements:	**Frame elements:**
ACCUSED	ACCUSED
ARRAIGN_AUTHORITY	PROSSECUTION_AUTHORITY
CHARGES	CHARGES

Lexical Units:	**Lexical Units:**
accuse.v, ⟶	Acusar.v
charge.n, ⟶	Acusarção.n
charge.v, ⟶	Acusar.v
indict.v, ⟶	Pronunciar.v
indictment.n ⟶	Pronúncia.n

Fig. 6. Equivalence of Notification_of_Charges and Charging frame

Frame Net   **Jur – Frame Net Brasil**

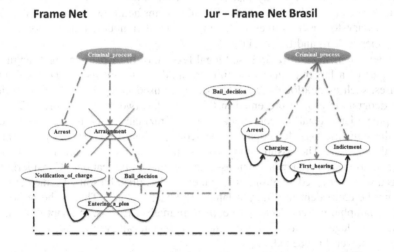

Fig. 7. Relations for Notification_of_charges and Charging frames

There are still frames that do not present any correspondence. This is the case of the FrameNet frame Arraignment. Arraignment frame describes a legal event that is typical of the American legal system, which is based on Common Law. Even the frame-evoking lexical units *arraign* and *arraignment* do not find an equivalent in Portuguese.

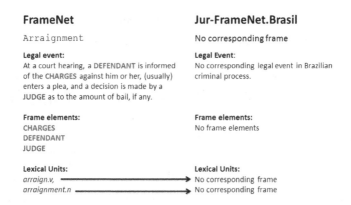

Fig. 8. Equivalence of Arraignment frame in the Brazilian legal system

8   Future Directions

This paper described the initial step to create a frame-based lexical resource for the Brazilian legal language. We have started developing this frame-based lexicon by the Criminal_process frame using the expansion methodology. As a first conclusion, it is possible to say that complex frames are difficult to expand to other languages, because of the differences of legal systems and Law in each country. Now, it is necessary to check frames that represent smaller nodes, that is, non-complex frames, such as Law and Legality.

The development of a frame-based legal lexicon of the Brazilian legal language is only a part of a larger project that aims to build legal knowledge and legal lexical databases, such as ontologies and lexicons, to be used in legal information retrieval. The lexicographic project presented in this paper has two main objectives. The first, developing a large legal lexical database of the Brazilian legal language. The second, using the semantic tags developed in the scope of this project for legal information retrieval proposes. The assumption here is that semantic tags are not completely applicable to other languages when used in automatic applications. Considering that Law as a social-oriented creation, the legal event described by some FrameNet frames could not be equivalent in different languages/legal systems. That is the reason why this lexicographic project decided to expand FrameNet frames, to adapt those frames that describe legal events similar to the Brazilian legal events, and to create new frames whenever it is needed.

Differently from WordNet-based databases, semantic relations between words the focus of a frame-based database. Therefore, a FrameNet-style database has different applications in natural language processing and information retrieval. This project aims to deliver in some years a large frame-based database of the Brazilian legal language and an annotated corpus of legal texts to be used as training corpus. The semantic tags could be used in a series of natural language processing applications, such as automatic legal decision summarization, legal information retrieval, and legal information extraction. Automatically annotating court decisions, it is possible to generate summaries of the decisions. These summaries are called in Brazil "ementa" and are used to inform those that search for court decisions of the general subject of the decision. Another possible use of frame semantics for legal information retrieval is to annotate in legal decision the participants of the legal event, such as, the *defendant*, the *judge*, the *attorney*, and the results of the legal event, such as the *findings*, and the *punishment*.

The work of describing the Brazilian legal language is still in the beginning of a long process. There are important procedures to be done yet. The first, expanding the number of frames to better represent the universe of the Brazilian legal language. The second, compiling a legal corpus to be used for semantic annotation. This corpus could be both a source of examples to the lexical database and a training corpus for automatic applications. The third, programming a database and a friendly internet interface to display this lexical database freely. The study of the Criminal_process frame presented here represents just the first stage of this lexicographic project focused on technological innovation in the Brazilian courts.

References

1. Benjamins, V.R., Casanovas, P., Breuker, J., Gangemi, A. (eds.): Law and the Semantic Web. LNCS (LNAI), vol. 3369. Springer, Heidelberg (2005)
2. Subirats, C.: Spanish FrameNet: A frame-semantic analysis of the Spanish lexicon. In: Boas, H.C. (ed.) Multilingual FrameNets in Computational Lexicography: Methods and Applications, pp. 136–162. Mouton de Gruyter, Berlin (2009)
3. Ohara, K.H.: Frame-based contrastive lexical semantics in Japanese FrameNet: The case of risk and kakeru. In: Boas, H.C. (ed.) Multilingual FrameNets in Computational Lexicography: Methods and Applications, pp. 163–182. Walter de Gruyter, Berlin (2009)
4. Salomão, M.M.M.: FrameNet Brasil: um trabalho em progresso. Calidoscópio 7(3), 171–182 (2009)
5. Schmidt, T.: The Kicktionary – A multilingual lexical resource of football language. In: Boas, H.C. (ed.) Multilingual FrameNets in Computational Lexicography: Methods and Applications, pp. 102–132. Mouton de Gruyter, Berlin (2009)
6. Burchardt, A., Erk, K., Frank, A., Kowalski, A., Padó, S., Pinkal, M.: Using FrameNet for the semantic analysis of German: annotation, representation, and automation. In: Boas, H.C. (ed.) Multilingual FrameNets in Computational Lexicography, pp. 209–244. Mouton de Gruyter, Berlin (2009)
7. Padó, S., Lapata, M.: Cross-lingual projection of role-semantic information. In: Proceedings of HLT/EMNLP 2005, pp. 859–866. Association for Computational Linguistics, Vancouver (2005)
8. Padó, S.: Cross-lingual Annotation Projection Models for Role-Semantic Information. PhD Thesis. Universität des Saarlandes, Saarbrücken (2007)
9. Tonelli, S., Pianta, E.: Frame Information Transfer from English to Italian. In: Proceedings of the Sixth International Language Resources and Evaluation (LREC 2008), pp. 28–30. European Language Resources Association (ELRA), Marrakech (2008)
10. Sagri, M.T., Tiscornia, D., Bertagna, F.: Jur-WorNet. In: Sojka, P., et al. (eds.) Second International Wordnet Conference, GWC 2004, pp. 305–310. Masaryk University, Brno (2003)

11. Sagri, M.T., Tiscornia, D.: Semantic Lexicons for Accessing Legal Information. In: Traunmüller, R. (ed.) EGOV 2004. LNCS, vol. 3183, pp. 72–81. Springer, Heidelberg (2004)
12. Miller, G.A.: WordNet: a lexical database for English. Communications of the ACM 11, 39–41 (1995)
13. Vossen, P.: Introduction to EuroWordNet. Computers and the Humanities 32(2-3), 73–89 (1998)
14. Cruse, A.: Lexical Semantics. Cambridge University Press, Cambridge (1986)
15. Fillmore, C.J., Baker, C.: A frames approach to semantic analysis. In: Heine, B., Narrog, H. (eds.) The Oxford Handbook of Linguistic Analysis, pp. 313–339. Oxford University Press, Oxford (2010)
16. Lönneker-Rodman, B.: Multilinguality and FrameNet. Technical report. TR-07-001. ICSI, Berkeley (2007)
17. Ruppenhofer, J., Ellsworth, M., Petruck, M.R.L., Johnson, C.R., Scheffczyk, J.: FrameNet II: Extended Theory and Practice. ICSI, Berkekey (2006)

Creative Commons and Grand Challenge
to Make Legal Language Simple

Matěj Myška, Terezie Smejkalová, Jaromír Šavelka, and Martin Škop

Masaryk University, Faculty of Law, Veveří 70,
611 80 Brno, Czech Republic
{matej.myska,terezie.smejkalova,jaromir.savelka,martin.skop}@law.muni.cz

Abstract. In this paper we analyse the Creative Commons computerized licensing system. We draw the attention to the fact that despite considerable efforts to make the complicated task of licensing work using so-called free license as simple as possible, the system is apt to give rise to countless ambiguities often leading to copyright infringements. We maintain that the phenomenon has been caused by the modifications of 'language' that facilitates the communication of the relevant section of law and consequent loss of vital context and structure in the framework of which the communication has to be perceived. We come to a conclusion that while context and structure preserving modifications should be regarded as the preferable method of simplifying legal language, its scope is too narrow to achieve the goal of making legal language easily understandable for a layperson. Unconstrained simplification is powerful enough to achieve the goal but entails a danger of driving a layperson, as well as a professional, into undesirable outcomes.

Keywords: legal language, simplification, legal certainty, copyright, Creative Commons.

1 Introduction

It has been a common cliché that legal language has been considered by the majority of society as unnecessarily complicated and even on the very edge of actual comprehensibility. (See [1], [2] and [3]) In English speaking countries legal language is often referred to as legalese which establishes a parallel between legal and foreign languages. (See [4] and [5]) Not surprisingly, it has been claimed that the situation has a negative impact on legal certainty which is undoubtedly an undesirable effect. Thus, tendencies to simplify legal language have naturally occurred, such as plain language campaigns[1] [6] or Creative Commons licensing scheme[2] [7]. However, we maintain that while the various efforts to simplify legal language are beneficial impulses for the development of both legal language and law, they have to be performed with high level of caution and awareness, since

---

[1] Addressing the problem of accessibility of legal language from various points of view.
[2] Thoroughly designed computer system providing assistance within the area of licensing copyrighted content.

M. Palmirani et al. (Eds.): AICOL Workshops 2011, LNAI 7639, pp. 271–285, 2012.
© Springer-Verlag Berlin Heidelberg 2012

in certain cases they are apt to lead to the actual decrease of legal certainty, i.e. the very opposite of the desired aim.

The following section provides the basic theoretical framework — the concept of legal certainty is established (although in substantially simplified form allowing to concentrate the focus exclusively on the impact of the communication of law) and the interaction between legal language and law is assessed. In the next section the theoretical framework is enriched by the elaboration on the impact of legal language simplification efforts on legal certainty — preciseness and comprehensibility (qualitative attributes of legal language) are introduced, their mutual dependency suggested and the role of context in uderstanding law assessed. The established theoretical framework is then employed in an analysis of the computerized Creative Commons (hereinafter abbreviated as "CC") licensing system.

2   Theoretical Framework

2.1   Legal Certainty

To allow a study of legal language from the perspective of its simplification's impact on legal certainty it is necesarry to introduce the notion of legal certainty. For the sake of simplicity the notion is developed only in the context of communication of law. Informally, it is an important aspect of legal certainty that a person — subject to the legal regulation — understands the regulation and considers its actual outcomes as predictable. Such is the reduced role of the notion of legal certainty within this paper. To explicate the notion in a more rigorous way it may be understood as a qualitative attribute of social reality referring to the ability of arbitrariness elimination from the domain of law. (cf. [8]) Furthermore, two instances of arbitrariness shall be distinguished — actual and perceived. To explain the meaning of these instances following example of a judge deciding individual cases by coin-flipping introduced by Bix [9] may be used. If the judge flips the coin in front of the public and decides upon the results of the flipping, the degrees of both instances would be very close to their maximum. Thus, the level of arbitrariness would be extremely high and legal certainty extremely low. In case the judge would flip the coin privately, decide upon the result, but publicly pretend to make the decision upon the merits of law, only the actual arbitrariness would be maximized while the perceived arbitrariness would be very low. The legal certainty in this case would be somewhat higher than in the previous case. Despite the fact that the decisions would often reach extremely dubious conclusions they would at least appear well-thought and supported with arguments. The very same conclusion applies for the case the judge would make the decision upon the merits of law while publicly pretending to decide upon the result of the coin flipping. The appearance of the process would surely raise much concern but strangely enough, on the long run, the outcomes of the individual decisions would be quite reasonable. Thus, the elimination of both components of arbitrariness produces the highest level of legal certainty.

2.2 Legal Language, Law and Their Interplay

It goes far beyond the scope of this paper to elaborate on the very nature of law and legal language. Nevertheless, we believe it is possible to assess their mutual relationship. Legal language, as the prominent means of capturing and consequently communicating law, is inevitably bound with law. Though the individual approaches and definitions may differ,[3] legal language may be described as a type of code intended and used as a means of capturing the complex of law, which has evolved hand in hand with it. The intuitive representation of the relationship between law and legal language may be compared to the Möbius strip: a three-dimensional object that appears to have two surfaces and two boundary components, while topologically having only one surface and only one edge. Similarly, law and legal language (the means of communication of law) are interconnected in such a way that it is impossible to think about one without the other. Not only is the legal language the means of communication of the legal rules, it also *"plays an important role in the construction, interpretation, negotiation and implementation of legal justice."* [11] Some scholars go even further by saying that law is not reducible to system of rules or policy choices but a that it is a language in its full sense. [12]

For the purpose of this paper, we maintain that despite their existential interconnectedness, it is still possible to distinguish between law and legal language. Whether this distinction is only a matter of appearance or not is another issue which shall not be discussed here. As in the case of the Möbius strip, when assessing one surface (e.g. legal language) there is always the other side (law); but when going further, it is possible to reach the other dimension without the need to change the surfaces. Whether perceived like this, or simply as two existentially-related elements of one system, their mutual cohesion is striking. However, it is essential to note that there is no causal link between them. Any consequences may be presupposed only on the level of probability. Similarly, this cohesion may be comparable to the relationship between a system and a chosen Level of abstraction on which the system is studied. (See generally [13]) As Floridi observes, "it makes no sense to wonder whether the system under observation is finite in time, space and granularity in itself, independently of the Level of abstraction at which it is being analysed". [14]

Seen from the point of view of an individual — an addressee of a legal rule — the above discussed relationship between law and legal language may be described in analogy to the operating system: between the user (the individual)

---

[3] There are of course more attitudes as to the nature of legal language whether be it considered a (sub-)variety of language or a language of its own. One of the definitions describes *"legal language as a type of register, that is, a variety of language appropriate to the legal situations of use."* [10].

and the operating system itself (law), there is an interface (the legal language) through which the system is perceived and controlled.[4]While retaining some of the characteristics of the operating system, it is mainly the interface that shapes the user's attitudes towards the system.

All this leads to the conclusion that extremely delicate and subtle interplay takes place between legal language and law. A question of key importance thus stands as to what is the nature of this interplay, its dynamics and towards what directions is it heading. To answer this question, it has to be taken into account that law and legal language are subsystems of extremely complex and almost ubiquitous social reality. At the same time they can be both considered as highly evolved and complex systems of their own right. When Kelly faced the question of what technology — system of high complexity — wanted, he has formulated, inspired by evolution of life, a list of attributes technology heads to. However, we believe that the list has much wider application and can be used more or less as a reference to any evolving system with sufficient degree of complexity, including both legal language and law. Thus, we propose that law and legal language simultaneously head towards efficiency, opportunity, emergence, complexity, diversity, specialization, ubiquity, freedom, mutualism, beauty, sentience, structure and evolvability. [16] With regard to their interconnectedness, the individual attributes of one grow in accordance with the other; e.g. the growing complexity of law entails the growing complexity of legal language. In this respect, law and legal language co-exist and evolve in harmony.

3   Making Legal Language Simple

Law and language have naturally reached a state at which their accessibility to a layperson has been disputed. The plain language movement has widely criticized the verbosity of legal language, unnecessarily complicated syntax or the overuse of the terms of art[5] [15], claiming its negative impact on legal certainty, and spreading a belief that *"legal documents can be expressed more or less in ordinary everyday English."* [16] To make legal language simpler means to modify it in order to enhance its comprehensibility. Various techniques can be employed to support the goal, e.g. abstraction (as an opposite to refinement), reduction (elimination of a component within a system) or certain impositions on structure (e.g. syntax in case of languages). Often the techniques are apt to change certain elements of legal language. Respectively, the modifications manifest themselves in revisions of the whole system. It should be taken into account that such revisions may eventually conflict with the natural evolution of the system; they

---

[4] This approach resembles, but not fully, the approach proposed by Floridi. (See generally [14] and [13]). For individual needs and requirements — and to target different groups of users — different interfaces may be adopted, based on choosing different variables to describe the system. [15] The choice of variables then shapes the nature of the interface and whether or not it is serving its purpose well.

[5] Such are some of the characteristics of legal language which were named by Mellinkoff in the 1960s.

may lead to slowing the speed of the evolution, stopping it altogether or even reversing it. However, it does not immediatelly follow that the revisions made to legal language are automatically projected to law. Quite on the contrary, they either have to be deliberately introduced — in case it is necessary to preserve the exact nature of the interaction between both systems — or it has to be accepted that the development of legal language had been altered while law has continued its natural evolution. Although not necessarilly, such state of affairs may cause an occurence of disproportion and tension between both systems. Consequently, the disproportion may result in errors in the communication of law. From the point of view of the user-interface-system metaphor, the interface has been changed in such a way that it fails to provide access to all the functions the operating system has to offer or perplex the user by offering those that do not exist within the system.

Assessing the impact of the tension between the systems on legal certainty is a rather difficult task. However, despite the fact that it would be extremely complicated to decide on the impact of legal language revisions on the levels of the individual instances of arbitrariness of law (actual and perceived), it seems likely that occurrence of the tension and raising disharmony between the systems is apt to increase both of them. Thus, the tension and discrepancies appearing as the consequence of legal language simplification efforts may compromise legal certainty. Paradoxically enough, legal certainty is the main reason why the legal language has been subjected to reform — simplification — in the first place. As Phillips maintains, *"it is certainly arguable that the use of a word-selection pre-programmed for plainness makes the complex message more difficult, not easier, to unravel."* [17] By using simplified legal language, while offering a more 'user-friendly' interface for law, important information contained within the system may be lost for the moment.

This phenomenon has been already noted by scholars studying legal language. Bhatia draws the attention to the issue when describing the relationship between integrity of legal documents and plain language. [18] It follows that it may be rather risky to simplify legal language with no regard to the interplay that takes place between legal language and law. Thus, it is worth considering directing the efforts on such modifications of legal language that would enhance its comprehensibility while not interfere with its composition and structure. In this respect, it is much more appropriate to talk about modifications aimed at matching the reading skills of the audience instead of simplification. Such an adjustment carried out in a way that *"the essential integrity of law is not sacrificed"* may be (after the fashion of Bhatia) called 'easification'. 'Easificated' legal language would provide more user-friendly interface by which the system of law may be accessed by a variety of its users (both lay and professional) while taking into account the delicate relationship between law and legal language. [19] However, 'easification' should not be understood as limited only to those techniques that do not interfere with the composition and structure of legal language at all — thus, eliminating the risk of creating the tension between the systems completely. Such an approach would inevitably prevent almost any non-trivial adjustment from being considered as an 'easificating' technique — even those adjustments that would actually fall within the definition provided above. The concept of 'easification' needs to be developed much further in order to be explained sufficiently. Such is the aim of the following subsections.

3.1 Legal Language Comprehensibility and Preciseness

On several occasions a concept of legal language comprehensibility has been mentioned. We understand it in such a way that the more effort is required for one to grasp the meaning of individual statements generated by the language the less comprehensible it is. Furthermore, a very important connection exists between the comprehensibility and the perceived arbitrariness (defined in subsection 2.1) — the less comprehensible legal language is, the higher the degree of this particular instance of arbitrariness.

At this point it is necessary to introduce a concept of preciseness. The more effort has to be put in understanding, assessing and describing law and individual legal problems, the less precise legal language is. The preciseness of legal language affects the actual arbitrariness (defined in subsection 2.1) in the same way the comprehensibility affects the perceived arbitrariness — the less precise legal language is, the higher the degree of the actual arbitrariness. Thus, the more precise and comprehensible the legal language is, the higher the legal certainty within any legal system.

The relationship between preciseness and comprehensibility of legal language is a very delicate one — increasing one of the elements would very often lead to the decrease of the other one. The relationship can be compared to that of inverse relationship only with great difficulties, yet in certain aspects it is very similar. For achieving maximum legal certainty, it is necessary to find an optimal point, equilibrium, where the sum total of the individual parameters of preciseness and comprehensibility is the highest. Unduly communicated law equals mere arbitrariness — Fuller refers to the minimum level of comprehensibility necessary for law to be legitimate. (See generally [20]) In this context it is only logical that when the comprehensibility of legal language approaches the minimum level, a tendency to reform or simplify the language appears. In this respect, we may formulate a fundamental constraint on 'easification' — it is possible to increase the comprehensibility of legal language as long as the preciseness is not decreased to such a level that the overall level of legal certainty would be compromised. This constraint is nothing else but a more rigorous expression of Bhatia's prohibition of sacrificing the essential integrity of law.

In the view of what has been said above, simplification of legal language that does not fall within the definition of 'easification' entails serious and most importantly unnecessary — as the overall level of legal certainty decreases — danger for the cohesion of the whole system; the only sensible way to proceed in making the legal language more comprehensible is the 'easification', a 'rational' adjustment of the interface. While any simplification leads to a higher level of comprehensibility of the legal language, serious errors in communication may occur because of the loss of preciseness in expression. Despite the fact that such legal language becomes more comprehensible, the overall legal certainty is compromised due to the tension between the natural evolution of law and a harshly/inconsiderately led adjustment of legal language.

3.2   Role of Context in Understanding Law

As has been shown above, the process of simplification may lead to the decrease of preciseness of legal language and consequently compromising legal certainty in the society. It has been claimed that adjustment of legal language in absence of reflecting the changes in the system of law may lead to a tension that can possibly cause undesirable consequences as regards legal certainty. The close connection of this phenomenon to the interplay between the comprehensibility and preciseness of legal language has been shown in the previous subsection. However, it remains to unravel yet another issue which is closely connected to the phenomenon — the issue of the context.

As expressed by Cao when paraphrasing Jackson, legal language may be understood only within the context of law: *"The words make sense within the context of legal system. Understanding an item of the legal system requires knowing the legal system."* [21] Understanding legal language does not equal understanding what the individual words mean, but also understanding their complex systemic consequences: the legal effect of a normative sentence cannot be simply read-off the surface of the text. [22] Therefore a layperson when reading a legal text *"may be quite oblivious to those systematic differences that give the same words a different meaning to the lawyer."* [23]

Understanding the complex of law and legal language thus requires certain type of knowledge. Reading legal texts (such as contracts) requires a 'competent reader'. This term, though used by Bourdieu in the context of art and its interpretation, is intriguingly fitting for the area of law as well. [24] A competent reader is the one who recognizes the system of law behind the interface of legal language and is able to make sense of it. A competent reader is legally literate.

In the process of increasing the comprehensibility of legal language, there is a tendency towards creating new shortcuts in terms of neologisms, symbols, or imagery; the use of images (pictures, pictograms, ideograms or icons) is perceived as useful in particular. [25] Some scholars claim the written text is not likely to lose its prominence in legal communication, [26] others go further and stress the growing importance of pictorial communication. (See [27] and [28]) At present law makes (however restricted) use of imagery, as is the case of various graphical forms (tables, charts) or informative pictograms and traffic signs.

The role of context in understanding law is analogical to the role of 'law' (concept) in the Piercean sense, where the symbol (language) is understood as *"a sign which refers to the Object that it denotes by virtue of a law, usually an association of general ideas, which operates to cause the Symbol to be interpreted as referring to that Object. It is thus itself a general type or law, that is, is a Legisign."* [29] To understand any symbol (may it be the word 'justice' or the image of a pair of scales), the reader has to know the 'law/legisign' that connects a Symbol to its Object.[6] Whether it is a traffic sign, a concept of contract or a pair of scales, a special kind of literacy is expected. The distinctive nature of legal language rests in the fact that it only obtains its legal meaningfulness against the matrix of a legal system. [30]

Based on what has been said above, an additional constraint on 'easification' must be formulated. It is possible to create shortcuts within the interaction between legal language and law in order to simplify legal language. However, these efforts have to preserve at least the neccesary level of context with regards to the audience, needs and requirements of such a simplification. Going back to Bhatia's prohibition of sacrificing the essential integrity of law this constraint goes beyond it, for it introduces subjective (audience) and circumstential (needs and requirements) elements. This brings us back to the previously mentioned Floridi's 'Method of the levels of abstraction' that may serve to explicate certain aspects of the relationship between law and legal language. For individual needs and requirements - and to target different groups of users - different interfaces may be adopted, based on choosing different variables to describe the system. (Floridi 319) The choice of variables then shapes the nature of the interface and whether or not it serves its purpose well. In the context of law, however, we believe that the full analogy to Floridi fails. The means of communicating law may not only reduce the perception of e.g. granularity of the system of law, but also provide additional misleading information of what can and cannot be done within the system. Legal language is not a mere way to describe or access the system; it is an integral part of the system itself to the extent that it itself shapes and creates the communication space of law,[7] while law in this sense cannot be reduced to a system unchanged by the way agents talk about it.

4   Creative Commons and Using Computers to Simplify Legal Language

4.1   The Idea of the Commons

The development of widespread internet use has led to the questioning of the contemporary paradigm of copyright law. [31] The critique has given rise to various initiatives, such as free software movement.[8] [32]

Obviously this idea proved to work extremely well in practice and was an enabler of what is nowadays being referred to as free or open source software (F/OSS). (see [34] and [35]) Furthemore, the same idea seems to be functioning with the same efficiency outside the domain of software, i.e. within the area of other types of copyrighted works such as literary, graphic or audio-visual.

---

[6] This relationship is basically the one of the triangle of signification: There is no direct relationship between a symbol/sign and its referent. This relationship is maintained only through the concept, which has to be learned. The relationship between the symbol may (or may not) be purely arbitrary.

[7] In this respect we maintain a position similar to White. [12]

[8] Backed up by Free Software Foundation since 1985. [33]

This initiative that is nowadays coordinated by the CC organization, which promotes its mission as follows: *"Our vision is nothing less than realizing the full potential of the Internet universal access to research and education, full participation in culture, and driving a new era of development, growth, and productivity."* [36] Despite the impulse this approach has brought into the domain of copyright law, it has also brought about surprisingly large amount of controversies and legal issues. (see [37] and [38]) A lot of them have risen out of the simple fact that the concept aiming to be embraced internationally has been designed within the framework of US system of law and was largely incompatible with civil law systems. On multiple occasions it has been already proven that this type of problem can be easily overcome.[9]

However, it has been the existence of a rather different class of problems caused by misunderstanding the whole concept by common laypersons that has been much more troubling and resulting in the lawyers' inability to assess and address the issue accordingly. These problems refer to various cases of errors in the process of placing works under one of the CC licenses leading to copyright infringements on a daily basis.[10] We believe that this issue can be directly attributed to the controversial popularization and promotion of the CC licenses that includes significant efforts of legal language simplification.

4.2 Simplification of Legal Language by the Commons

As has been already suggested earlier, the CC community employs rather controversial methods of legal language simplification. This simplification is based upon a neatly designed computerized system, which represents the individual license agreements in three different layers. All of the layers are stored in the source code, which is a combination of a mark-up language[11] accompanied with simple scripts[12] but these expressions usually remain hidden from the users sight. It is the interpretation of the code provided by a web browser that is primarily accessible to the users. However, this should not be considered a problem since from the legal point of view the information provided by each expression are equal; unlike the expression displayed by the browser the source code includes tags and scripts intended to instruct the browser about the way how to display the content and the functions it should perform upon various actions done by the user. On the other hand, what can be considered a problem is the difference among the individual layers, which is rather extensive.

---

[9] E.g. the incorporation of the section 46 para 5 to the Czech Copyright Act enabling the use of free licenses. If the concept is valuable, works fine in other jurisdictions and offers significant advantages for the development of current information and knowledge economy it would be futile to resist it continuously on the grounds of current legal provisions instead of reinterpreting or amending them in order to harness the potential benefits of the free licensing scheme.

[10] See subsection 3.3 of this paper.

[11] HTML 4.0.

[12] Mostly JavaScripts.

The first layer consists of a grey array with small pictograms distinguishing the type of CC license. The first pictogram is usually the CC pictogram informing the user that the license belongs to the CC licensing scheme and that it may be followed by various combinations of non-commercial, share-alike or no-derivatives pictograms. The badge itself serves as a link to the second layer and is often, but not absolutely necessarily, accompanied with the text informing the user about the same facts as the badge and sometimes provides certain additional information. These may selectively include the format of the work, its title, attribution, URL of the work, source work URL and URL of the webpage containing information regarding other permissions not covered by the CC license. It is also a very important function of the layer (perhaps the most important one) to associate the work with the individual license; this can be done by including the optional URL of the work and its title and further supported by appropriate placement of the badge.

The second layer, often referred to as the Commons Deed, is represented by neatly looking webpages summarizing the individual CC licenses. These are located at the CC website and are general, i.e. there is no association between them and the work (only the link connecting the first layer with the second one). Thus, once the user has entered the second layer she no longer has any information regarding the work the license is associated with; all she can do is to use the back function of the web browser to return to the first layer in which the information is (or at least should be) accessible. Every individual webpage belonging to the second layer of CC licensing scheme consists of a header informing the user about the fact that the page belongs to the CC system and about the type of license that is currently displayed. The body of the documents consists of three parts what is one free to do with the work, under what conditions and with what understanding. The permissions and conditions have a form of simple lists of activities and limitations accompanied with the similar pictograms that have been already included in the badge within the first layer while the understanding takes form of a short list of sentences informing the user about certain features of the license. At the very bottom there is information that the website is a human readable summary of the Legal Code (the full license), a link to the third layer the actual text of the license, and a disclaimer explaining the non-legal nature of the summary.

The third layer is represented by the websites containing the full text of the individual CC licenses. As in the case of the second layer, the licenses are general and no association between the license and the work exists at this particular level. All the websites belonging to the third level are equipped with the very same header as those that belong to the second layer, followed by a short disclaimer regarding the position of the CC organization within the process and provisions of the license. At the very bottom there is a notice expressly stating the non-involvement of the CC organization in the legal relationship that have been established by the license and imposing restrictions on use of the CC logo.

4.3   Outcomes of the Creative Commons Simplification

This subsection seeks to demonstrate the actual impact of simplifying the 'interface' of law by the CC licensing scheme. The idea of reducing a large number of copyright licensing nuances and niches to a few pictograms has resulted in a plethora of practical problems. Even though CC do provide a licensor with an extensive FAQ and help section on what aspects should be taken into account before using their license, [39] it may be claimed that a large number of problems result from misunderstandings and consequent incorrect application of the CC licenses.

It is quite interesting that serious flaws appear frequently even with respect to the most fundamental aspects of the legal relationships established by the CC licensing scheme. Such is the case of incorrect licence association, which can be observed mostly at personal blogs, where various types of works are posted. Often an author states that the whole blog is licensed under the CC licence without referring to particular works that should be subsumed under the licensing conditions. Subsequently, the visitors may tend to re-use the allegedly licenced materials in good faith and thus infringe the rights of the original right-holder. Another issue is related to the proper attribution of the author.[13] In accordance with the CC BY 3.0 Unported License, the work must be attributed in the manner specified by the author or licensor. Thus, a question remains how to attribute the content properly, in case the author does not provide the necessary information or in case the author's identification is missing completely. As there is no standard[14] for correct identification of the author, this problem still causes much uncertainty.

Table 1. Summary of the brief internet survey

Country	Work clear	Author identifiable	Infringment suspicion
Australia	29/50 (58 %)	35/50 (70 %)	7/50 (14 %)
Czech Republic	25/50 (50 %)	40/50 (80 %)	8/50 (15 %)
Germany	31/50 (62 %)	38/50 (76 %)	5/50 (10 %)
USA	33/50 (66 %)	44/50 (88 %)	15/50 (30 %)
TOTAL	**82/200 (41 %)**	**43/200 (21.5 %)**	**35/200 (17.5 %)**

In May 2011 we have carried out a brief internet survey the sole purpose of which has been to shed some light on the above mentioned phenomenon and estimate how commonly do these mistakes actually appear. In total 200 webpages chosen in accordance with the agreed methodology have been assessed. Three aspects have been evaluated — whether it has been clear what work has been covered by the license; whether the author has been identifiable; and whether a suspicion of copyright violation has existed. The summary of the internet survey is presented in the Table 1. (for detailed information on the survey see [42])

---

[13] This issue manifests itself mainly at the popular photo sharing sites such as Flickr. [40].

[14] Although basic guidelines are provided by the CC. [41].

It is difficult to identify the cause of such disturbing results. However, it seems that even concepts as fundamental as 'work' or 'author' do cause trouble within the framework of CC licensing scheme.[15]

A number of court disputes related to the inappropriate use of the CC licenses have arisen as well. In the *Lichdmapwa v. L'asbl Festival de Theatre de Spa* case,[16] a Belgian folk band sued a theatre company for using a snippet of their CC BY-NC-ND licensed song in a radio advertisement without proper attribution. During the trial the theatre company claimed that it had been unaware of the existence of such a license and had believed that the song could have been used freely since it had been available online for free download. The court rejected the defense and awarded the claimant damages of 4500 EUR (1500 EUR for every term of the license violated).[17]

In 2006 the tabloid Weekend published a set of photos taken by Adam Curry that he uploaded on the popular photo sharing site Flickr. These were marked as public, while still licensed under the CC BY-NC-SA 2.0 License. The license was indicated by the appropriate graphical icon. Despite the fact that the photos were published with the appropriate copyright notice (i.e. © Adam Curry) the court found the publishing house liable for copyright infringement.[18] As a professional party, the publishers of Weekend were obliged to inquire the license to the respective work thoroughly. The mark 'public' simply means that these photos can be viewed by anyone, not used freely. In this particular case, such use of the photos did not comply with non-commercial term of the license.

In *Chang v. Virgin Mobile* an Australian telecommunications operator used a CC-BY licensed photo depicting a 15-year old girl in a national advertisement campaign. The family of the girl found such use libelous and filed a complaint against the operator and the creators of the CC licenses. The latter had allegedly the obligation to educate and warn the user of the CC license *"of the meaning of commercial use and the ramifications and effects of entering into a licence allowing such use"*.[19] Even though the case was finally dismissed for the lack of personal jurisdiction, it has still demonstrated the potential interpretation problems related to the CC licensing scheme.

To sum up, in this subsection it has been shown that even when used with good intention the wrongful use of CC licences can occure easily. Consequently,

---

[15] Not to mention other complicated aspects as e.g. the possibility to prohibit the use of work for commercial purposes and the option to ban the re-use of the licensed work - as these two terms are extremely ambiguous they allow a lot of space for possible interpretation as to what falls within the definition of commercial.

[16] For details of the case refer to: http://wiki.creativecommons.org/
09-1684-A_%28Lich%C3%B4dmapwa_v._L%27asbl_Festival_de_Theatre_de_Spa%29

[17] Full decision of Le Tribunal de Premiere Instance de Nivelles in French available at: http://wiki.creativecommons.org/images/f/f6/
2010-10-26_A%27cision-trib.-Nivelles-Lichodmapwa.pdf

[18] Full decision of the Court in English translation available at: http://wiki.creativecommons.org/images/3/38/Curry-Audax-English.pdf

[19] Full text of the complaint available in English at: http://lessig.org/blog/complaint.pdf

when unclear about her granted rights, the user will rather opt for a defensive approach in order to avoid the legal actions of the entitled copyright holder and will restrain herself from the creative re-use of the unclearly licensed content. However, such a state of affairs is contradicotry to the basic principle of the CC movement, which is to engage as many people as possible in a hassle-free creative sharing of content without the need to seek a professional legal advice.

5   Conclusions

The Creative Commons have provided wide public with the carefully designed computerized system that is able to facilitate an extremely demanding process of licensing a work within the free licensing scheme. Even a skilled lawyer would have to employ a significant effort to provide the client with the equivalent service. However, experience shows that certain constituents of the system should be carefully assessed and perhaps even redesigned since its usage leads to copyright infringements on an almost regular basis. We maintain that in the core of the problem lies the incorrect comprehension of the tool by its users.

The legal language simplification efforts of CC consist mainly of representation of the complicated license agreements by a set of pictograms and short statements divided between the CC badge and the Commons deed. This fact alone would not cause any trouble if only the representation would adhere to the described principles of 'easification'. However, it seems that the license contains a large number of highly relevant information that remain hidden to the ordinary user. Furthermore, it uses highly complicated general terms — such as 'work', 'attribution' or 'non-commercial' — in situations requiring a higher level of granularity and preciseness. In context of the theoretical framework developed in sections 2 and 3, it may be asserted that the simplification of legal language provided by the CC licensing platform goes too far in increasing the comprehensibility of legal language — this is done at the expense of the preciseness of legal language so the overall level of legal certainty may be compromised. Furthermore, it also seems that to a certain degree the way in which the simplification is done does not take proper account of the audience for which it has been employed at the first place. It is possible the ordinary user of the CC licensing platform lacks the literacy that is needed to use the system correctly. To paraphrase the above said within the framework of the user-interface-system metaphor — the interface (legal language) that is used to communicate with the system (law) does not represent the system accurately, allows the user — who is not fully aware of the fact — to use only a small portion of its functions and often encourages her to employ operations she is in fact not allowed access to. The main problem is that the described employment of operations the user is not allowed access to has a very specific meaning when speaking about the domain of law — namely infringments of law followed by normative consequences.

In conclusion to the elaboration on the 'Method of levels of abstraction' Floridi asks a very serious question: *"Can a complex system always be approximated more accurately at finer and finer levels of abstraction, or are there systems which simply cannot be studied in this way?"* He provides no answer to the question. [43] In the view of our findings it seems sensible to ask in a similar way whether the task of simplifying legal language in such a way the CC organization pursues can ever be acomplished. As the answer necessarilly cannot be given at this stage of the CC licensing scheme's development it is only possible to formulate the principles to which all the efforts have to adhere. Some of them have been identified in this paper as the constraints imposed on the 'easification'. Identification of the others shall be the subject of future research.

References

1. Hager, J.W.: Let's Simplify Legal Language. Rocky Mountain Law Review 32, 74–88 (1959)
2. Phillips, A.: Lawyers Language. How and why legal language is different. Routledge, London (2003)
3. Butt, P., Castle, R.: Modern Legal Drafting. A Guide to Using Clearer Language, p. 18. Cambridge University Press, Cambridge (2006)
4. Bhatia, V.K., Candlin, C.N., Engberg, J. (eds.): Legal Discourse across Cultures and Systems. Hong Kong University Press, Hong Kong (2008)
5. Goldstein, T., Lieberman, J.K.: The Lawyer s Guide to Writing Well. University of California Press, Los Angeles (2002)
6. Plain English Campaign, http://plainlanguagecampaign.com/
7. Creative Commons, http://creativecommons.org
8. Raitio, J.: Legal Certainty,
   http://ivr-enc.info/index.php?title=Legal_Certainty
9. Bix, B.: Law, Language, and Legal Determinacy, p. 106. Clarendon Press, Oxford (1993)
10. Cao, D.: Translating Law, p. 9. Multilingual Matters, Clevedon (2007)
11. Bhatia, V.K., Candlin, C.N., Engberg, J.: Concepts, Contexts and Procedures in Arbitration Discourse. In: Bhatia, V.K., Candlin, C.N., Engberg, J. (eds.) Legal Discourse across Cultures and Systems, p. 9. Hong Kong University Press, Hong Kong (2008)
12. White, J.B.: The Legal Imagination. p. xiii. University of Chicago Press, London (1985)
13. Floridi, L.: The Philosophy of Information. Oxford University Press, Oxford (2011)
14. Above 13, p. 318
15. Floridi, L.: The Method of Levels of Abstraction. Minds & Machines, p. 319. Springer Science+Business Media (2008)
16. Kelly, K.: What Technology Wants, pp. 269–274. Penguin Group, New York (2010)
17. Above 2, p. 40
18. Above 11, p. 4
19. Bhatia, V.K.: Simplification v. Easification - The Case of Legal Texts. Applied Linguistics 4, 42–54 (1983)
20. Fuller, L.L.: Positivism and Fidelity to Law a Reply to Professor Hart. Harvard Law Review 71, 644 (1958)
21. Above 10, p. 17
22. Hutton, C.: Language, Meaning and the Law, p. 65. Edinburgh University Press, Edinburgh (2009)
23. Above 10, p. 17

24. Bourdieu, P.: The Rules of Art: Genesis and Structure of the Literary Field, p. 286. Stanford University Press, Stanford (1995)
25. Boehme-Nessler, V.: Pictorial Law. Modern Law and the Power of Pictures, p. 52. Springer, Heidelberg (2010)
26. Above 19
27. Above 28
28. Above 28, pp. 10–151
29. Peirce, C.S.: Logic as Semiotics: The theory of Signs. In: Buchler, J. (ed.) Philosophical Writings of Peirce, p. 102. Dover Publications, Mineola (1955)
30. Above 10, p. 16
31. Lessig, L.: The Future of Ideas: the Fate of the Commons in the Connected World, pp. 4–5. Random House, New York (2001)
32. Lessig, L.: Free Culture, pp. xv–xvi. The Penguin Press, New York (2004)
33. Free Software Foundation, http://fsf.org
34. Stallman, R., Gay, J.: Free Software, Free Society: Selected Essays of Richard M. Stallman. Free Software Foundation, Cambridge (2002)
35. Raymond, E.S.: The Cathedral and the Bazaar: Musings on Linux and Open Source by an Accidental Revolutionary. O'Reilly Media, Sebastopol (1999)
36. Above 7
37. Dulong de Rosnay, M.: Creative Commons Licenses Legal Pitfalls: Incompatibilities and Solutions. Institute for Information Law, Amsterdam (2010)
38. Angelopoulos, C.J.: Creative Commons and Related Rights in Sound Recordings: Are the Two Systems Compatible? Institute for Information Law, Amsterdam (2009)
39. Creative Commons FAQ, http://wiki.creativecommons.org/Before_Licensing/
40. Sullivan, D.: Flickr's Big Fail On Creative Common's Attribution Guidelines, http://daggle.com/flickr-fail-on-creative-commons-attribution-691
41. Above 42
42. Koscik, M., Savelka, J.: Dangers of over-enthusiasm in licensing under Creative Commons (unpublished manuscript), For private use only the survey is accessible, http://is.muni.cz/www/134449/Koscik_Savelka-DangersOfCCAnnex.pdf
43. Above 13, p. 326

From User Needs to Expert Knowledge: Mapping Laymen Queries with Ontologies in the Domain of Consumer Mediation

Meritxell Fernández-Barrera[1] and Pompeu Casanovas[2]

[1] Cersa,CNRS-Université Paris2, 10,
Rue Thénard, 75005-Paris, France
Meritxell.Fernandez@cersa.cnrs.fr
[2] Universitat Autònoma de Barcelona
Institute of Law and Technology UAB-IDT,
Cerdanyola del Vallès, 08290 Spain
pompeu.casanovas@uab.cat

Abstract. We believe that an important element towards the improvement of online legal services lies in the ability to identify legally relevant textual segments in users queries. In this line, this paper presents a case study in the processing of citizens queries in the domain of consumer justice from the point of view of legal domain semantics. The analysis is made in the framework of the ONTOMEDIA project, which aims at the design of a semantic platform enabling users and professional mediators to meet in a community-driven Web portal. The paper first presents the characteristics of the platform and its requirements; then reports the term extraction methodology from a corpus of consumer queries, and draws some conclusions on the double nature of laymen representation of legal problems: both terminological-conceptual and script-like (in the form of storytelling narratives). The paper further proposes an annotation structure based on these dimensions, which can be exploited to train machine learning algorithms to automatically recognize legally relevant text fragments. The paper concludes with the discussion of open issues for future work.

Keywords: Legal Electronic Institutions (LEI), mediation, user-generated content, term-extraction, domain ontologies, semantic mapping.

M. Palmirani et al. (Eds.): AICOL Workshops 2011, LNAI 7639, pp. 286–308, 2012.

1 The Catalan White Book on Mediation (CWBM), Legal Electronic Institutions (LEI), and the Ontomedia Project

The CWBM is a large research project (2008-2010)[1] aiming at the implementation of mediation as defined by the EU Directive 52/2008[2]. One of its most surprising findings is that at present near 20% of the population in Catalonia (7,5 million people) has pending cases in the Spanish Courtrooms (18% in 2008). Heavy caseloads and chronic shortage of judges and magistrates, on the one side, and increasing social problems on the other (especially large immigration rates and the emergence of all kind of violence in families, schools, hospitals and institutions) have fostered the need to draw a map of dispute resolution techniques in the country, before drafting a general statute. It is worthwhile taking into account that from 2000 to 2010, more than one million people have landed in Catalonia (15.9% of the population are newcomers, according to the 2010 census). Therefore, we conceived mediation not only as an Alternative Dispute Resolution (ADR) device, but as a set of tools operating near the communities, Courts and Administrations. In this way, mediation as institution may be adapted to the nature of conflicts arising within the different environments, contexts and settings (neighbourhoods, colleges, hospitals, administrations etc).

To apply technology to mediation, we followed a twofold strategy leading to two separate models: (i) building mediation as a Legal Electronic Institution (LEI)[3]; and (ii) setting up a general platform for citizens, administrations,

---

[1] All the results of the Catalan White Book (Department of Justice, 2010-2011) are available at http://www.llibreblancmediacio.com in both languages, Catalan (1186 pp.) and Spanish (1206 pp.). The Spanish version contains the programming and a computer prototype of mediation as electronic institution.

[2] Art. 3.a. Mediation means a structured process, however named or referred to, whereby two or more parties to a dispute attempt by themselves, on a voluntary basis, to reach an agreement on the settlement of their dispute with the assistance of a mediator ; art. 3.b. Mediator means any third person who is asked to conduct a mediation in an effective, impartial and competent way, regardless of the denomination or profession of that third person in the Member State concerned and of the way in which the third person has been appointed or requested to conduct the mediation. It is worth to mention R. (9): This Directive should not in any way prevent the use of modern communication technologies in the mediation process.

[3] Electronic Institutions (EIs) organize interactions by establishing a restricted environment where all interactions take place (e.g. e.commerce, e-learning, or ODR). They create a virtual environment where interactions among agents in the real world correspond with illocutions exchanged by agents within this restricted environment. When an EI is entitled to perform legal acts, or at the end of successive steps may produce a result with legal value, or an agreement that can be alleged in Court or before other appropriate ruling institutions, we face a Legal Electronic Institution (LEI) See [19]. See also http://e-institutions.iiia.csic.es. See for a more detailed analysis [20]; for a comparison of the grounds of LEI and Ontomedia [7]; for the state of the art of the ODR existing platforms, [28]. The LEI software code for mediation [21] is available at http://www.llibreblancmediacio.com (Spanish version).

institutions and professionals. The first strategy (LEI) models the performative structure of mediation as a set of procedural rules. The second one (ONTOME-DIA) allows users and professional mediators to meet in a community-driven Web portal (in which contents are provided by users and annotated by the ODR web platform).

ONTOMEDIA is, then, a semantic platform for relational justice. It has been conceived as a bus of services to offer to both citizens and mediators a kit of tools and services to facilitate a better access to justice. Attention is focused on the development and synergy between different technologies stemming from Web Services (WS), the Semantic Web (SW), Social Networks (SN), Multi-agents Systems (MAS), Computer Vision (CV) and legal applications. LEI and ON-TOMEDIA are orthogonally related, according to the original James Hendlers diagram on the link between Web 2.0 and the emergent Web 3.0 (Fig. 1):

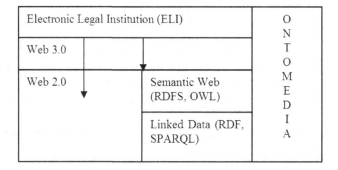

Fig. 1. Use of Hendlers diagram (with permision). **Source:** [7].

The sections of ONTOMEDIA are tailored to the domains previously identified within the CWBM: commercial and business disputes, consumer complaints, labor conflicts, family, restorative justice (adult and juvenile mediation in criminal issues), community problems, local administration, health care, environmental management, and education (Fig. 2).

We have planned a lifecycle of five years to the full development of all the functionalities. We chose the consumer domain, first, to implement some of them specifically addressed to citizens. We made this decision because we had a good description of all the procedures and the precise workflow of pre-mediation, mediation and post-mediation stages [1]. Moreover, as it will be shown later, the Catalan Consumer Agency would give us access to more than 30,000 complaints and information requests to work with.

As a result of gathering consumer mediation related resources, a relational schema for a database was proposed as well. This database is a critical component of the platform's data tier. However, the proposed relational schema is only a little portion of it, storing entities and relations involving national regulations, regional regulations, soft-law, consumer offices, and so on. The database contains so far information on 19 Spanish regions, 892 towns and 52 provinces, holding

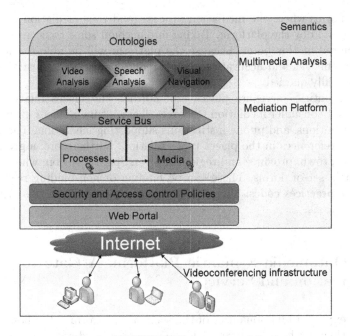

Fig. 2. ONTOMEDIA layered architecture

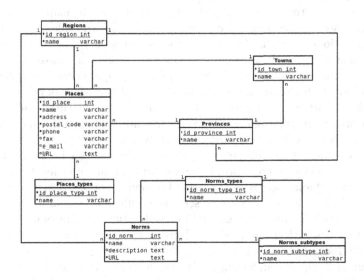

Fig. 3. Database diagram. **Source**: [15].

1,264 consumer mediation resources (which include consumer offices and other public and private institutions) and 75 different regulations (Fig. 3).

The idea behind this schema is to provide basic legal and judicial resources to citizens involved in consumer mediation processes, including users in conflict starting scenarios where the mediation process (and thus mediation resources) may be suggested by the platform. Taking into account some basic geographical data is required at this point, because the platform will be able to locate the user and an efficient and accurate response requires norms and institutions to be geographically queried.

The basic entities here are places and norms. The places relation stores registers about consumer mediation resources, like consumer offices, private consumer organizations, and public institutions supporting mediation. These types of places are references in the places types relation. Furthermore, a given place is located in a town, province and region. A given Spanish region, where a place belongs, has a set of norms. These can be binding or non-binding regulations, such as best practices codes. These distinctions are made in norms types and subtypes relations.

2   The Ontomedia Semantic Platform: A Gate to Information and Services

One of the expected functionalities of the semantic platform is to allow citizens to present their problem in natural language and to redirect them either to relevant information already available online or to the suitable state agency. The assumption at the basis of this process is being able to map two different conceptual systems: the user representation/section of a problem in the form of concrete actions, actors and contexts (non-expert model); and the regulative representation of the problem usually in the form of general classes of actions, actors and normative provisions (expert model). We propose here the existence of a middle-level which corresponds to the practices and know-how of professionals.

Professionals are indeed frequently in charge of reformulating regulative information into more comprehensible texts that are subsequently published in the form of electronic leaflets in institutional websites, and they are usually as well the ones interacting directly with citizens. Thus it can be assumed that theirs is an intermediary conceptual system, bridging abstract legal provisions with concrete conflictive situations presented by non-expert citizens. Our conception of domain knowledge can thus be seen as a multidimensional figure, which, vis--vis flat knowledge models, takes into consideration elements such as different domain actors (citizen, professional, legislator) and communicative contexts (information request, complaint).

The technical aspects underlying this functionality are related to the automatic classification of consumer queries according to a conceptual scheme which models citizens problems or conflictive situations on the basis of available institutional structures and procedures. This model has been described thoroughly in the Catalan White Book of Mediation [8] and an ontological representation of mediation expert knowledge has been proposed in the Mediation Core Ontology, available in OWL-DL [28]. A further representation of the domain of consumer mediation is provided by the consumer mediation ontology [29].

We have a diachronic corpus of around 10,000 questions and 20,000 complaints which have been addressed by consumers to the Catalan Consumer Agency[4] from 2007 to 2010. The difference between queries and complaints relies on the fact that while queries are mere requests of information, complaints are meant to initiate an administrative process of mediation between the consumer and the seller or service provider[5]. A further distinction relevant to characterize our corpus is the input language. Indeed, since both Catalan and Spanish are official languages in Catalonia and thus citizens are entitled to address state agencies in both languages, a previous step for treating automatically our corpus has been to classify documents according to their language.

At this initial stage we have decided to concentrate exclusively on queries expressed in Spanish, corresponding to the year 2010. The subset of queries of 2010 has been used to extract representative terminology from subsets of consumer questions classified by topic (Internet service providers, travel agencies, vehicles), and the extracted terminology has been linked to the available ontological domain models (See Section 5).

However, as highlighted in Section 6 user questions are not mainly terminological or conceptual. They do not present definitions, but they describe contextually situated stories that are analysed by domain experts and given legal interpretations which entail certain institutional reactions. An annotation structure that captures both the terminological and the narrative structure of citizens questions is proposed in Section 6.2.

Section 3 discusses the technical challenges of an intelligent platform able to process citizens queries and presents the model that will be used in our case study. Section 4 details the process of terminology extraction from a set of consumer queries; Section 5 describes the extension of the available formal ontologies with consumer terminology through a has lexicalisation property. Section 7 discusses the main contributions of the paper and identifies the issues that require being dealt with in the follow-up of the ONTOMEDIA project.

3 Bridging the Gap between Knowledge in Action and Theoretical Legal Knowledge: Web 2.0 vs. Web 3.0

Enabling the intelligent processing of non-expert generated content is strongly connected with the problem of interfacing Web 2.0 with Web 3.0. Indeed, with the advent of Web 2.0, semantic technologies face a new challenge: the processing of heterogeneous non-standardized knowledge, with unknown producers and with the absence of explicit terminological and conceptual harmonization. This

---

[4] The mission of the Catalan Consumer Agency is to defend citizen's rights as consumers, and thus on the one hand it provides information regarding consumer affairs and on the other it has a role in the resolution of conflicts between consumers and companies through mediation and arbitration. http://www.consum.cat/qui_som/index_en.html

[5] One of the requirements for being able to initiate a mediation process is to have previously contacted the seller or service provider.

problem was already highlighted in connection to the need of conceiving artificial intelligence as the ability to cope with heterogeneous and disperse data, based on different ontologies, instead of focusing on highly axiomatised and unified ontological models [20, 13, 12].

Coping with this challenge implies finding a way to bridge Semantic Web data structures, such as formal ontologies expressed in RDF or OWL, with unstructured implicit ontologies emerging from user-generated content. Sometimes these emergent lightweight ontologies take the form of unstructured lists of terms used for tagging online content by users. Accordingly, some works have dealt with this issue especially in the field of social tagging of web resources in online communities. More concretely, different works have proposed models for making compatible the so-called top-down metadata structures (ontologies) with bottom-up tagging mechanisms (folksonomies)[6]. Some authors, such as [37], point out that the emergent problem of linking Web 2.0 and Web 3.0 lies on the way in which emergent collective rationality of the Web 2.0 relates to the proposed connective rationality of Web 3.0.

The possibilities range from transforming folksonomies into lightly formalized semantic resources [19, 17] to mapping folksonomy tags to the concepts and the instances of available formal ontologies [36, 24]. At the basis of these works we find the notion of emergent semantics [19], which questions the autonomy of engineered ontologies and emphasizes the value of meaning emerging from distributed communities working collaboratively through the web. An important element in the model proposed by many of those works is the actor, who tags a specific resource with a particular tag.

This is not the case in our corpus, since users in our case study simply provide input texts describing their problems and asking for institutional assistance. Thus we do not have a ready-made folksonomy created collaboratively by users. In this context the implicit ontology is understood as the set of linguistic structures on which users rely to represent the concepts of the domain. Thus in this framework a further challenge consists in being able to extract recurrent linguistic structures from non-normalized texts. Indeed, while texts following the standards defined by a particular community of experts lend naturally to the extraction of patterns, this task becomes much less obvious with regard to texts which do not necessarily conform to pre-established guidelines.

This way, the terminological, argumentative and semantic patterns of texts following certain standards in the legal community (i.e. bills, acts, judgments, legal expert files) has been deeply studied (see state-of-the-art on legal ontologies, legal argumentation models and XML models for legal documents in [32]), while the analysis of recurrent structures in the way citizens express their legal problems has been paid less attention. In the domain of semantic technologies for

---

[6] It should be highlighted that the terms top-down and bottom-up are here used as referring to the participants in the construction of the resource: while in the first case the resource is the result of an agreement on a world model reached by the members of a particular community, in the second case the resource emerges from the distributed tagging activity of a big number of anonymous users.

the legal domain efforts have indeed mostly concentrated on making explicit formal ontologies deriving from normative sources and from legal expert texts [33]. This work has lead to the creation of several domain ontologies but so far explicit mappings between these formal ontologies and implicit ontologies emerging from the citizens representation of particular legal problems are not available. This implies an important drawback for legal web-based services, since the linguistic and conceptual schemas used by citizens in the expression of their needs are not taken into account. The improvement of such services requires taking into account the particularities of non-expert common discourse and above all, to connect it to specialised legal discourse.

As a first effort in this direction, this paper presents a case study in the consumer law domain. We propose to reuse the available (a) Mediation-Core Ontology (MCO) and (b) Consumer Mediation Ontology (COM) as anchors to legal, institutional and expert knowledge, and therefore as entry points for the queries posed by consumers in common language. We will follow the approach proposed by [24] and enrich the available ontologies with the terminology appearing in the consumer corpus. For so doing, Owl classes and instances will be complemented with a has_lexicalization property linking them to consumer terms.

Our methodology is thus based on the following steps: i. extraction of relevant terminology from the consumer queries on the basis of morphological tagging; ii. enrichment of the ontological resources with consumer terminology through a has_lexicalization property.

4   NLP Extraction of Consumer Terminology

Since the corpus of consumer queries has not been previously annotated or semantically tagged there is no available semantic representation of consumer knowledge. This is why it has been decided to semi-automatically extract a list of representative terms through NLP techniques. The goal was to see whether despite the fact that producers are unknown and do not follow explicit guidelines in the construction of their message common lexical patterns emerge.

Firstly, the set of queries of 2010 was manually classified into subsets according to a list of topics used by the Catalan Consumer Agency [8] in order to enable the extraction of contextually-related groups of terms. The topics defined by the Agency are: *commerce, e-commerce, electrical appliances, housing, hotel industry, finance industry and insurers, services, professional services*[7]*, supplies, telephone, passenger air transport, transport, vehicles and travel agencies.* Fig. 4 reports the number of queries corresponding to each topic as well as their average length in number of tokens.

---

[7] Professional services refer mostly to the services provided by liberal professionals such as lawyers, doctors, dentists. On the other hand, services refer in general to services such as sport facilities, hairdressers, cultural shows (theatres, cinemas) or educational services (for instance e-learning).

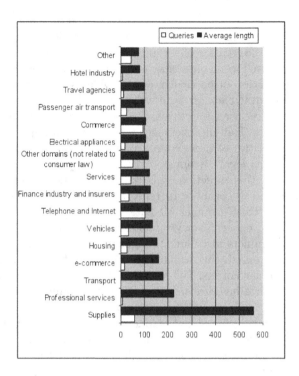

Fig. 4. Number of queries and query average length per domain

It can be observed that the topics concentrating the highest number of queries are, in this order, telephone supply, commerce, other supplies (gas, electricity, water,), and services. The average question length varies according to the domain. For instance, the average length in the domain of supplies (560 tokens) is five times that of queries in the domain of commerce (104 tokens). This fact can be explained due to the characteristics of the problems arising in the domain of supplies, which are usually connected to technical failures which require a certain degree of precision to determine where responsibility lies (nature of electric installation, power of the electric circuit,). On the other hand, queries in the domain of commerce require usually only a few details to present the situation (such as place of purchase or guarantee length) and therefore tend to be shorter. These are important aspects to take into account in the process of terminology extraction.

Another relevant feature to be highlighted is that a high number of queries belong to other domains which go beyond the competences of the Consumer Agency and which mainly belong to other areas of law such as private law (i.e. disputes between tenant and landlord; private deals) or administrative law (i.e. tax paying; appeals to speeding tickets or to penalties for drunk driving). This reinforces the need of a semantic platform able to classify and distribute citizens queries to the state agencies which are able to provide useful information and assistance to solve the conflict.

Next, the questions were tagged and lemmatized using Tree-tagger. Tree-tagger [34] is a probabilistic morphosyntactic tagger and lemmatiser which estimates transition probabilities on the basis of a binary decision tree in order to avoid the limitations of probabilistic taggers based on Markov Models.

By using the make_separate.pl module we created an XML version of the tagged documents and imported them into the NooJ platform. NooJ is a platform that enables the linguistic processing of texts at different levels (i.e. morphological, syntactic, semantic) with the aid of different types of grammars (among which, inflexion grammars, morphological grammars and syntactic grammars) [35]. In order to enable NooJ to recognise tree-tagger tags as morphosyntactic annotations we adapted the original XML element and attribute to NooJ standards[8]. NooJ offers the possibility of analyzing morphologically any input text but it does not offer in-built disambiguation grammars, so whenever there is ambiguity all the possible syntactic categories will be maintained. This would have created a lot of noise in the subsequent search of morphosyntactic patterns, so we decided to rely on the probabilistic tagging of Tree-tagger, which provides disambiguated morphosyntactic tags with a low level of error.

Once the queries had been imported into the NooJ platform we extracted first simple terms and then multiword terms. Firstly, regarding simple terms, we follow the traditional trend in terminological studies that states that the most common linguistic unit carrying conceptual meaning is the noun. In this line, we extracted through the NooJ function Locate pattern all the nouns of our corpus. Nevertheless we do not rule out the possibility of extending term extraction to other linguistic units such as predicates in the future, since recent works have highlighted that units of specialized knowledge can take different syntactic forms[9].

Table 1 reports some of the simple terms (nouns) extracted in each subset of questions. Some of the extracted terms are recurrent in different topic subsets and therefore we can consider that they belong to the general domain of consumer queries. They denote: the seller, such as firm (empresa); the contractual binding between consumer and seller, such as contract (contrato), invoice (factura), guarantee (garantía); or the amount paid by the consumer, such as money (dinero), euros, amount (importe); the cause of the conflict, such as problem (problema), abuse (abuso), failure (fallo); and the expectations of the consumer, such as return-refund (devolución).

Other terms seem to be topic-specific. They denote either the actors of specific domains such as real state agency (inmobiliaria), in the domain of housing; campsite (camping), camper (campista), in hotel industry; bank (banco, entidad), insurer (correduría), in finance industry and insurers; lawyer (abogado, letrado), dentist (dentista), psyschiatrist (psiquiatra), hospital (hospital), in professional

---

[8] In other words, the original element and attribute "$< TOKENtag =>$" have been transformed into "$< LUcat =>$".

[9] More concretely, [3] highlight that units of specialised knowledge can be: morphological units (morphemes); one word units; syntagmatic units, that is to say, multiword units and phraseological units; and phrasal unit.

services; customer (abonado), telephone operator (operador), in the domain of telephone and the Internet services; taxi driver (taxista) in transport; garage (taller), car dealer (concesionario), in vehicles.

Actors are often denoted in the corpus through named entities, specially in the domain of telephone and Internet, such as Jazztel, Telefónica, Vodafone; and the domain of passenger air transport, such as Iberia, Easyjet, Aerlingus. They will be integrated into the available ontologies as concept instances.

Other nouns denote actions and states which are typical of particular domains, such as sales (rebajas), in commerce; activation (activación), prepaid card (prepago), permanence clause (permanencia), contract cancelation (baja), portability (portabilidad), in the domain of phone and Internet services; technical maintenance (mantenimiento), supply (suministro), in the domain of supplies.

Finally, some nouns denote typical objects of certain domains: appliance (aparato), washing machine (lavadora), computer (ordenador), microwave (microondas), in the domain of electrical appliances; meter (contador), boiler (caldera), heating (calefacción), in the domain of supplies.

The analysis of extracted terms according to semantic categories paves the way for their insertion into the available domain ontologies. This task will be described in Section 5.

Secondly, in order to extract complex terms, we applied a series of morphosyntactic grammars to the annotated corpus. The grammars correspond to patterns which are recurrently carriers of conceptual meaning in specialized discourse and, more concretely, in legal discourse. In the line of the approach followed for the extraction of simple terms, the grammars we chose are all syntagmatic units with a noun header[10]. By applying the grammars to our corpus we observed that not all the patterns were suitable for non-specialized discourse, specially the most complex patterns with embedded noun phrases (such as N+PREP+ART+N+ADJ) and those containing a syntactic inversion (ADJ+N, or ADJ+N+PREP+N).

The patterns that were finally applied are summed up in Table 4 with their corresponding examples. Similarly to simple terms, multiword terms denote either domain actors (air company -compañía aérea-, motorcycle insurer -empresa aseguradora de motos-, phone company -compañía de telefonía, compañía de teléfono-, voice over IP operator -operador de voz por ip-, debt collector -empresa de gestin de cobros-); events giving place to the conflict between seller and consumer (unexpected flight connection -escala imprevista-, undue charging -cobro indebido-, erroneous fee -error de tarificacin-, damages on a wall -desperfectos en una pared-, uninhabitable house -inhabitabilidad de la vivienda-); or events creating a contractual relation (deed signature -firma de la escritura-, purchase deposit -firma de las arras-).

Some extracted terms deserve a particular attention. This is the case of volcano cloud (nube volcánica) and Icelandic volcano (volcán islandés). A priori and out of context these terms do not belong to the domain of consumer law, but to geologic phenomena. However they appear repeatedly in consumer queries as a

---

[10] The set of grammars was built on the basis of a legal corpus in [12].

Table 1. Sample of extracted terms (N) by topic

Commerce	tienda, euros, dinero, producto, empresa, devolucin, problema, servicio, cámara, garantía, vale, rebajas, reclamación, tarjeta, fabricante, importe
e-commerce	tarjeta, cargo, teléfono, cuenta, garantía, devolución, producto, precio, reclamación, estafa, factura, paquete, transporte, calidad
Electrical appliances	servicio, reparación, cambio, garantía, lavadora, tienda, marca, ordenador, reclamación, tele, ACER, aparato, avería, denuncia, establecimiento, fallos, microondas
Housing	piso, arras, casa, problema, puerta, vecino, contrato, inmobiliaria, vivienda, cliente, empresa, propietario, alquiler, ascensor, reparacin, comunidad, parking, edificio, fianza
Hotel industry	hotel, tarjeta, reserva, nieve, noche, importe, viaje, camping, campista, agosto, autopista, cancelacin, caravana, Llagostera, PortAventura, restaurante, recargo
Finance industry and insurers	abogado, abuso, banco, entidad, cargo, cobertura, coche, complemento, compraventa, conductor, correduría, deuda, dinero, escritura
Services	contrato, curso, dinero, empresa, autoescuela, bono, cambio, casa, honorario, factura, gimnasio, guardería, enseñanza, estudio, fotografía, formación, gestora
Professional services	banco, gestoría, gestión, abogado, dentista, psiquiatra, medicación, dentadura, hospital, letrado
Supplies	gas, factura, luz, suministro, consumo, contador, contrato, agua, euros, servicio, Endesa, vivienda, mantenimiento, domicilio, reclamación, canon, electricidad, abuso, aparato, caldera, inspección, subida, teléfono, recibo, suministradora, apagón, calefacción, cuota, facturación, lampista
Telephone and Internet	contrato, factura, teléfono, abonado, abuso, acceso, activación, servicio, llamada, permanencia, baja, Internet, portabilidad, operador, línea, Vodafone, Jazztel, Adel, sms, penalización, cuota, telefonía, móvil, conexin, contratación, contraoferta, prepago, blackberry, Movistar
Passenger air transport	vuelo, billete, compañía, reclamación, aeropuerto, avión, destino, retraso, salida, billete, maleta, reserva, pasajero, easyjet, Aerlingus, compensación, Iberia, espera, indignacin, viaje-circuito, volcán
Transport	cinturón, taxi, taxista, trayecto, minusválidos, peaje
Vehicles	moto, taller, coche, concesionario, vehículo, garantía, fallo, problema, dinero, marca, vendedor, reparación, freno, motor, pieza, taller, airbag, fabricante, motocicleta, avería, centralita, distribuidor, Honda, caravana, carburador, ciclomotor, coche, embrague, homologación, Suzuki, volante, válvula
Travel agencies	viaje, dinero, euros, reserva, crucero, devolución, hotel, importe, reembolso, adelanto, anulación, compañía, reclamación, tour, agencia, alquiler, reembolso, Tailandia

source of conflict in the domain of air passenger transport. This makes evident that once general normative provisions materialize in real facts, the control over concepts and vocabulary becomes more and more sophisticated, because there is no predefined domain restriction.

Table 2. Sample of multiword terms

Pattern	Examples
N+ADJ	compañía aérea, vuelo regional, escala imprevista, nube volcánica, volcán islandés, cobro indebido
N+ADJ+PREP+NC	acción redhibitoria por vicios, placa identificativa de voltaje, empresa aseguradora de motos
N+PREP+N	compañía de telefonía, compañía de teléfono, contrato de Adsl, error de tarificación, fecha de activación
N+PREP+N+PREP+N	fecha de fin de permanencia, operador de voz por ip, empresa de gestión de cobros
N+PREP+ART+N	desperfectos en una pared, firma de la escritura, inhabitabilidad de la vivienda, firma de las arras

The levels of precision of patterns vary, but they are mostly situated between 50% and 60% of precision. The percentage of precision shown in Fig. 5 has been calculated as an average of the precision of each pattern per domain, so all patterns did not have the same performance in all domains. For instance, the pattern with a higher average precision, N+Prep+N (62%), presented a considerably lower level of precision in the domain of Transport (50%), while in the domain of electrical appliances the precision reached 75%. Similarly, the pattern N+Adj, with an average precision of 60%, has a precision of 56% in transport and of 70% in Supplies. This fact might be related to the length of the corpus Transport (which is the shortest with around 400 tokens). The levels of performance of grammars will be studied in a detailed way per domain in further research.

Furthermore, in the follow-up of the project we plan to add statistical measures to reduce the levels of noise, as proposed by the most efficient current terminology extractors ([2],[25]). It is further to be noted that at this initial stage we did not set up a threshold of occurrence in the corpus, so we included all candidate terms even if they were hapaxes. As it will be detailed in Section 7 one of the core issues of the ONTOMEDIA platform is to evaluate the terminological content of user-generated text vis--vis text produced by domain experts. Both morphosyntactic patterns and statistical measures currently applied to the detection of domain terms in a specialized text will have to be tuned to the characteristics of user-generated corpora. We plan to apply the results of the analysis of our corpus to the design of a new set of NLP tools tailored to the nature of user queries.

As an initial step in this direction, however, we consider that the results obtained are rich enough to support lexically the available ontologies. This is shown in the next section.

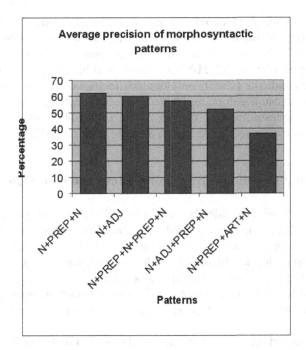

Fig. 5. Precision of morphosyntactic patterns

5   Laymen Knowledge through Expert Lenses

As mentioned in Section 3, a possible approach to deal with user-generated data from a domain perspective is to rely on available domain ontologies. This section explores available domain ontologies in the domain of consumer justice and mediation and explores the mapping of extracted vocabularies to ontological classes.

5.1   The Available Domain Ontologies

The Mediation-Core Ontology [28] contains the basic concepts of the domain of mediation, since it is aimed at providing the conceptual anchors for the set of domain mediation ontologies that will be developed in the ONTOMEDIA platform. This way, its top classes denote the agents involved in the mediation process (MediationAgent), any information source used in the process (MediationInformationSource), the mediation process according to the domain (MediationProcess) (Fig. 6), the different phases of the process (MediationProcessStage), the sessions of the mediation process (MediationSession), the roles that actors might play in the mediation process (MediationRole) and the domains in which mediation can intervene (MediationTopic). It may be noticed that MCO is a structured general ontology that focuses on the mediation system, while the second one (COM) is a

domain ontology especially focused on legal institutional features [5]. The underlying conceptual structure of MCO points to the social, political and economic
features of ADR, ODR and relational justice processes including negotiation,
Victim-Offender Mediation (VOM-) and transitional justice.

On the other hand, the (CMO) ontology [28] focuses on the particularities of
mediation in the consumer domain. Its main classes denote the parties involved
in the conflict (PartiesinConflict), the regulation applicable to the conflict (Regulation), the geographic area (Territory), and the type of conflict.

5.2   Integrating Consumer Terminology into the Ontologies

Among the lists of topics provided by the Core-Mediation Ontology as subclasses
of the MediationTopic top class we find ConsumerTopic. One possibility would
thus be to link all the extracted terms to this class through a has_lexicalization
property. However, this would imply the loss of the fine-grained classification
of terms by topic presented in the previous section. This is why we decided
to create 14 OWL subclasses of the class ConsumerMediation corresponding to
each of the domains and to link to each of the subclasses the terms belonging to
each domain.

Once the extracted terms were mapped to the newly created OWL subclasses
we linked the terms to the COM, this time according to their semantic nature
and not to the topic they belong go. We do not reproduce here all mapped terms.
The main semantic typologies of extracted terms were presented in Section 4.

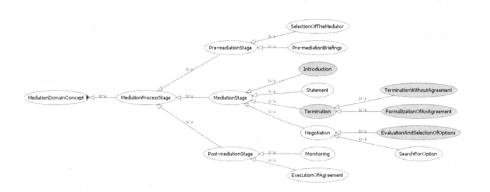

Fig. 6. Fragment of the Mediation-Core Ontology. **Source:** [28].

As an example we show in Fig. 7 how we linked to the class PartiesinConflict,
and more concretely, to its subclasses Consumer and Seller, respectively, some
of the terms we identified as being actors in different consumer domains. The
figure shows as well the introduction of some named entities as instances of the
class Seller.

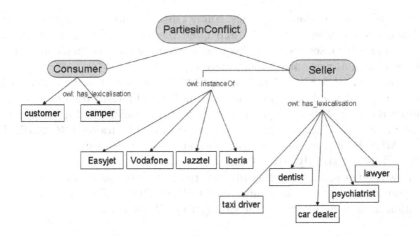

Fig. 7. Integration of consumer terminology into consumer mediation domain ontology

6 Laymen Narratives and Script-Like Structures

By linking the most representative terms of the user queries to ontological classes (Section 5) domain semantics is partially embedded in laymen discourse. Nevertheless, common-sense discourse is not mainly conceptual. In other words, since it is not technical discourse, there is not necessarily a shared underlying conceptual model to all user questions and therefore single lexical units are not the main carriers of meaning. What is shared, on the contrary, is the eventive structure that describes legal conflicts. This structure can be formalized in terms of contextual frames with certain participants and modalities (past tense, rights, obligations,). In this section we describe the narrative structure of user questions and propose an annotation structure for capturing it.

6.1 Narrating Legal Conflicts

Narratives and storytelling have been highlighted as basic mechanisms of cognition used by laymen in the conceptualization of legal conflicts. Story construction has been for instance considered an important element of juror decision making [26, 27, 17], and more generally of legal cognition [31]. A lot of research has concentrated on storytelling at trial, but as highlighted by [31] story construction is part of the legal case even from its very beginning (the client tells a story to the lawyer, the lawyer presents it to the court, the judge makes a decision which is based on a story which has been built through the evidence presented at the trial). Here we analyze the narrative structure of consumers questions in describing a conflict in the domain of consumer justice.

We observed a series of recurrent narrative patterns in consumer questions that have a particular domain semantic value, that is to say, that are given a domain interpretation by experts. Each frame contains certain recurrent frame elements. The frame CONFLICT, for instance, appears in all user questions. It describes the situation that lead to the conflict and therefore has as participants the Consumer, the Seller, some Place and Time, the concerned Object (which can be either a good or a service according to consumer law) and certain contractual conditions (such as price or payment method).

The frame ARGUMENT-SELLER identifies the opinion that the Seller holds on the conflict at hand and some justification for it. Its elements are a certain instantiation of the Seller, the argument content, its backing and in some cases a point in time in which the opinion was expressed. The frame ARGUMENT-CONSUMER is similar, but from the perspective of the consumer. The frame REQUEST is not strictly part of the story built by the consumer, but is the element which transforms the story in a question, either by asking for information on steps to take to solve the conflict (REQUEST-PROCEDURAL) or on the regulation in force applicable to the case (REQUEST-INFO).

CONFLICT [Consumer, Seller, Place, Time, Object, Contractual conditions, Breach]
ARGUMENT-SELLER [Seller, Argument-Content, Backing, Time]
ARGUMENT-CONSUMER [Consumer, Argument-Content, Backing, Time]
NEGOTIATION [Consumer, Seller, Third-parties, Content, Result, Time]
REQUEST [Consumer, Content]
REQUEST-INFO [Consumer, Content-What]
RESQUEST-PROCEDURAL [Consumer, Content-How]

Fig. 8. Sample of consumers' narrative frames

Figure 8 provides a sample of consumer narrative frames identified in the corpus (translated from Spanish). Figure 9 reports a user question annotated with several narrative frames.

[START CONTRACT[Consumer] I ordered [Object] a new mobile phone from [Seller] Vodafone [Contractual conditions] in exchange for points.] [CONFLICT [Breach] I never received it.] [ARGUMENT-SELLER [Seller] They say [Argument-Content] they delivered the phone to my neighbour] but my neighbour says he does not have it. [ARGUMENT-SELLER [Seller] Vodafone says [Argument-Content] that I should sue my neighbour]. [ARGUMENT-CONSUMER [Consumer] I think [Argument-Content] that if I have to sue someone it's Vodafone].
[REQUEST-PROCEDURAL What can I do?] [ARGUMENT-SELLER I was told by [Seller] Vodafone [Argument-Content] that they would not hesitate to impose a financial penalty for breach of permanence].

Fig. 9. Frame-based annotation of a user question

It is interesting to note that the frame-based structure is complementary to the annotation with ontological domain concepts.

6.2 A Domain-Driven Script-Like Structure for the Annotation of the User Corpus

In order to capture both the conceptual meaning and the frame-based structure of consumer questions we propose a two-layered annotation structure. The goal of the annotation structure proposed in this paper is to identify from the text of the user questions the elements that would be relevant from the perspective of the domain expert. The first level, namely, the level of concepts is only partially informative (enabling to identify, for instance, the topic of the query and the regulated objects or subjects the consumer, the seller-). The second level provides factual information relative to the interpretation of the case.

Up to now we have identified seven frames[11] with different combinations of Frame Elements. The initial annotation task has shown that the instantiation of the frames takes heterogeneous morphosyntactic patterns. In some cases, certain Frame Elements remain implicit and are not lexicalised in separate lexical units (for instance the Consumer and Seller Frame Elements, specially in the ARGUMENT-SELLER[12] and ARGUMENT-CONSUMER[13] frames). This is due to the characteristics of the Spanish language, in which Subjects are only lexicalised to add an emphatic value to the sentence. In some other cases the lexicalisation takes place through pronouns (ex. me) and verb endings. Similarly, time usually remains implicit and emerges from verbal tense and from the immediate textual context. Furthermore it should be noted that the annotation can be done at different levels of syntactic depth. In order to deal with the usually non-lexicalised frame elements we foresee a general annotation category that would enable the expert annotator to add any relevant information inferred from the text although not instantiated.

We believe that we have identified frames which are recurrent in user-questions and that are meaningful from a domain perspective. In other words, we believe that these frames select pieces of text relevant for the expert interpretation and evaluation of the case and we expect them to be a source for machine-learning algorithms.

Our ultimate goal is to annotate 200 user questions with the aid of a team of experts from the Catalan Consumer Agency. We are currently working with the Agency in order to identify a team of annotators and we are designing a set of guidelines in order to ensure the objectiveness of the annotations. We foresee as well to set up an online annotation environment in order to simplify the annotation task, and in order to enable the measure of the degree of agreement between the different expert annotations. The level of agreement between the annotators will enable us to either confirm or adapt the annotation structure before this allows the training of machine learning algorithms. Annotation guidelines will be provided to the annotators, indicating them to identify fragments carrying

---

[11] Start-Contract, Conflict, Argument-Seller, Argument-Consumer, Negotiation, Request-Procedural, Request-Info.

[12] For instance, "they say I have to pay the price anyway".

[13] For instance, "I already told them to fix it".

information they consider relevant for case interpretation, and describing the structure and values of the different frames.

We conceive this first stage of annotation as an iterative process by which experts will highlight their difficulties and disagreements with the proposed frame-structure. This process will lead to the refinement of the annotation model.

In order to enable the posterior processing of the annotated corpus with different statistically-based textual analysis tools (such as Alceste, Lexico, Textometrie and Sato) we plan to translate the annotations into the XML-TEI based exchange format proposed by [10, 11].

7   Conclusions and Further Work

This paper has described the initial steps taken in the ONTOMEDIA project in order to align the knowledge and linguistic structures used by citizens to represent their conflicts in the domain of consumer justice with the expert and institutional knowledge of the domain. More concretely, the goal of the paper was to propose a strategy for identifying in a set of user queries the domain relevant information that would be used by an expert to give an interpretation of the case. As explained later on, results lead us to a double annotation strategy.

A first hypothesis tested in this paper was the terminological-conceptual character of user queries. Results are counterintuitive since we observed that a considerable number of lexical units or phrases lent naturally to a mapping to ontological domain concepts through different semantic links (lexicalisation, sub-class, instance of). However we noted that the nature of user questions is only partially terminological (contrarily to expert discourse) since an important part of their informative content lies in the narrative structure.

Indeed, expert interpretation of laymen case narration will rely on words that evoke domain concepts (such as consumer, seller, contract, and Named Entities evoking instances of domain concepts), but as well on larger textual structures denoting procedural aspects. An example of the later would be the NEGOTI-ATION frame, which describes an attempt to solve the conflict between the seller and the consumer. In presence of a textual instantiation of the NEGOTI-ATION frame the expert would probably recommend different actions than if no negotiation had already been attempted.

In accordance to this double interpretation of consumer questions this paper has proposed a two-layered annotation structure that combines terminological-conceptual knowledge and script-like information which has a substantive and procedural value in the interpretation of the case by the domain expert.

Further work includes the use of a sample of annotated examples with machine learning algorithms.

Moreover, the paper has provided some hints on the theoretical issues that underlie the Ontomedia project. More precisely, this research opens a Pandoras Box in terms of automatic processing of user-generated content. Indeed, several issues in the domain of Natural Language Processing will have to be tackled in the follow-up of the Ontomedia project in order to ensure the efficiency of the semantic platform.

First of all, as mentioned above, an in-depth analysis of the notion of term is required. Term has been traditionally defined as a linguistic unit carrying conceptual meaning in a particular domain. The morphosyntactic characteristics of terms have been widely studied with regard to technical texts, but research on the linguistic form taken by terms in common-language discourse are much less common. In this paper we provided an analysis of user-generated corpora on the basis of morphosyntactic grammars previously designed for the processing of legal texts. We saw that not all of them were reusable and this indicates that a more detailed analysis of the linguistic characteristics of user-generated content in the domain of consumer law is required. Aspects such as unithood, that is to say, the level of stability of syntagmatic combinations [2] and termhood, the extent to which terms are representative of a domain will have to be re-explored in user-generated texts.

On the basis of these observations, one of the hypothesis on which our future work will rely is that term is any linguistic unit which is carrier of concepts relevant for the description of any type of conflictive situation in the domain.

We expect this provisional definition to enable us to render more objective the task of annotating the relevant domain terms in a user-generated corpus. This will furthermore enable us to measure recall and thus to overcome one of the limitations of our current approach (since we were only able to measure precision and could not estimate the number of potential terms left out by our grammars).

Secondly, in our future work we will have to deal with orthographic errors and common abbreviations in short online messages (i.e. cía instead of compañía company-). We will have to deal as well with language mixture in some queries, since both Catalan and Spanish being official some citizens mix both languages in their message (this occurs specially when they are using reported speech and literally quoting what was said by the seller or by another state agency in Spanish).

Thirdly, we observed two potentialities in our corpus that will have to be exploited in the future. The first one refers to the presence of terms in more than one topic subset. Exploiting this multiple occurrence as links between terms expressed in the form of graphs might give us an idea of the semantic relations between different consumer topics. The second one refers to the presence of a large number of expressions denoting psychological states (powerlessness -situacin de impotencia-, leg-pull -tomadura de pelo-) which give clues to the domain expert about the characteristics of the conflict and the most convenient resolution mechanisms. The construction of a database of those expressions might be useful in other of the mediation platform.

In terms of ontological models, it should be noted that we were able to find anchors in the available domain ontologies for linking the terms extracted from the user-generated corpus. This indicated that even if domain ontologies are difficult to use in an open environment, in a relatively restrained legal-institutional environment we build on them, because citizens, in a way, are already adapting their discourse to what they believe are the available institutional mechanisms.

Fourthly, we might consider adding to the available domain ontology an ontological representation of the workflow of treatment of queries and complaints by the Agency, and of the specific services dealing with them in order to enhance the semi-automatic redirection of questions.

Acknowledgments. ONTOMEDIA: Platform of Web Services for Online Mediation, Spanish Ministry of Industry, Tourism and Commerce (Plan AVANZA I+D, TSI-020501-2008, 2008-2010); ONTOMEDIA: Semantic Web, Ontologies and ODR: Platform of Web Services for Online Mediation (2009-2011), Spanish Ministry of Science and Innovation (CSO-2008-05536-SOCI).

References

1. Barral, I., Suquet, J.: La mediación en el ámbito de consumo. In: Casanovas, P., Magre, J., Lauroba, M.E. (Dir.) Libro Blanco de la Mediación en Cataluña, Generalitat de Catalunya, Dpt. de Justícia, Cap. 3, Barcelona, p. 7 (2010-2011)
2. Cabré, M.T., Estopà, R., Vivaldi, J.: Automatic term detection: a review of current systems. In: Recent Advances in Computational Terminology, pp. 53–87. John Benjamins, Amsterdam (2001)
3. Cabré, R., Estopà, R.: Unidades de conocimiento especializado, caracterizacin y topología. In: Cabré, M.T., Bach, C. (eds.) Coneixement, Llenguatge i Discurs Especialitzat, pp. 69–94 (2005)
4. Casanovas, P., Poblet, M.: Concepts and Fields of Relational Justice. In: Casanovas, P., Sartor, G., Casellas, N., Rubino, R. (eds.) Computable Models of the Law. LNCS (LNAI), vol. 4884, pp. 323–339. Springer, Heidelberg (2008)
5. Casanovas, P., Poblet, M.: Esquema general de los conceptos y ámbitos de la justicia relacional. In: Casanovas, P., et al. (eds.) Materiales del Llibro Blanco de la Mediación en Cataluña. La mediación: conceptos, ámbitos, perfiles, indicadores. Generalitat de Catalunya, Departament de Justícia, Centre d'Estudis Jurídics i Formació Especialitzada, Justícia i Societat, vol. I(32), pp. 21–33 (2009)
6. Casanovas, P.: The Future of Law: Relational Law and Next Generation of Web Services. In: Fernández-Barrera, M., de Filippi, P., Nuno Andrade, N., Viola de Azevedo Cunha, M., Sartor, G., Casanovas, P. (eds.) The Future of Law and Technology: Looking into the Future.Selected Essays. Legal Information and Communication Technologies Series, vol. 7, pp. 137–156. European Press Academic Publishing, Florence (2009)
7. Casanovas, P., Pagallo, U., Sartor, G., Ajani, G.: Introduction: Complex Systems and Six Challenges for the Development of Law and the Semantic Web. In: Casanovas, P., Pagallo, U., Sartor, G., Ajani, G. (eds.) AICOL-II/JURIX 2009. LNCS, vol. 6237, pp. 1–11. Springer, Heidelberg (2010)
8. Casanovas, P., Magre, J., Lauroba, M.E. (Dir.): Llibre Blanc de la Mediació a Catalunya. Generalitat de Catalunya, 1184 p. (2010), http://www.llibreblancmediacio.com
9. Casanovas, P., Magre, J., Lauroba, M.E. (Dir.): Libro Blanco de la Mediación en Cataluña. Generalitat de Catalunya, 1206 p. (with the code of LEI for mediation in Annex) (2011), http://www.llibreblancmediacio.com

10. Daoust, F., Marcoux, Y.: Logiciels d'analyse textuelle: vers un format XML-TEI pour l'échange de corpus annotés. In: Proceedings of the 8th International Conference on the Statistical Analysis of Textual Data (JADT 2006), Besanç, April 19-21 (2006)

11. Daoust, F., Marcoux, Y., Viprey, J.M.: Lannotation structurelle. In: Bolasco, Chiari, Giuliano (eds.) Proceedings of 10th International Conference Journées dAnalyse Statistique des Données Textuelles, June 9-11. Sapienza University of Rome (2010)

12. D'Aquin, M., Motta, E., Sabou, M., Angeletou, S., Gridinoc, L., Lopez, V., Guidi, D.: Toward a new Generation of Semantic Web Applications. IEEE Intelligent Systems, 20–28 (May/June 2008)

13. Fensel, D.: STI Technical Report 2008-01-10, STI Innsbruck (2008), http://www.sti-innsbruck.at/fileadmin/documents/SemanticTechnology.pdf

14. Fernández-Barrera, M.: From specialised legal knowledge to user-generated knowledge through legal ontologies: paving the way towards a Semantic Web 2.0. PhD dissertation, European University Institute, Florence, Italy (forthcoming)

15. González-Conejero, J., Meroño, A.: Mediation Tools for eGovernment: the Medi-Web and MediApp applications (2011) (unpublished paper)

16. Hendler, J.: Web 3.0 emerging. IEEE Intelligent Systems, 88–90 (January 2009)

17. Lempert, R.: Telling tales in court: Trial procedure and the Story Model. Cardozo Law Review 13, 559–576 (1991)

18. Lux, M., Dsinger, G.: From folksonomies to ontologies: Employing wisdom of the crowds to serve learning purposes. International Journal of Knowledge and Learning (IJKL) 3(4/5), 515–528 (2007)

19. Mika, P.: Ontologies Are Us: A Unified Model of Social Networks and Semantics. In: Gil, Y., Motta, E., Benjamins, V.R., Musen, M.A. (eds.) ISWC 2005. LNCS, vol. 3729, pp. 522–536. Springer, Heidelberg (2005)

20. Motta, E., Sabou, M.: Next Generation Semantic Web Applications. In: Mizoguchi, R., Shi, Z.-Z., Giunchiglia, F. (eds.) ASWC 2006. LNCS, vol. 4185, pp. 24–29. Springer, Heidelberg (2006)

21. Noriega, P.: Regulating Virtual Interactions. In: Casanovas, P., Noriega, P., Bourcier, D., Galindo, F. (eds.) Trends in Legal Knowledge, the Semantic Web and the Regulation of Electronic Social Systems. Papers from the B-4 Workshop on Artificial Intelligence and Law, May 25-27, 2005. XXII World Congress of Philosophy of Law and Social Philosophy, IVR 2005, Granada, May 24-29, 2005, pp. 55–77. European Press Academic Publishing, Florence (2007)

22. Noriega, P., López, C.: Towards a platform for Online Mediation. In: Poblet, M., Shield, U., Zeleznikow, J. (eds.) Proceedings of the Workshop on Legal and Negotiation Support Systems 2009, in Conjunction with the 12th International Conference on Artificial Intelligence and Law (ICAIL 2009). IDT Series, Barcelona, June 12, vol. 5, pp. 67–75 (2009), http://www.huygens.es/site/service4.html

23. Noriega, P., Lóopez de Toro, C.: Software de Desarrollo. In: P. Casanovas, J. Magre, E. Lauroba (Dirs.) Libro Blanco de la Mediación en Cataluña (2011), http://www.llibreblancmediacio.com

24. Passant, A.: Using Ontologies to Strengthen Folksonomies and Enrich Information Retrieval in Weblogs. In: Int. Conf. on Weblogs and Social Media (2007)

25. Pazienza, M.T., Pennacchiotti, M., Zanzotto, F.M.: Terminology Extraction: An Analysis of Linguistic and Statistical Approaches. STUDFUZZ, vol. 185, pp. 255–280 (2005)

26. Pennington, N., Hastie, R.: A cognitive theory of juror decision making: The Story Model. Cardozo Law Review 13, 519–557 (1991)

27. Pennington, N., Hastie, R.: Reasoning in explanation-based decision making. Cognition 49, 123–163 (1993)
28. Poblet, M., Casellas, N., Torralba, S., Casanovas, P.: Modeling Expert Knowledge in the Mediation Domain: A Mediation Core Ontology. In: Casellas, N., et al. (eds.) 3rd Workshop on Legal Ontologies and Artificial Intelligence Techniques Joint with 2nd Workshop on Semantic Processing of Legal Texts, LOAIT 2009. IDT Series, Barcelona, vol. 2 (2009)
29. Poblet, M., Casanovas, P., López-Cobo, J.M.: Online Dispute Resolution for the Next Web Decade: The Ontomedia Approach. In: Journal of Universal Computer Science, Proceedings of the 10th International Conference on Knowledge Management and Knowledge Technologies, Graz, Austria, pp. 117–125 (2010)
30. Poblet, M., Noriega, P., Suquet, J., Gabarró, S., Redorta, J.: Capítulo 16: Tecnologías para la mediación en línea: estado del arte, usos y propuestas. In: Casanovas, P., Magre, J., Lauroba, E. (Dirs.) Libro Blanco de la Mediación en Cataluña (2011)
31. Rideout, J.C.: Storytelling, narrative rationality, and legal persuasion. Journal of the Legal Writing Institute 14, 53–86 (2008)
32. Sartor, G., Casanovas, P., Casellas, N., Rubino, R.: Computable Models of the Law and ICT: State of the Art and Trends in European Research. In: Casanovas, P., Sartor, G., Casellas, N., Rubino, R. (eds.) Computable Models of the Law. LNCS (LNAI), vol. 4884, pp. 1–20. Springer, Heidelberg (2008)
33. Sartor, G., Casanovas, P., Biasiotti, M.A., Fernández-Barrera, M. (eds.): Approaches to Legal Ontologies. Theories, Domains, Methodologies. LGT Series, vol. 1, pp. 1–15. Springer, Heidelberg (2011) ISBN: 978-94-007-0119-9
34. Schmid, H.: Probabilistic Part-of-Speech Tagging Using Decision Trees. In: Proceedings of the International Conference on New Methods in Language Processing, pp. 44–49 (1994)
35. Silberztein, M.: NooJ Manual (2003), http://www.nooj4nlp.net/
36. Specia, L., Motta, E.: Integrating Folksonomies with the Semantic Web. In: Franconi, E., Kifer, M., May, W. (eds.) ESWC 2007. LNCS, vol. 4519, pp. 624–639. Springer, Heidelberg (2007)
37. Spivack, N.: The Semantic Web, Collective Intelligence and Hyperdata (2007), http://novaspivack.typepad.com/nova_spivacks _weblog/2007/09/hyperdata.html

Author Index

Agnoloni, Tommaso 93
Ambrossio, Agustín 189
Araszkiewicz, Michał 33

Bertoldi, Anderson 256
Boella, Guido 131
Boer, Alexander 235
Bourcier, Daniele 73

Casanovas, Pompeu 171, 286
Ceci, Marcello 116, 245
Chishman, Rove Luiza de Oliveira 256

De Filippi, Primavera 73

Fernández-Barrera, Meritxell 286
Francesconi, Enrico 147, 162

Humphreys, Llio 131

Laukyte, Migle 204
Lesmo, Leonardo 245

Martin, Marco 131
Mazzei, Alessandro 245
Mendoza, Leandro 189
Myška, Matěj 271

Pagallo, Ugo 48
Palmirani, Monica 116, 245
Peruginelli, Ginevra 162
Plaza, Enric 171

Radicioni, Daniele P. 245
Rossi, Piercarlo 131
Rotolo, Antonino 189
Ruyter, Jelle de 106

Sagri, Maria-Teresa 93
Sartor, Giovanni 1
Šavelka, Jaromír 271
Škop, Martin 271
Smejkalová, Terezie 271
Smith, Clara 189

Tiscornia, Daniela 93

van der Torre, Leendert 131
van Engers, Tom 235
Vincent, Andrew 217

Weng, Yueh-Hsuan 61
Winkels, Radboud 106

Zhao, Sophie Ting Hong 61